Cell-Based Meat in the European Union and Beyond

Giulia Formici • Maria Cecilia Mancini •
Lucia Scaffardi
Editors

Cell-Based Meat in the European Union and Beyond

An Interdisciplinary Study

Springer

Editors
Giulia Formici
Department of Law, Politics and International Studies
University of Parma
Parma, Italy

Maria Cecilia Mancini
Department of Economics and Management
University of Parma
Parma, Italy

Lucia Scaffardi
Department of Law, Politics and International Studies
University of Parma
Parma, Italy

ISBN 978-3-032-14623-6 ISBN 978-3-032-14624-3 (eBook)
https://doi.org/10.1007/978-3-032-14624-3

The publication of this book is financed by the Research Project of Relevant National Interest (PRIN) 2022 "The Future of Food, the Food of the Future. Novel Food, Innovation, Sustainability and Legal Issues", funded by the European Union – Next Generation EU (Mission 4, Component 2, CUP D53D23007160006).

This Springer imprint is published by the registered company Springer Nature Switzerland AG
The registered company address is: Gewerbestrasse 11, 6330 Cham, Switzerland

Foreword

As a proposed solution for an imminent global crisis to satisfy the growing demand for animal proteins, cultivated meat has reached a stage where marketing approvals in several countries have been slow but, in the end, granted. This sets the stage for large-scale production and consumption. Indeed, a handful of companies are scaling production to do so. During the decade-long development of cultivated meat, it has become clear that there are many societal ramifications of this technology, including legal, regulatory, social, market, and economic ones, many of them strongly interacting.

This volume provides an integrated, neutral, and highly academic exposé of the most pressing societal issues, with a strong focus on the European situation, which is arguably the most complex political and social landscape in the world. It is also the region where the concept of cultivated meat originated. Cultivated meat thrived through private money invested in startup companies and was relatively late followed by much-needed public funding. The present volume is the outcome of such public funding and as a result has a refreshing neutral and academic approach to the issues discussed. Each of the contributions is from world class academic experts in the field.

In the new academic field of cultivated meat, most papers and books cover scientific and technological aspects in addition to consumer acceptance studies. None of those span the entire societal panorama exposing and discussing the interaction between, for instance, the regulatory framework, nomenclature, policy, understanding of cultural background, and consumer behavior. This book is unique in providing that overview and understanding of the interactions. It is a highly recommended read for anyone interested in the societal issues of the upcoming protein transition.

Maastricht University, Department of Physiology
Maastricht, The Netherlands

Mark Post

Preface

This book is one of the main outputs of the Research Project of Relevant National Interest (PRIN) 2022 "The Future of Food, the Food of the Future. Novel Food, Innovation, Sustainability and Legal Issues," funded by the European Union—Next Generation EU (Mission 4, Component 2, CUP D53D23007160006). Coordinated by Professor Lucia Scaffardi (Principal Investigator) and involving three Research Units (University of Parma, University of Milan, and Roma Tre University), the project hosted several conferences and seminars dedicated to cell-based meat, a highly innovative and controversial Novel Food. The complex scientific, regulatory, and economic challenges of this emerging product—as well as the vibrant debate it has sparked—inspired this Volume.

Bringing together contributions from scholars affiliated with Italian and foreign Universities and Institutions, the present work offers an interdisciplinary analysis that draws on diverse expertise and research experience. It is intended as a resource for a vast audience, including policymakers, public authorities, food business operators, researchers, students, and citizens. The aim is to provide the clear, research-based insights needed to navigate an intricated topic and to counteract disinformation and purely ideological narratives.

The Editors thank all the Authors for their constant collaboration and precious contributions. The Editors would also like to thank Helen Ampt for revising the English, Maria Giulia Corazza for her meticulous formal review and enthusiastic participation in this editorial journey, the Springer team for embracing our book proposal and supporting its realization, as well as the anonymous reviewers for their constructive and valuable suggestions.

Finally, the Editors express their gratitude to the Department of Law, Politics and International Studies of the University of Parma for institutional support in the broader framework of the "Food for Future" Project (Department of Excellence 2023–2027), as well as to the Italian and EU Institutions that made the PRIN 2022 project possible, enabling us to publish this book as open access.

Parma, Italy

Giulia Formici
Maria Cecilia Mancini
Lucia Scaffardi

Acknowledgment

The publication of this book is financed by the Research Project of Relevant National Interest (PRIN) 2022 "The Future of Food, the Food of the Future. Novel Food, Innovation, Sustainability and Legal Issues," funded by the European Union—Next Generation EU (Mission 4, Component 2, CUP D53D23007160006).

Contents

Socio-Economic and Cultural Issues

Contributors

Giancarlo Anello Department of Humanities, Social Sciences, and Cultural Industries, University of Parma, Parma, Italy

Antonella Baldi Department of Veterinary Medicine and Animal Sciences, University of Milan, Milan, Italy

Garrett M. Broad Ric Edelman College of Communication, Humanities and Social Sciences, Rowan University, Glassboro, NJ, USA

Katharina Burdorf Hydrosol GmbH & Co. KG, Ahrensburg, Germany

Federica Cheli Department of Veterinary Medicine and Animal Sciences, University of Milan, Milan, Italy

Maria Giulia Corazza Department of Law, Politics and International Studies, Department of Excellence 2023–2027 "Food for Future," University of Parma, Parma, Italy

Chiara Dall'Asta Department of Food and Drug, University of Parma, Parma, Italy

Giacomo Degli Antoni Department of Law, Politics and International Studies, Department of Excellence 2023–2027 "Food for Future," University of Parma, Parma, Italy

Marco Faillo Department of Economics and Management, University of Trento, Trento, Italy

Giulia Formici Department of Law, Politics and International Studies, Department of Excellence 2023–2027 "Food for Future," University of Parma, Parma, Italy

Carlotta Giromini Department of Veterinary Medicine and Animal Sciences, University of Milan, Milan, Italy

Hilva Gjoni Department of Food and Drug, University of Parma, Parma, Italy

Leland Glenna Department of Agricultural Economics, Sociology, and Education, College of Agricultural Sciences, Pennsylvania State University, University Park, PA, USA

Davide Lanzoni Department of Veterinary Medicine and Animal Sciences, University of Milan, Milan, Italy

Nick Lin-Hi Faculty I: Management of Social Services, University of Vechta, Vechta, Germany

Maria Cecilia Mancini Department of Economics and Management, University of Parma, Parma, Italy

Cinzia Piciocchi Faculty of Law, University of Trento, Trento, Italy

Simone Pitto Department of Law, University of Genoa, Genoa, Italy

Mark Post Department of Physiology, Maastricht University, Maastricht, The Netherlands

Mosa Meat, B.V., Maastricht, The Netherlands

Silvia Rolandi Department of Agriculture, Food and Environment, University of Pisa, Pisa, Italy

Lucia Scaffardi Department of Law, Politics and International Studies, Department of Excellence 2023–2027 "Food for Future," University of Parma, Parma, Italy

Eleonora Sirsi Department of Law, University of Pisa, Pisa, Italy

Tamil Selvi Sundaram Department of Veterinary Medicine and Animal Sciences, University of Milan, Milan, Italy

Gabriella Tedeschi Department of Veterinary Medicine and Animal Sciences, University of Milan, Milan, Italy

Bernd van der Meulen European Institute of Food Law, Amstelveen, The Netherlands

Abbreviations

ANPR	Advance Notice of Proposed Rulemaking
BOAH	Board Of Animal Health
BSE	Bovine Spongiform Encephalopathy
CAFOs	Confined Animal Feeding Operations
CAP	Common Agricultural Policy
CFR	Code of Federal Regulations
CJEU	Court of Justice of the European Union
CM	Cell-based Meat
CRS	Congressional Research Service
CSR	Corporate Social Responsibility
DARPA	Defense Advanced Research Projects Agency
DOE	Department of Energy
EC	European Community
EESC	European Economic and Social Committee
EFSA NDA	EFSA Panel on Nutrition, Novel Foods and Food Allergens
EFSA	European Food Safety Authority
EPRS	European Parliamentary Research Service
ESCs	Embryonic Stem Cells
EU	European Union
F2F	EU "Farm to Fork" Strategy
FAO	Food and Agriculture Organization
FBOs	Food Business Operators
FBS	Fetal Bovine Serum
FDA	Food and Drug Administration
FFDCA	Federal Food, Drug, and Cosmetic Act
FMIA	Federal Meat Inspection Act
FPLA	Fair Packaging and Labelling Act
FRMD	Food Risk Management Department
FSA	Food Standards Agency
FSANZ	Food Standards Australia New Zealand
FSIS	Food Safety Inspection Service

FSSA	Food Safety and Security Act
FSVO	Food Safety and Veterinary Office
GAO	Government Accountability Office
GAP	Good Agricultural Practice
GEM	Genome-scale Model
GFI	Good Food Institute
GFSI	Global Food Safety Initiative
GFL	General Food Law
GHG	Green House Gas
GLEAM	Global Livestock Environmental Assessment Model
GMOs	Genetically Modified Organisms
GMP	Good Manufacturing Practices
GRAS	Generally Recognized as Safe
HACCP	Hazard Analysis Critical Control Points
HHS	Department of Health and Human Services
IIA	Israeli Innovation Authority
IPCC	Intergovernmental Panel on Climate Change
iPSCs	induced Pluripotent Stem Cells
ISO	International Organization for Standardization
LCA	Life Cycle Assessment
LCAs	Life Cycle Analysis
LEAP	Livestock Environmental Assessment and Performance
MAHA	Make America Healthy Again
MEWR	Ministry of the Environment and Water Resources
MSE	Ministry of Sustainability and the Environment
NACMCF	National Advisory Committee on Microbiological Criteria for Foods
NAMs	New Approach Methodologies
NASA	National Aeronautic Space Administration
NFC	Novel Food Committee
NFS	National Food Service
OECD	Organisation for Economic Co-operation and Development
OLFs	Other Legitimate Factors
OLS	Ordinary Least Squares
PAFF	Standing Committee on Plants, Animals, Food and Feed
PHPFL	Public Health Protection Food Law
PPFNL	Protection of Public Health Food and Nutritional Labelling
PPIA	Poultry Products Inspection Act
SD	Standard Deviation
SDGs	Sustainable Development Goals
SFA	Singapore Food Agency
SFAA	Singapore Food Agency Act
STS	Science and Technology Studies
TEA	Techno-Economic Assessment
TFEU	Treaty on the Functioning of the European Union
TRIS	Technical Regulation Information System

UAE	United Arab Emirates
UK	United Kingdom
UN	United Nations
UNEP	United Nations Environment Programme
UNGA	United Nations General Assembly
USA	United States of America
USC	United States Code
USCA	United States Cattlemen's Association
USDA	United States Department of Agriculture
USDA-FSIS	United States Department of Agriculture—Food Safety and Inspection Service
VAT	Valued Added Tax
WHO	World Health Organization
WMO	World Meteorological Organization
WTO	World Trade Organization

Cultivating a Revolution: Legal and Policy Trajectories for a Sustainable Protein Transition

Lucia Scaffardi

1 Beyond the Lab: The Legal Dimension of Cultivated and Alternative Proteins

The growing interest in cell-based meat is no longer simply a matter of scientific curiosity or cultural provocation. It now stands as a tangible and symbolic frontier of the agri-food transition, a "technological promised land"[1] capable of reshaping the relationship between production, consumption and sustainability in all its dimensions.[2] Cell-based meat therefore occupies a particularly emblematic position

[1] As explored in this book, at the time writing, specific categories of cell-based meat (mainly chicken, beef and duck) can be marketed in Singapore, Israel, the United States of America and recently in Hong Kong. In the European Union (EU), two applications for market authorisation under the Novel Foods Regulation—Regulation (EU) 2015/2283 of the European Parliament and of the Council of 25 November 2015 on Novel Foods, amending Regulation (EU) no 1169/2011 of the European Parliament and of the Council and repealing Regulation (EC) no 258/97 of the European Parliament and of the Council and Commission Regulation (EC) no 1852/2001—were submitted in July 2024 and January 2025. The authorisation procedure is therefore in its initial stage.

[2] According to the well-known definition of the World Commission on Environment and Development's 1987, Brundtland report "Our common future" (World Commission on Environment and Development (1987)), sustainable development is "development that meets the needs of the present without compromising the ability of future generations to meet their own needs" and can be considered in three dimensions: environmental, social and economic. See Bifulco and D'Aloia (2008), Fracchia (2010), Atkinson et al. (2014), Porena (2017), and Cuocolo (2025). Focusing on the agri-food sector, on a world scale, we are now witnessing a shift from a "farm to fork" paradigm to a "factory to lab" one: this warrants thorough examination not only of environmental sustainability but also of social and economic sustainability. On this point see Cerini (2025).

L. Scaffardi (✉)
Department of Law, Politics and International Studies, Department of Excellence 2023–2027 "Food for Future", University of Parma, Parma, Italy
e-mail: lucia.scaffardi@unipr.it

G. Formici et al. (eds.), *Cell-Based Meat in the European Union and Beyond*,
https://doi.org/10.1007/978-3-032-14624-3_1

due to the promises it offers, the resistance it encounters, and most significantly, the breadth of issues it engages.[3] Hence the present scholarly volume is conceived for international open access readership, to foster inclusive and critical reflection on these pivotal themes at the crossroads of our shared future.

Given this premise, the approach of this book is necessarily interdisciplinary.[4] It reviews public comparative, constitutional, agri-food and even multicultural law, while maintaining a sustained dialogue with sciences like biology, economics, agro-economics and food sociology. The aim is not merely to chart the current legal landscape, but rather to construct a coherent and reflective framework that captures the many dimensions at stake.[5] It is now commonly held that only through an integrated approach—what comparatists might call inter- and transdisciplinary—can we grasp the complex implications arising from scholarly engagement with these evolving situations.[6] Such an approach enables us to move beyond the ideological simplifications that too often accompany narratives on innovation.[7]

Faced with this complexity, the task of this book is to demonstrate that any serious engagement with cultivated meat requires more than a scientific or industrial analysis: it demands a broader inquiry into cultural, legal and socio-economic implications.[8]

The introductory remarks proposed here provide some preliminary insights into the role of innovation in the agri-food sector as well as the urgent ongoing call for a legal transition from reactive norms to proactive governance. These are necessary premises to better explain the reasons behind this book, its structure and the approaches and key findings characterizing its interlinked and articulated chapters.

Moreover, it is worth taking into account the ongoing debate on a potential fourth dimension (or pillar) of sustainable development: cultural sustainability. This aspect is particularly relevant for the specific topic of our book: in fact, issues related to the preservation of cultural identities—including those connected to food and culinary traditions—have increasingly featured in the political, social and legal discussions surrounding cell-based meat, as underlined in several chapters of this book (in particular see *Cell-Based Meat and Food Policy* by Mancini; *Food Justice in Religious Laws: Hypotheses on Cultivated Meat* by Anello; *Who Hates Cultured Meat? Examining Public Opposition to Cellular Agriculture* by Broad and Glenna, as well as *Navigating Consumer Acceptance of Cultured Meat: Insights from Social Science Research* by Lin-Hi and Burdorf). On the concept of cultural sustainability, see Meireis and Rippl (2018) and Sabatini (2019).

[3] For an overview of these profound and diverse issues and debate, see Lanni (2025).

[4] On the need to apply an interdisciplinary method, especially when dealing with topics linked to sustainability, see Borghi (2023).

[5] Di Lauro (2025).

[6] Saija (2025).

[7] Paganizza (2025).

[8] Formici (2024).

1.1 *Interlocking Crises and the Call for a Legal Transformation with Regard to Food Systems*

Humanity is faced with an intricate set of challenges, many arising from the complex interdependence that has evolved in the last century, primarily as a consequence of globalised trade.[9] Additional pressures on ecosystems, particularly those from intensive agriculture and industrial animal farming, have accelerated global phenomena such as biodiversity loss, climate change and soil degradation. At the same time, growing nutritional inequalities, systemic food waste and widespread food insecurity underscore the fundamental unsustainability of our prevailing models.[10] In our endeavour to ensure a sufficient and resilient supply of food for a growing global population, we can be likened to sailors on different ships, all bound for the same destination.[11] Traditional food systems are increasingly unable to respond to climate emergencies,[12] and to conflicts and disruptions caused by new fiscal and trade measures. Natural resources have been consumed disproportionately, also due to exponential growth of world population in the last 25 years.

While landmark initiatives such as the United Nations 2030 Agenda and its Sustainable Development Goals (SDGs)[13] have laid out a bold and near-universal vision, unfolding global events make it increasingly clear that these goals are unlikely to be met by the designated deadline,[14] or even to have partial success in the near term.[15]

---

[9] "On the other hand, for some time there was an illusion that the multiple phenomena of economic interdependence generated or encouraged by more "open" trade could somehow replicate the virtuous mechanisms, that have generally characterized the experience of European integration, on a global scale. The redistribution of wealth that followed the creation of a single market in Europe, combined with aspects of (limited) mutual dependence of Member States, led to an undeniable period of peace, prosperity and cooperation among EU countries. However, it has proven substantially impossible to reproduce the virtuous aspects of that experience in the international context. Rather, productive specialization and interdependence of states and economies have become the foundation for the increasingly frequent emergence of true coercive dynamics", Borghi (2023), p. 249. Translation.

[10] FAO (2018), Scaffardi (2023), Pitto (2023), Iacovino and Randazzo (2023), and Pitto (2024).

[11] According to the Food and Agriculture Organization (FAO), by 2050 the world's population will reach 9.1 billion (FAO 2009).

[12] EFSA (2020) and Gornall et al. (2010).

[13] UN (2015).

[14] According to the 2023 United Nations report, over 50% of the targets show weak or insufficient progress, while 30% have experienced stagnation or even a reversal of trends due to factors such as the pandemic, climate change, and geopolitical conflicts; see https://www.cfr.org/councilofcouncils/global-memos/sdgs-are-not-track-here-what-world-should-do?utm_source=chatgpt.com. Accessed 10 May 2025. See also Shepherd (2023), who pointed out that, halfway through the timeline, the SDGs were far from a success, with many goals off track and progress either slowed or reversed.

[15] See Sustainable Development Goals Report 2024 at https://unstats.un.org/sdgs/report/2024/The-Sustainable-Development-Goals-Report-2024.pdf. Accessed 10 May 2025. This document underlines that, although some progress has been made, most of the goals remained far from being

In continental Europe, the European Union (EU) has sought to address the intersection of environmental crises and agri-food sustainability through flagship strategies such as the Green Deal and the Farm to Fork initiative.[16] These programmes have placed transformation of the food system at the centre of broader sustainability efforts.[17] Yet for all their symbolic and strategic value, implementation has been uneven and hindered by the absence of a strong binding legal framework capable of compelling coordinated action across member states.[18]

It is now increasingly evident—as stressed in the Food 2030 Pathways for Action 2.0 report[19]—that such multifaceted challenges can no longer be met with sector-specific approaches. The EU has accordingly begun to promote a more systemic and integrated understanding of food systems, capable of simultaneously addressing environmental, health, social and economic concerns. In this emerging paradigm, sustainability is thus taking shape as a transverse legal principle, embedded in European and global policy frameworks. Crucially, as underlined with reference to the specific case of cell-based meat, it should not be reduced to environmental protection alone: it must also include economic viability across supply chains and equal access to food as a condition of social inclusion and intergenerational justice.

1.2 Law and Innovation: From Reactive Norms to Proactive Governance

If sustainability is the regulatory and policy objective, food innovation increasingly emerges as one of its principal instruments. Emerging technologies such as cultivated meat, alongside other promising developments like precision fermentation and molecular farming, offer multifaceted responses to pressing global challenges:

achieved. Critical areas such as the fight against hunger and the protection of biodiversity were—and still are—of particular concern, a relevance further heightened by the thematic focus of the volume in which this chapter appears.

[16] Communication from the Commission to the European Parliament, the Council, the European Economic and Social Committee and the Committee of the Regions—A Farm to Fork Strategy for a fair, healthy and environmentally-friendly food system, COM(2020) 381 final, part of the European Green Deal, COM(2019) 640 final; Proposal for a Framework for Sustainable Food Systems (2023) https://food.ec.europa.eu/horizontal-topics/farm-fork-strategy/legislative-framework_en. Accessed 10 May 2025. The Proposal, which represented one of the so-called flagship initiatives of the European Commission, failed to be concretely adopted. About the evolution of EU law and policies related to the promotion of sustainability in the agri-food sector, see Holle (2025).

[17] Schebesta affirms that "Among the 27 specific legal action points identified, the proposal to create a framework for sustainable food systems (FSFS) stands out as an overarching legal instrument that should promote policy coherence, mainstream sustainability in all food-related policies, and strengthen the resilience of food systems." See Schebesta (2023), p. 232.

[18] On this point see Alessandrini et al. (2024).

[19] European Commission: Directorate-General for Research and Innovation (2023).

from reducing emissions associated with animal protein production to minimising land and water usage and enhancing animal welfare.[20]

While cell-based meat continues to occupy a paradigmatic role in the transition to alternative proteins,[21] not only due to its symbolic resonance as an alternative to conventional animal farming, but also because of the wide-ranging and intersecting issues it raises, *precision fermentation* is gaining momentum, distinguished by its use of biotechnologically modified microorganisms that can synthesize complex molecules, including milk[22] and egg proteins.[23] In particular, "this technique allows us to produce high-value compounds with precision and consistency, with cost savings, less environmental impact and improved product quality. It has been applied for years in biopharmaceutical production (e.g. insulin, vaccines, gene therapies, etc.), and is currently used in food production to obtain specific "animal-free" components (e.g. collagen, lactoferrin, ovalbumin)."[24] In the EU context, these products are Novel Foods pursuant to Regulation (EU) 2015/2283, which places particular emphasis on food safety, transparency and ingredient traceability, as underlined in the pages of this volume.[25]

Another emerging technology is *molecular farming*, which employs transgenic plants as biofactories for the production of recombinant proteins.[26] Although still in the early stages of application, molecular farming holds considerable promise in terms of sustainability, accessibility and environmental impact reduction. However,

[20] European Parliamentary Research Service (EPRS) (2024).

[21] For an in-depth analysis of the role of cell-based meat and its positioning within the broader transformation of protein production, see *What Is Cell-Based Meat? Food Safety and Sustainability Aspects* by Cheli et al.; *Cell-Based Meat in the European Union: A Regulatory Crossroads* by Corazza and Formici, as well as *Cell-Based Meat and Food Policy* by Mancini, all in this volume.

[22] de Grahl and Treuil (2024).

[23] FAO (2023).

[24] *Safety Assessment of Cell-Culture-Derived Food: The Role of the European Food Safety Authority* by Dall'Asta and Gjoni in this volume. See also FAO (2025).

[25] See also Canfora and Leccese (2024).

[26] "Molecular farming can be defined as the production of recombinant proteins in plants, where the aim is to recover and utilize the protein product rather than the plant itself. The target protein is either extracted and purified or used as part of a crude extract, whereas the plant is merely a host and is either discarded and destroyed at the end of the process or utilized as a separate side-stream. Molecular farming often involves the use of whole terrestrial plants such as tobacco and cereals, but the technology also encompasses other plant-based systems, including plant cell and tissue cultures, aquatic plants, moss, algae, and even in vitro transcription and translation systems derived from plant cells. The diversity of these plant-based systems means that molecular farming comprises a range of different platforms that have the potential to compete in many different markets, ranging from technical enzymes and research reagents that are typically produced in bacteria and yeast, to biopharmaceutical proteins that are usually produced in mammalian cells, particularly Chinese hamster ovary (CHO) cell lines", Fischer and Buyel (2020), p. 1. Molecular farming is also studied in vaccine research and for metabolites of nutritional or pharmacological interest.

it requires a robust and nuanced regulatory framework due to the hybrid nature of the products.[27]

What makes these technologies particularly noteworthy is the complexity and breadth of the legal challenges they pose. They prompt urgent reconsideration of traditional legal categories, many of which were conceived with conventional production systems in mind. Only a few years ago, the legal debate on Novel Foods centred mostly on edible insects;[28] today, as shown by this volume on cultivated meat, the horizon has already broadened significantly.[29]

Indeed, definition uncertainties—particularly regarding what is "food", "natural" or "traditional"—call for regulatory attention that must also tackle issues related to consumer transparency, genetic traceability and consistency across food, environmental and even pharmaceutical regulations.[30] These technologies raise fresh legal questions concerning authorisation, labelling, transparency and more generally the balance between precaution and innovation. The question of regulatory harmonisation at EU and international levels is also important to avoid technical barriers to trade and to ensure a coordinated, forward-looking approach to legal governance.

In this light, the law must not merely react to innovation but also anticipate emerging asymmetries. Science and technology must serve the pursuit of systemic sustainability—but this is only possible in a normative framework that is not just reactive, but actively anticipatory and capable of steering change.[31] In such a scenario, legal intervention cannot be reduced to technical codification; rather it must construct flexible and inclusive regulatory frameworks that integrate the ethical, environmental, social and economic dimensions of food innovation.

Framed within this context, cultivated meat is therefore not only a technological challenge but also a legal and cultural one. It demands critical reflection on the law's capacity to govern this transformation effectively.

---

[27] This is a promising technique with over a decade of research. However, from a legal perspective it encompasses a variety of different cases, adding an additional layer of complexity. For instance, the use of Genetically Modified Organisms implies the application of Directive 2001/18/EC of the European Parliament and of the Council of 12 March 2001 on the deliberate release into the environment of genetically modified organisms and repealing Council Directive 90/220/EEC, and in certain cases, also pharmaceutical regulations, thereby making the authorisation process particularly complex.

[28] Formici and Scaffardi (2022).

[29] It is therefore of paramount importance to deepen and articulate the ongoing debate on the normative treatment of these emerging food technologies at European Union level. The potential benefits and risks inherent in their marketing should be critically evaluated with a view to determining whether a dedicated regulatory framework is warranted. Particular attention should be paid to labelling regimes capable of clearly indicating the product's origin and production method, thereby safeguarding the consumer's right to informed and autonomous choices, an approach already adopted in the contexts of genetically modified organisms and edible insects. See Albisinni (2024).

[30] Di Lauro (2025).

[31] Cerini (2025).

2 The Structure of the Volume and the Chapters' Main Findings

As indicated by these introductory remarks, this volume thoroughly explores innovation in the agri-food sector and the innovative foods at the centre of lively political, legislative, economic, scientific and ethical debate. Focusing on cultured meat, it aims to offer a comprehensive and updated view of the issues surrounding this product intended for human consumption, and of its potential.

An interdisciplinary approach is used, given the diverse yet closely interconnected questions surrounding cell-based meat. The contributing authors are scholars from a range of academic disciplines and geographic regions, who in recent years have been involved with the multifaceted challenges posed by cell-based meat, including the legal, socio-economic, cultural and scientific dimensions.

The book is organized in three parts that reflect this approach: Part "Scientific and Sustainability Challenges" is dedicated to a presentation of the scientific and sustainability challenges of cell-based meat, with specific attention to food safety questions and the role played by the European Food Safety Authority (EFSA). Part "Legal and Policy Debate" proposes an in-depth analysis of the regulatory and policy debate on this innovative food, taking a comparative perspective inside and outside European Union borders. The rules and policies on the marketing and labelling of cell-based meat are explored, underlining different approaches and different ways of balancing fundamental rights and interests. This part reveals the complex, heterogeneous and rapidly evolving legislative scenario, where some states are enhancing investments in future foods while others are banning production or sale of cell-based meat. Part "Socio-Economic and Cultural Issues" explores cultural and economic aspects, such as consumer perceptions, new consumption models and religious concerns.

These three parts should be read as closely interconnected. While the ultimate purpose of the book is to offer a broad understanding of the challenges surrounding cell-based meat—drawing on both state-of-the-art knowledge and future perspectives—the necessary starting point is a clear grasp of what cell-based meat actually is from a scientific perspective. For this reason, and in order to guide the reader step by step through the diverse set of topics related to cell-based meat, the first two chapters are devoted to its scientific notions. This initial focus is essential, since the legal frameworks, policy questions and political debates that follow are all deeply dependent on scientific definitions, processes and unresolved questions.

Building on this foundation, the reader could then move on to Part "Legal and Policy Debate", where all the chapters are united by their exploration of potential legal solutions, ongoing discussions that remain open to scholarly and policy debate.

Once these fundamental insights are established, the book continues by broadening the perspective by investigating cultural and socio-economic aspects.

Although each chapter can be read as a stand-alone contribution, they are best appreciated in relation to one another, sharing the same overarching purpose of contributing to a holistic view of cell-based meat. The interconnectedness of the different parts of this book ideally enables the reader to perceive the full complexity of this emerging and multifaceted topic.

In fact, the added value of the volume lies in the integration of comprehensive legal analysis with socio-economic and scientific perspectives. Most existing books on cultured meat focus on environmental and food safety issues. While these profiles are of key importance, the present volume promotes a strong and careful dialogue between studies on law, food safety, food security and sustainability (intended in its three dimensions, economic, environmental and social): only through such a challenging yet stimulating dialogue is it possible to fully understand the complex elements of the ongoing regulatory debate and policy strategies. Shedding light on the key regulatory aspects, from authorisation procedures to labelling provisions, from animal welfare to religious precepts, is crucial to grasp how legal decisions can affect the market, private and public investments and, more vastly, the entire future of cell-based meat production.

2.1 Scientific and Sustainability Challenges

Moving to a brief overview of the contributions, Part "Scientific and Sustainability Challenges" begins with an investigation of the scientific challenges related to cell-based meat, focusing specifically on safety and sustainability considerations and on the role of EFSA in the risk assessment phase.

The chapter "What Is Cell-Based Meat? Food Safety and Sustainability Aspects" is written by experts from the University of Milan (Cheli, Tedeschi, Sundaram, Lanzoni, Baldi and Giromini). It offers a first definition of cultured meat and a detailed analysis of the production process. The authors identify correct knowledge of the different interrelated stages as the necessary basis for detecting food safety risks at any step of the process. Similar evaluations of security, safety and sensory/nutritional properties are in turn essential for informed regulatory choices. With specific attention to the role played by cutting-edge omics approaches and machine learning applications, the contribution also explores environmental sustainability aspects, investigating data on water and soil consumption, greenhouse gas emissions, energy efficiency and waste management. The authors conclude describing potentialities and shortcomings, highlighting the need for real-world data and comprehensive long-term studies: such research is crucial to "ensure that cell-based meat can be a reliable and sustainable alternative in the global food system".

In line with these scientific reflections, in the chapter "Safety Assessment of Cell-Culture-Derived Food: The Role of the European Food Safety Authority", Dall'Asta and Gjoni examine the vital role of EFSA in evaluating cell-based food products under the Novel Food Regulation, with the final aim of ensuring the safety of food marketed in the European Union. Emphasizing the importance of transparency, rigorous assessment and science-based regulatory decision-making in fostering consumer acceptance and promoting public safety and trust, the authors clarify EFSA's role in the risk assessment of cell-based food. They explore its scientific and procedural frameworks, as well as key challenges and potential hazards associated with the risk assessment process. The contribution also provides insights into how EFSA can effectively navigate the delicate balance "between promoting innovation and safeguarding public health in a rapidly evolving food landscape", by focusing on how EFSA tackles delicate and complex concerns regarding allergenic potential and nutritional equivalence. This analysis allows readers to reflect on the difficult task attributed to this independent authority in a highly politicized context "in which EFSA's scientific opinion must remain objective, while mediating between innovation, safety and competing interests".

2.2 *Legal and Policy Debate*

Part "Legal and Policy Debate" of this volume offer insights into the debate on legal and policy aspects of cell-based meat. The innovative nature of this vast analysis lies in different aspects: firstly, the comparative approach used to extend investigation beyond European Union borders in order to explore different regulatory and policy solutions. Secondly, the ability to look at ongoing and rapidly evolving issues and discussions in an attempt to promote reflection on the state-of-the-art as well as on possible future developments and legal scenarios. A third element of innovation characterizing Part "Legal and Policy Debate" lies in the many aspects investigated, from authorisation procedures and the rules governing the marketing of cell-based meat, to labelling, investment strategies, policies, and the impact of legislative decisions on animal welfare.

The chapter "Cell-Based Meat in the European Union: A Regulatory Crossroads" by Corazza and Formici discusses the normative situation in the European Union, starting with the Novel Foods Regulation. The authors observe what they call "centrifugal" and "centripetal forces", namely national provisions banning the marketing of cell-based meat on one hand, and on the other, calls for reform of the current supranational legislative framework. They discuss the ongoing debate and political initiatives in place in several Member States, together with the stand taken by different EU Institutions (in particular, the European Commission and the Parliament). On this basis they investigate future scenarios and challenges with specific attention to the need for regulatory solutions that not only address food safety evaluations but also socio-economic, ethical and environmental concerns. According to the authors, "promoting a regulatory discourse capable of engaging with, and possibly untangling, all these multifaceted challenges is not only vital for the destiny of cell-based meat but also for the attractiveness of the EU market, the resilience of its agri-food sector, the protection of consumer freedom of choice and the achievement of long-term sustainability goals".

Complementing the EU regulatory perspective, the chapter "The Legal Debate on Cultured Meat Around the World: Comparative Insights from the United States, Singapore and Israel" provides a comparative legal analysis of the different laws on the marketing of cell-based meat in three jurisdictions, where the sale of this product has already been authorised: the United States, Singapore and Israel. In this contribution, Pitto analyses several aspects of the regulatory framework in those countries, looking not only at legal definitions and pre-market approval mechanisms, but also at the role of public authorities in licensing and oversight. The author, therefore, examines "institutional competences, legal classifications and political dynamics". Interestingly, specific attention is also paid to recent developments: from growing opposition at State level in the US and the recently enacted 2025 Food Safety and Security Act in Singapore, to public funding initiatives in Israel. In comparing these complex scenarios, Pitto shows the common features and differences of the three case-studies.

The chapter "The Regulatory Landscape of Cultured Meat and Fish: Terminology and Consumer Information Provisions" by Sirsi and Rolandi examines another crucial legal aspect: that of the terminology and labelling of cultured meat and fish. The authors explore the terminology and the information provisions of Business to Business and Business to Consumer dynamics, with the premise that "informed choices rely on appropriate marketing communication terminology, product names defined according to the regulatory context and providing correct information". Using comparative analysis, they look at the European Union context, where several member states have already attempted or manifested willingness to adopt specific national rules, and then turn to the US and Singapore. Based on their findings, the authors highlight the growing importance of having a clear reference point and guidance for terminology, nomenclature and information, that could also derive from an agreement within the Codex Alimentarius. The establishment of such rules is not to be seen as a "stylistic exercise but concerns commercial opportunities and consumer protection".

The chapter "Cell-Based Meat and Food Policy" by Mancini sheds light on the crucial dimension of food policies concerning cell-based meat, or cell-based foods more generally. The contribution investigates the role of public policy in boosting cultivated meat innovation, focusing on global policy approaches and economic instruments in different countries of the world. The author also shows two other interrelated and seriously problematic socio-economic aspects related to the shift to alternative protein sources: market concentration and social inequalities. These elements allow Mancini to reflect on the importance of inclusive governance and food policies able to provide a "unified, long-term strategy that aligns regulatory policies, public investments and sustainability goals". The study, in fact, affirms the pivotal role of adaptive policymaking, present and future, in ensuring that innovation is accompanied by safety, inclusivity and equal market access, emphasizing the close link between food policies and regulatory solutions and the way fragmented legislative scenarios could ultimately create a "barrier to the harmonised advancement of these technologies".

The chapter "Animal Welfare as a Constitutional Value: The Debate in Comparative Law and the Emerging Role of Cultivated Meat" by Piciocchi addresses the legal and ethical topic of recognition of animal welfare as a constitutional value. In examining the process of legal recognition of animal rights, the author provides an in-depth comparative analysis of different regulatory approaches to the topic, starting with an evaluation of important constitutional case law and constitutional provisions in different countries. The author pays specific attention to the dignitarian and welfarist approaches, exploring their relevance to the debate on cultivated meat. Although certain aspects (such as the use of fetal bovine serum) could be in conflict with animal welfare, the production of cell-based meat reflects clear attention to the reduction of animal suffering and enables a quite unexpected "alliance" between animal rights groups and scientific research. In conclusion, Piciocchi reflects on the way classic themes of constitutional law are re-proposed with regard to cultivated meat and stresses the importance of evaluating the potential impact of cultivated meat on animal rights through the lens of constitutional law.

2.3 Socio-Economic and Cultural Issues

Part "Socio-Economic and Cultural Issues" of the volume is dedicated to other key cultural and socio-economic factors that warrant careful consideration in an interdisciplinary analysis of cell-based meat: firstly, religious rules that can affect and regulate food consumption; secondly, socio-economic considerations mainly regarding consumer acceptance and public opposition based on cultural tradition and techno-economic evaluations.

The chapter "Food Justice in Religious Laws: Hypotheses on Cultivated Meat" by Anello explores the vast set of religious laws, including dietary laws and specific prohibitions, that can significantly influence meat production and consumption across various cultures. The contribution offers a comparative examination of the idea of "meat security": moral issues linked to the act of "taking a life" and recognition that an animal's life belongs to God lead to eco-religious considerations on animals, linking religious rules with their ecological implications. The author then shows that cell-based meat is raising new and unprecedented issues and questions. While religious leaders will likely await further development before establishing rulings, the contribution provides a clear picture of the ongoing religious debate elicited by this innovative product across faiths and cultures, particularly with regard to Judaism, Islam, Hinduism, Buddhism and Christianity.

The chapter "Who Hates Cultured Meat? Examining Public Opposition to Cellular Agriculture" considers so-called oppositional coalitions composed of diverse stakeholders "motivated by a heterogeneous set of sometimes contrasting ideological principles and personal incentives". Broad and Glenna delve into the variety of actors, arguments and socio-technical ideologies against cell-based meat. They apply the conceptual principles and methods of consumer psychology, marketing, social studies, communications, media and the anthropology of food. The authors argue that in order to fully understand people's attitudes to new foods it is necessary to look at individual preferences and attitudes determined by collective forms of communication (in physical and digital contexts), influenced by cultural attachment to specific traditions, socio-technical imaginaries, and in some cases *conspirituality*. Careful analysis of who "hates" cell-based meat and their reasons is therefore essential for promoting participatory dialogue based on a correct understanding of the risks and potential of this innovative product.

While the chapter "Who Hates Cultured Meat? Examining Public Opposition to Cellular Agriculture" looks at factors associated with rejection, the last two chapters examine factors that can influence consumer acceptance. Burdorf and Lin-Hi propose a synthesis of the existing literature, categorizing and exploring three key factors: personal (including demographic traits, psychological characteristics and individual awareness), product-related (including health concerns, taste expectations, labelling and terminology) and external factors (such as environmental concerns, trust in the food industry and social influences). The authors also reflect on data limits, motivating and discussing the necessary evolution of research efforts, especially with regard to the importance of promoting evolving rather than static studies that examine the ways different factors "combine and influence one another". The evaluations in the first part of the chapter "Navigating Consumer Acceptance of Cultured Meat: Insights from Social Science Research" lead to an overview of the most significant practical implications for food businesses, while emphasizing the fundamental nature of interdisciplinary research and dialogue for consumer acceptance and the future of the cell-based meat market.

The chapter "Socio-Demographic and Behavioural Determinants of Consumer Attitudes to Cell-Based Meat and Insect-Flour-Based Products: An Empirical Study of Italians' Propensity to Accept New Foods for Human Nutrition and Pet Food" by Degli Antoni and Faillo represents a concrete and practical application of the research questions and methodologies emerging from previous contributions. They examine the results of empirical analysis of original data collected between October and November 2024, based on online surveys completed by 2102 participants, with a sample representing the adult Italian population in terms of age groups, gender and macro-geographical area of residence (north-west, north-east, centre and south). Socio-demographic and behavioural factors affecting Italian consumer attitudes to cell-based meat and insect-based flour are presented and explained, highlighting the important role of drivers of acceptance such as age, gender, education, political views, technological beliefs and trust in science. Based on these findings, the authors state the need for future research to "identify the determinants of acceptance of different Novel Foods in a comparative perspective", which could be crucial for the design of targeted policies.

3 Illuminating Complexity: Interdisciplinary Approach to Emerging Scientific Challenges

Technological and scientific innovation clearly highlights profound challenges for legal scholars, policymakers and legislators—namely, the pressing need to foster a strong connection and dialogue between science, law and society. In other words, "we must work to inform the public of what sciences is, that it is often full of uncertainty, and prudent decision-making requires acting on the available information."[32]

This fundamental and unavoidable task entails a twofold effort. First, it requires learning to speak "interdisciplinary". If, as has been said, "in inter-disciplinarity what is needed is the presence of people capable to ask questions",[33] then in this book we have aimed, first and foremost, to ask questions, to underline doubts, uncertainties and challenges across a range of different but interconnected domains: scientific, socio-economic, cultural, legal and ethical. Second, communicating innovative and highly scientific matters to the public means stepping out of the comfort zone and embracing "techniques of creativity and legal innovation".[34] This involves critically assessing existing legal framework and designing new regulatory solutions capable of shaping a legal landscape that can properly discipline cutting-edge technologies, new products and production methods.

With this twofold effort in mind, the present volume provides a 360-degree study on cell-based meat as a paradigmatic and compelling case-study, offering useful resources to propel informed debate on a topic that is gaining increasing momentum.

Through accessible science-based analysis of legal, ethical, scientific and economic profiles, the book therefore ensures a systematized and comprehensive study. It is grounded in the recognition of the vital role played by the academic community not only in generating knowledge but also in promoting public understanding and supporting evidence-based choices at every level of society, from individual consumers and industry players to institutional actors. This is especially necessary in a context where disinformation, fake news and ideological discourses often accompany the spread of potentially revolutionary and unprecedented discoveries.

Ultimately, the goal of this volume is, on the one hand, to bring together diverse voices in dialogue around a shared challenge; on the other, to raise awareness among different stakeholder. It aims to keep the debate on cell-based meat—as well as, more broadly, on innovation in the agri-food sector—alive and to inspire innovative answers to urgent and complex questions on the future of food systems.

References

Albisinni F (2024) Un laboratorio di innovazione permanente. Rivista di Diritto Alimentare 1:1–3

Alessandrini M, Cazzini F, Rolandi S (2024) The future of food law after and beyond the farm to fork strategy. Eur J Risk Regul Special Issue 15(2):215–221

Atkinson G, Dietz S, Neumayer E, Agarwala M (eds) (2014) Handbook of sustainable development, 2nd edn. Edward Elgar, London

Bifulco R, D'Aloia A (eds) (2008) Un diritto per il futuro. Teorie e modelli dello sviluppo sostenibile e della responsabilità intergenerazionale. Jovene, Napoli

[32] Fortin (2025), p. 153.

[33] Goldoni (2019), p. 3.

[34] Pascuzzi (2019), p. 18.

Borghi P (2023) "La" sostenibilità e "le" sostenibilità. L'UE, la sicurezza alimentare e il senso di un sistema agroalimentare più sostenibile. In: Rossi Dal Pozzo F, Rubino V (eds) La sicurezza alimentare tra crisi internazionali e nuovi modelli economici. Cacucci Editore, Bari, pp 245–260
Canfora I, Leccese VS (2024) Social sustainability as the milestone for a sustainable food system: the essential role of people working in agriculture. Eur J Risk Regul 15(2):253–264
Cerini D (2025) "From farm to fork vs. From factory to lab": Riflessioni su sostenibilità sociale e benessere animale a partire dalla L. 172/2023 in materia di alimenti e mangimi da colture cellulari. Rivista di Diritto Alimentare 1:76–103
Cuocolo L (2025) Diritto dello sviluppo sostenibile. Il Mulino, Bologna
De Grahl A, Treuil E (2024) Regulatory situation in the EU on lab-grown ingredients. Focus on ingredients imitating milk proteins. Eur Food Feed Law Rev 19(2):72–76
Di Lauro A (2025) Carne "artificiale" e meatsounding: un orizzonte cannibale possibile e la retorica del "senza". Rivista di Diritto Alimentare 1:4–16
EFSA (2020) Climate change as a driver of emerging risks for food and feed safety, plant, animal health and nutritional quality. EFSA Supporting Publications 17(6):1881E
EPRS (2024) Alternative protein sources for food and feed. https://www.europarl.europa.eu/RegData/etudes/STUD/2024/757806/EPRS_STU(2024)757806_EN.pdf. Accessed 10 May 2025
European Commission: Directorate-General for Research and Innovation (2023) Food 2030 – pathways for action 2.0 – R and I policy as a driver for sustainable, healthy, climate resilient and inclusive food systems. https://data.europa.eu/doi/10.2777/365011. Accessed 10 May 2025
FAO (2009) How to feed the world in 2050. https://www.fao.org/fileadmin/templates/wsfs/docs/expert_paper/How_to_Feed_the_World_in_2050.pdf. Accessed 10 May 2025
FAO (2018) Sustainable food systems. Concept and framework. https://openknowledge.fao.org/server/api/core/bitstreams/b620989c-407b-4caf-a152-f790f55fec71/content. Accessed 10 May 2025
FAO (2023) Food safety aspects of cell-based food. Rome. https://doi.org/10.4060/cc2353en. Accessed 10 May 2025
FAO (2025) Precision fermentation – with a focus on food. Rome. https://openknowledge.fao.org/items/527bf0bc-8d7e-4fef-8ce0-cad2c3b01a18. Accessed 10 May 2025
Fischer R, Buyel JF (2020) Molecular farming – the slope of enlightenment. Biotechnol Adv 40:107519
Formici G (2024) Un assaggio di futuro? La carne a base cellulare e il vivace dibattito regolatorio dentro e fuori i confini dell'Unione europea. BioLaw J 4:165–197
Formici G, Scaffardi L (eds) (2022) Novel foods and edible insects in the European Union – an interdisciplinary analysis. Springer, Cham
Fortin N (2025) Can science be a neutral arbiter in adjudicating trade disputes? In: Costato L, Albisinni F, Georgopoulos T (eds) European and global food law, 3rd edn. Wolters Kluwer, Milan, pp 143–154
Fracchia F (2010) Lo sviluppo sostenibile. La voce flebile dell'altro tra protezione dell'ambiente e tutela della specie umana. Editoriale Scientifica, Napoli
Goldoni M (2019) Agricultural and food law: new challenges for legal scholars. In: AIDA-IFLA (ed) Innovation in agri-food law between technology and comparison. Wolters Kluwer, Milan, pp 3–4
Gornall J, Betts R, Burke E, Clark R, Camp J, Willett K, Wiltshire A (2010) Implications of climate change for agricultural productivity in the early twenty-first century. Philos Trans R Soc Lond Ser B Biol Sci 356(1554):2973–2989
Holle M (2025) Sustainability in European agri-food law. In: Costato L, Albisinni F, Georgopoulos T (eds) European and global food law, 3rd edn. Wolters Kluwer, Milan, pp 409–419
Iacovino A, Randazzo L (2023) Climate change and food security in the One Health perspective. BioLaw J 2:293–314
Lanni S (2025) Le riflessioni sulla carne coltivata. Rivista di Diritto Alimentare 1:1–3
Meireis T, Rippl G (eds) (2018) Cultural Sustainability: perspectives from the humanities and social sciences. Routledge, London

Paganizza V (2025) Carne coltivata: una comparazione tra Australia e Unione europea, passando da Singapore. Rivista di Diritto Alimentare 1:62–75
Pascuzzi G (2019) The creativity of agri-food lawyers. In: AIDA-IFLA (ed) Innovation in agri-food law between technology and comparison. Wolters Kluwer, Milan, pp 17–25
Pitto S (2023) Cambiamento climatico e sicurezza alimentare: Dall'approccio One Health ai modelli olistici del Global South. BioLaw J 2:315–347
Pitto S (2024) La rilevanza costituzionale e globale della sicurezza alimentare. Una lettura olistica e comparata della food security. Editoriale Scientifica, Napoli
Porena D (2017) Il principio di sostenibilità. Contributo allo studio di un programma costituzionale di solidarietà intergenerazionale. Giappichelli, Torino
Sabatini F (2019) Culture as fourth pillar of sustainable development: perspectives for integration, paradigms of action. Eur J Sustain Dev 8(3):31. https://doi.org/10.14207/ejsd.2019.v8n3p31
Saija R (2025) Carni e pesci in vitro. Quali regole in vista della sostenibilità? Rivista di Diritto Alimentare 1:33–44
Scaffardi L (2023) La sicurezza alimentare ovvero come il "diritto a togliersi la fame" evolve in un mondo che cambia. DPCE Online 59(2):2167–2193
Schebesta H (2023) How to save the farm to fork strategy: a two phased approach. Eur Food Feed Law Rev 4:231–238
Shepherd A (2023) The road to 2030: is inclusive, sustainable economic transformation achievable? ODI Global. https://odi.org/en/insights/achieving-the-2030-agenda-is-inclusive-sustainable-economic-transformation-achievable/. Accessed 10 May 2025
UN (2015) Transforming our world: the 2030 agenda for sustainable development. https://sustainabledevelopment.un.org/content/documents/21252030%20Agenda%20for%20Sustainable%20Development%20web.pdf. Accessed 10 May 2025
World Commission on Environment and Development (1987) Report: our common future. https://sustainabledevelopment.un.org/content/documents/5987our-common-future.pdf. Accessed 10 May 2025

Scientific and Sustainability Challenges

What Is Cell-Based Meat? Food Safety and Sustainability Aspects

Federica Cheli, Gabriella Tedeschi, Tamil Selvi Sundaram, Davide Lanzoni, Antonella Baldi, and Carlotta Giromini

Abstract Food security faces several challenges now and in the future. As the global population is projected to reach 9–11 billion by 2050, more sustainable protein production is essential in the next 30 years. Biotechnology can play a critical role in providing humanity with nutritious, safe and healthy food while minimizing use of resources, such as energy, water and land. Cellular agriculture produces animal-based foods directly from animal cells and offers an ethical and sustainable alternative to traditional animal protein sources. In 2013, Professor Mark Post created the first cell-cultured hamburger, which was later cooked and tasted live in London. Since then, discussions on its pros and cons have become increasingly widespread, engaging researchers across various sectors. A key concerns in this debate are the safety and sustainability of cell-based meat. Food safety is a prerequisite regardless of how food is produced. Understanding the cell-based food production system is crucial for safety assessment. This chapter provides an overview of the cell-based meat production process, its challenges and safety considerations. Risks at each production stage, the role of omics technologies in research and development, and environmental sustainability concerns related to water, soil and energy use are also explored.

Keywords Cell-based meat · Cell-based meat production process · Foodomics · Food safety · Environmental sustainability

F. Cheli (✉) · G. Tedeschi · T. S. Sundaram · D. Lanzoni · A. Baldi · C. Giromini
Department of Veterinary Medicine and Animal Sciences, University of Milan, Milan, Italy
e-mail: federica.cheli@unimi.it; gabriella.tedeschi@unimi.it; tamil.sundaram@unimi.it; davide.lanzoni@unimi.it; antonella.baldi@unimi.it; carlotta.giromini@unimi.it

G. Formici et al. (eds.), *Cell-Based Meat in the European Union and Beyond*,
https://doi.org/10.1007/978-3-032-14624-3_2

1 Introduction

The United Nations (UN) and the Food and Agriculture Organization (FAO) estimate that the global population will reach 9 to 11 billion by 2050.[1] Consequently, the demand for livestock food products, including meat, milk and eggs, is expected to double by 2050. Additionally, factors such as increasing urbanization and rising average incomes in developing countries are also driving the global rise in *per capita* meat consumption and overall meat production.[2] By delivering 15% of global per capita calories and 31% of protein through meat, milk, and eggs, the livestock industry remains a cornerstone of food supply and security, although contributions differ by region.[3] A recent 2023 FAO report highlights that global meat production surged from 124 million tonnes in 2000 to 357 million tonnes in 2021, a 53% increase. Notably, meat production grew by 4% between 2020 and 2021, marking the fastest annual growth rate in the past decade.[4] Current trends in livestock production forecast significant pressure on land, soil and water resources. Escalating greenhouse gas (GHG) emissions, climate change and the growing risk of antibiotic-resistant infections also pose significant challenges to global health, food security and the sustainability of current livestock production systems.[5] To sustainably meet the nutritional requirements of a growing population, research in cellular agriculture is being conducted to develop various protein-based foods as alternatives to conventional animal proteins. This includes the development of cell-based meat (CM) and other protein-based food products, ingredients and additives derived from the cells of animals, plants and microorganisms. Over the last decade, numerous CM industries have emerged worldwide, producing CM at laboratory scale and working towards large-scale commercialization.[6] As CM production expands, it is crucial to identify potential hazards, assess risks and establish regulatory guidelines to ensure the safety, quality and consumer acceptance of CM and other alternative protein sources. This chapter provides an overview of the general processes involved in CM production, including safety considerations, environmental sustainability challenges and the use of omics technologies in its research and development. Biotechnological and technical challenges and the evaluation of security, safety and sensory/nutritional properties of CM along its entire production chain must necessarily consider the regulatory aspects for Novel Food authorisation in different countries. How CM is regulated will be a determining factor in the success and acceptance of the product. The legislative context of CM is beyond the scope of this

[1] See FAO and WHO (2023) and FAO (2009).

[2] Godfray et al. (2018).

[3] Godde et al. (2021); FAO (2017), p. 3.

[4] FAO (2023a), p. 16.

[5] Committee on Considerations for the Future of Animal Science Research et al. (2015).

[6] Kumar et al. (2021) and Jahir et al. (2023).

chapter, however those interested can find more specific information in Part "Legal and Policy Debate" of the present volume.[7]

2 Cell-Based Meat: Origin and Future

Livestock meat is a vital source of high-quality dietary protein for humans, providing all the essential amino acids with enhanced digestibility, bioavailability and nutritional value.[8] Cell-based meat, also referred to as *in vitro,* cultured or lab-grown meat, is edible meat produced in the laboratory by an innovative method that involves tissue engineering and bioprocess technologies. It can potentially reduce or eliminate the need to slaughter livestock. In other words, CM is produced by growing different types of meat cells, mainly muscle satellite cells, adult muscle cells and adipocytes, on edible three-dimensional scaffolds. These scaffolds are supplied with nutrient-rich culture media and maintained in bioreactors under precisely controlled culture conditions. The carefully optimized physicochemical environment in the bioreactors facilitates tissue growth that closely resembles the structure and composition of conventional meat.[9] The concept of CM was originally envisioned by a French chemist, Pierre Eugen Marcellin Berthelot in 1894.[10] Over more than a century, this idea has evolved from speculative theory to astounding fact. In 1931, former British Prime Minister, Winston Churchill pictured a future where "we will escape the absurdity of growing a whole chicken in order to eat the breast or wing, by growing these parts separately under a suitable medium."[11] This idea laid the groundwork for Dutch researcher, Willem van Eelen, who in the 1950s became one of the "godfathers" of CM development. His experiences during World War II, including periods of starvation, fuelled his interest in finding a sustainable way to produce meat without slaughtering animals.[12] In 2004, American businessman Jon F. Vein secured a patent for the production of tissue-engineered meat for human consumption, proposing the cultivation of muscle and fatty tissue together to create

[7] In particular, see *Cell-Based Meat in the European Union: A Regulatory Crossroads* by Corazza and Formici; *The Legal Debate on Cultured Meat Around the World: Comparative Insights from the United States, Singapore and Israel* by Pitto; *The Regulatory Landscape of Cultured Meat and Fish: Terminology and Consumer Information Provisions* by Rolandi and Sirsi; and *Animal Welfare as a Constitutional Value: The Debate in Comparative Law and the Emerging Role of Cultivated Meat* by Piciocchi in this volume. For an in-depth analysis of policy solutions see also *Cell-Based Meat and Food Policy* by Mancini in this volume. See also Cheli et al. (2024) and Lanzoni et al. (2024).

[8] Kaur et al. (2022).

[9] Post (2012), Post and Hocquette (2017), and Hocquette (2016).

[10] As carefully explained by Kirsch et al. (2023).

[11] Churchill (1931).

[12] Chase (2020).

animal meat.[13] In 2001, Willem Van Eelen, businessman Willem Van Kooten and dermatologist Wiete Westerhof collectively filed applications for patents all over the world for a specific process to produce cultured meat. Their method involved seeding a collagen matrix with muscle cells and supporting it with nutrient-rich solution to promote cell division.[14] About the same time, the US National Aeronautics and Space Administration (NASA) began exploring CM as a potential food source for astronauts on long missions, examining the feasibility of cultivating goldfish and turkey cells to produce CM in space.[15] In 2013, Prof. Dr. Mark Post from the University of Maastricht developed the world's first cultured beef burger. The burger weighed 85 g and was composed of approximately 10,000 thin strands of muscle tissue, as well as breadcrumbs, salt and natural colouring agents. It cost over $325,000 and took nearly 2 years to produce. It was publicly cooked and tasted in London, marking a significant milestone in demonstrating the possibility of CM production.[16] In the same year, Singapore became the first country to approve the sale of cell-based chicken. Between 2010 to 2021, close to 100 CM industries and start-ups emerged worldwide, with the aim of bringing CM to market. New Harvest, Memphis Meats, JUST, Modern Meadow, Aleph Farms, Mosa Meat are just some that work in collaboration with research institutions.[17] In July 2024, the French start-up Gourmey became the first company to submit a Novel Food application to the European Commission regarding the production of CM, specifically their cultivated duck *foie gras*.[18] In January 2025, Mosa Meat lodged the second application to EFSA for the cultivation of cell-based meat (beef fat) in the EU.[19] The results of the authorization procedure as well as the risk assessment evaluation provided by the European Food Safety Authority (EFSA) could have a significant impact on these companies and, in case of a positive evaluation, could bring them closer to introducing their innovative products to the European market.

---

[13] Vein (2004).

[14] Van Eelen et al. (1998).

[15] Norton (2021), p. 164.

[16] Post (2014), p. 1039.

[17] On this point, see *Cell-Based Meat and Food Policy* by Mancini in this volume. Kumar et al. (2021), Jahir et al. (2023), Cheli et al. (2024), and Lanzoni et al. (2024). The themes explored in the previous pages are based on a review and critical analysis of the academic literature as well as on the results of research activities as part of a Project funded under the National Recovery and Resilience Plan (NRRP), Mission 4, Component 2, Investment 1.1, Call for tender No. 104, published on 2.2.2022 by the Italian Ministry of University and Research (MUR), funded by the European Union—NextGenerationEU—Project Title "Cellular agriculture for sustainable and innovative food production (CELLtoFOOD)"—CUP G53D23003940006—Grant Assignment Decree No. 1048 adopted on 7.14.2023 by the Italian Ministry of Ministry of University and Research (MUR).

[18] European Institute of Innovation and Technology FOOD (EIT Food) (2024).

[19] See *Cell-Based Meat in the European Union: A Regulatory Crossroads* by Corazza and Formici in this volume. See also Ford (2025).

3 Approaches to Cell-Based Meat Production: Application of Omics Technologies

The research on CM relies on knowledge and techniques from cell and molecular biology, tissue engineering, food engineering, bioprocessing, animal and omics sciences. The term "omics" refers to sciences, the main goal of which is to identify, describe and quantify all biomolecules and molecular processes that contribute to the structure and function of cells and tissues. They are based on technologies that are now cornerstones of modern biotechnology, and offer insights into the molecular components and processes that make up a living cell.[20] Omics science includes a range of disciplines and tools focused on exploring the following aspects of cells:

- genomics is the study of genomes and the complete DNA sequences of organisms;
- transcriptomics focuses on RNA (messages encoded by DNA in the coding regions of genomes) and patterns of gene expression;
- proteomics concerns the entire set of proteins translated by RNA, fundamental for understanding cell functions occurring essentially at proteome level and enabling identification and quantification of all the proteins in a sample;
- peptidomics is a subset of proteomics aimed at identification and quantification of all the endogenous peptides in a sample and analysis of their possible biological activity;
- metabolomics is analysis of the entire set of metabolites, largely derived from the functioning of enzymes and proteins. In the context of metabolomics, specific molecules such as lipids can also be considered (lipidomics);[21]
- multiomics and systems biology are integrations and combinations of other omics technologies, providing a comprehensive view of biological systems and detailed insights into the complexities of cell function and organisms.[22]

Mass spectrometry, a sensitive, widely used, up-to-date analytical technique, encompasses four of the previously mentioned fields: proteomics, peptidomics, metabolomics and lipidomics. The application of omics techniques in food science led to coining of the term "foodomics" in 2009. Foodomics is a discipline that explores the nutritional and health aspects of food by integrated application of advanced omics technologies, in order to improve consumer well-being, health and confidence.[23] For more than 20 years, omics technologies have been extensively used to assess meat quality, including research into: (1) the mechanisms underlying meat quality; (2) the biochemical pathways behind the conversion of muscle into fresh meat; and (3) the biochemical pathways associated with manufacturing

[20] Dai and Shen (2022).
[21] *Ibid*; Balkir et al. (2021).
[22] Su et al. (2024).
[23] Cifuentes (2017), p. 1; Rocha and Silva (2023).

processes and their effects on meat products.[24] Likewise, in-depth study of the production processes of CM, including critical points and production chain safety, as well as profiling of nutritional and functional characteristics, should employ cutting-edge omics approaches that offer a high degree of precision and sensitivity.[25] Their application spans the entire production pipeline of CM, from donor animal selection and cell line development to technical, economic and social barriers to scaling and commercialization, while also minimizing the carbon footprint. Some major aspects of their practical applications are listed below and represented in Fig. 1.

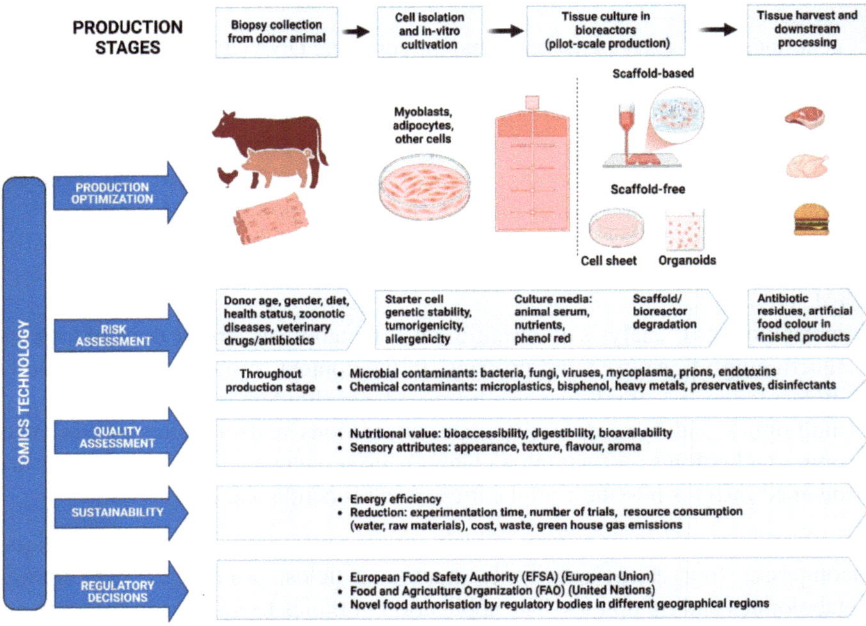

Fig. 1 Scheme of the key stages in cell-based meat production, highlighting potential risk factors at each stage and omics technology used throughout the production pipeline for optimization, risk assessment, quality control, sustainability evaluation and regulatory decisions (Figure created with BioRender.com)

[24] Gagaoua (2022).

[25] Tomiyama et al. (2020).

3.1 General Process of Cell-Based Meat Production

The production of CM typically involves four key stages that include selecting suitable cells, designing culture media and scaffolds to support tissue development, growing the tissues in bioreactors, harvesting the mature tissues and processing into various meat products, as outlined below and in Fig. 1.

3.1.1 Cell Sourcing

The isolation and selection of cells that proliferate vigorously, self-renew and differentiate are the first critical steps in designing CM production. Edible meat is about 90% muscle fibre, 9% connective and fatty tissue and 1% blood cells.[26] The major cell components of meat include myoblasts, stem cells, adipocytes, fibroblasts and hematopoietic cells. Over the years, experimental models for CM production have used different cell types, including adult stem cells, adult progenitor cells (such as satellite cells), embryonic stem cells (ESCs), induced pluripotent stem cells (iPSCs) tissue specific stem cells and primary or immortalised cell lines.[27] Stem cells are preferred for their robust self-renewal capacity and ability to differentiate into specialized cell types corresponding to their tissue or organ of origin,[28] while also producing sufficient cell densities for pilot-scale CM production. Cell lines can be obtained from cell banks, while primary cells are isolated from animals raised for meat through minimally invasive biopsy or sourced from carcasses.[29] Cells sourced for meat production can differ vastly in growth behaviour and medium requirements. Genomics, proteomics and other omic techniques are now becoming consolidated practice for mapping and characterizing these cells to achieve genetic stability, optimal growth, texture and nutrition, while also allowing standardization of the production process.[30] Genomic genotype profiling promotes understanding of genetic stability and the scalability potential of selected cells. Genomic analysis also detects any genetic drift or mutations, ensuring that the cells used in production retain the genetic traits of the original tissue.[31] On the other hand, transcriptomics, proteomics and metabolomics allow phenotype profiling to understand how genes are expressed under different conditions and to identify the underlying signalling pathways.[32] Based on these studies, it is possible to select or engineer cells with desirable traits such as faster growth rates, improved metabolic and nutrient profiles and specific textures.

---

[26] Listrat et al. (2016).

[27] Reiss et al. (2021), Lanzoni et al. (2022), and Specht et al. (2018).

[28] Su et al. (2024).

[29] Lanzoni et al. (2022) and Specht et al. (2018).

[30] Martins et al. (2024).

[31] Bonilla et al. (2024).

[32] Dai and Shen (2022) and Bonilla et al. (2024).

3.1.2 Development of Culture Media

The cell culture medium is the most important factor for maintaining cells *ex vivo*, making the development of suitable growth media critical for CM production. Muscle fibres that contribute to typical meat texture are formed by an intricate developmental process called myogenesis. Myogenesis involves two major cell growth phases: proliferation and differentiation. During proliferation, satellite cells divide into mononucleated myocytes until sufficient cell density is reached. Then they fuse with adjacent cells and differentiate into multinucleated myotubes, which eventually develop into myofibers.[33] Since cells are cultured without *in vivo* biochemical and mechanical signals from surrounding tissues, identifying and optimizing the components of the culture medium that best support these growth phases is critical. In general, the culture media contain adequate amounts of essential nutrients (glucose, proteins, amino acids, vitamins), growth factors (Insulin-like Growth factor (IGF), Fibroblast Growth Factor (FGF), Transforming Growth Factor-β (TGF-β)) and animal serum-based growth supplements (fetal bovine serum (FBS), horse serum (HS)) to stimulate *in vitro* myogenesis.[34] However, due to ethical, safety, variability, economic and environmental sustainability concerns, animal serum is increasingly replaced with commercially available serum-free media (FBMTM, TesRTM, Essential 8TM, Beefy9) or formulations containing nutrients derived from plants, microbes or food by-products.[35] All omic techniques can be used to characterize cells grown in different media to identify the optimal combination of nutrients and conditions. The final goal is selection of a medium that promotes efficient cell growth while also being sustainable, safe and cost-effective, making the process more economically viable. Metabolomic approaches in particular, allow us to understand cell health, growth rate and nutrient utilization in order to adjust culture media for efficient metabolite production. Characterization of protein profiles by transcriptomics and proteomics makes it possible to obtain optimal growth media for consistent product quality in terms of taste, texture and nutritional properties.[36]

3.1.3 Scaffold Bioengineering

Cells sourced for meat cultivation are subsequently seeded into porous two- or three-dimensional scaffolds, that provide a structural framework for cell adhesion, proliferation, muscle fibre formation and alignment into muscle tissue. Scaffolding technology plays a central role in recreating the sensory and nutritional attributes of conventional meat, including its appearance, texture and flavour. The porous

---

[33] Alheib et al. (2023).

[34] Sundaram et al. (2025), Soleymani et al. (2024), and Dayton and Hathaway (1990).

[35] Hadi and Brightwell (2021), Stout et al. (2022, 2023), and Dolgin et al. (2025).

[36] Swartz (2019).

network mimics the extracellular matrix, facilitating the exchange of nutrients, oxygen and waste within the scaffolds, which are critical for cell metabolism and for preventing the formation of necrotic cores. The polymers used for scaffolding must meet criteria such as biocompatibility, edibility, biodegradability and digestibility.[37] They are commonly derived from a wide range of natural and synthetic sources, including plants (proteins, polysaccharides), animals (collagen, gelatine, whey), marine polysaccharides (alginate, chitosan), fungi (mycelium) and other improvised biomaterials.[38] Over the years, a number of scaffolding techniques such as electrospinning, 3D bioprinting, microcarrier cultures, freeze-drying, solvent casting and plant tissue decellularization, have been increasingly explored for CM development.[39] Each technique imparts distinct characteristics to the scaffold, which in turn influences the properties of finished CM products. For instance, 3D bioprinting fabricates scaffolds with channelled structures by a precise layer-by-layer printing method. This approach is advantageous for establishing co-culture tissues that closely resemble conventional meat layers.[40] On the other hand, electrospinning creates fibre structures with diameters ranging from 10 nm to several microns, offering a high surface-to-volume ratio for achieving high cell density.[41] Recently, electrospun silk fibroin scaffold demonstrated high cytocompatibility, enhanced cell migration, scalability and cost-effective expansion of bovine satellite cells for cultivated beef production.[42] Alternatively, scaffold-free CM meat can be established using methods such cell sheet technology, which produce tissues with a thickness of 1.3–2.7 mm by stacking cultured myoblast cell layers.[43] Omics strategies, mainly transcriptomics and proteomics, can be applied in the development of edible scaffolds made from different materials. A recent study, for example, provided insights into the characteristics of a marine biopolymer-based scaffold, using a transcriptomics approach that revealed significant differences in the gene expression profiles of myoblasts grown on microstructured versus flat scaffolds. The results highlighted the crucial role of topography in cell growth and differentiation.[44] Using gene ontology and gene network modelling, they also unveiled important aspects involved in filament sliding, muscle contraction and the organization of sarcomeres.[45] In line with this, analysing and characterizing protein interactions to improve muscle fibre formation using proteomics and transcriptomics has now become a common practice in CM production.

[37] Alam et al. (2024) and Seah et al. (2022).

[38] *Ibid.*

[39] Singh et al. (2023) and Zandonadi et al. (2025).

[40] Dong et al. (2023).

[41] Alam et al. (2024) and Seah et al. (2022).

[42] Dagès et al. (2025).

[43] Tanaka et al. (2022).

[44] Bezjak et al. (2024).

[45] *Ibid.*

3.1.4 Tissue Growth in Bioreactors

The next stage of CM production involves the stimulation of tissue growth in cell-laden scaffolds or scaffold-free cell cultures through nutrition, aeration and temperature conditions in a sterile environment. This process is crucial as it enables the proliferation, maturation and development of cells into muscle tissue. Cell cultures (~10^4 cells) are initially expanded into standard culture vessels until the desired density (~10^{13} cells) is achieved, after which they are transferred to bioreactors for pilot-scale production.[46] Bioreactors are specialized automated systems, designed to provide controlled environments for stimulating cell and tissue growth. They precisely regulate key parameters, including temperature, pH and O_2/CO_2 levels, and can establish dynamic culture conditions as in the biological environment. This includes the application of chemical, electro-magnetic and mechanical stimulation to enhance tissue growth. Bioreactors are equipped to maintain a steady nutrient supply and remove metabolic waste, such as lactate, which could inhibit cell growth.[47] Different types of bioreactors, such as stirred tank, rocking bioreactor, rotating wall vessel, spinner flask and flow perfusion, are commonly employed in the bioprocess industries for cell culture applications.[48] In particular, stirred tank bioreactors, which can operate in batch, fed-batch and continuous modes, are more suitable for scaling up (~20,000 L) CM production. They are advantageous for achieving time-averaged mixing by mechanical, pneumatic or hydraulic agitation, which is necessary for uniform mass transfer and nutrient transport throughout the cell-laden scaffolds, as well as homogenous tissue development.[49] Bioprocess operations should be optimized for scaling and commercialization of the products. Omics data plays a major role in fine-tuning bioreactors and continuously monitoring growth conditions in order to provide real-time data on cell health and metabolism. This enables process adjustments to ensure consistent quality and yield. For instance, cell density is a crucial operational target, as it directly impacts bioreactor yield and running costs.[50] It is therefore essential to use metabolomics to monitor the production of growth-inhibiting metabolites, such as ammonia and lactate, derived from glucose and amino acid metabolism, and to develop media formulations that maintain rapid growth while minimizing metabolite production.[51] Furthermore, Genome-Scale Models (GEM) specific to each breed of interest are particularly promising for analysing cell metabolism in fed-batch culture experiments.[52] They are based on computational modelling to identify the essential genes to consider in order to probe and optimize cell metabolism. Unfortunately, since no

---

[46] Post et al. (2020).

[47] De Carvalho et al. (2024).

[48] Oragui et al. (2011) and Rosser and Thomas (2018).

[49] Allan et al. (2019).

[50] Martins et al. (2024).

[51] *Ibid.*

[52] Romero and Boyle (2023).

cell-line-specific metabolic models for CM are publicly available,[53] it is necessary to exploit closely related models. Future efforts should focus on filling this gap in the CM field by leveraging the data obtained by omics approaches.

3.1.5 Tissue Harvest and Downstream Processing

The tissue is harvested once it has matured and reached quantities sufficient for downstream processing into various meat products. Harvesting is carried out under aseptic conditions to prevent any contamination and maintain food safety standards.[54] Depending on scaffolding technique and material, tissue constructs may need to be separated from the scaffolds or incorporated along with the tissue mass in the finished products.[55] In scaffold-free cultivation systems, such as cell sheet technology, stacked cell layers can be harvested directly from culture plates or substrates. In scaffold-based cultivation, tissue can be harvested by different methods, including mechanical removal, enzyme digestion, or by thermal or pH modification of scaffold materials with specific properties. These materials allow tissue detachment by inducing a reversible structural change in the scaffolds through variation in temperature or pH.[56] Harvested tissue may be washed to remove any residual culture medium before processing to enhance sensory attributes such as appearance, taste, texture and flavour. This step includes shaping, seasoning, cooking and blending with fats, myoglobin, or other additives to produce commercially viable meat products such as minced meat, nuggets or hamburgers.[57] The processing phase should essentially conclude with packaging, labelling, quality and safety assessments and pre-market authorization in line with the Novel Food guidelines set by the EFSA, FAO and relevant authorities in different countries and economic zones, before the products can be distributed to markets.[58]

---

[53] *Ibid.*

[54] Ben-Arye and Levenberg (2019).

[55] Smith-Uchotski and Wanjiru (2023).

[56] *Ibid*; Kadim et al. (2015).

[57] Kadim et al. (2015) and Siddiqui et al. (2022).

[58] See FAO and WHO (2023), FAO (2009), Cheli et al. (2024), Lanzoni et al. (2024), and EFSA (2024). See also *Safety Assessment of Cell-Culture-Derived Food: The Role of the European Food Safety Authority* by Dall'Asta and Gjoni in this volume; for an in-depth analysis of the regulatory aspects related to the authorisation process, see *Cell-Based Meat in the European Union: A Regulatory Crossroads* by Corazza and Formici and *The Legal Debate on Cultured Meat Around the World: Comparative Insights from the United States, Singapore and Israel* by Pitto, both in this volume.

3.2 *Safety and Quality Challenges Associated with Cell-Based Meat and Its Production*

Authenticity, safety and quality are the three key food attributes to consider, whether the food is cultured or conventional. There are significant gaps in our understanding of the risks that may arise in the different stages of CM production, as outlined in the previous section. Factors such as culture medium supplements (animal serum), antibiotics, chemicals, additives, genetic modification of cell lines, pathogen exposure and scaffolding materials can all introduce health hazards. These must be carefully evaluated before commercialization to ensure the safety and quality of CM products.

3.2.1 Risk Factors in Cell Sourcing

When obtaining biopsies to establish primary cell cultures for CM production, it is crucial to consider the age, gender and diet of the donor animals. These factors are reported to significantly influence the density of stem cells, such as satellite cells, in muscle tissue. Viral contaminants and prions in donor animals can also be transmitted to humans through cell cultures. Most viruses are reportedly host-specific with limited possibility of causing cross-infection, however they can mutate and infect different hosts.[59] The overall health status of the animals and the presence of any zoonotic risk factors or veterinary drugs, including antibiotics, must therefore be monitored in donor animals, biopsy samples and the final CM products.[60]

Cell lines established through genetic reprogramming are increasingly explored for CM production. These modified cells can divide indefinitely without the need for frequent tissue sampling. The technology of iPSCs is an emerging method for reprogramming somatic cells to make them pluripotent through genetic modification by means of viral vectors, chemicals or episomal reprogramming.[61] Note, however, that these modifications can involve unintended risks, including possible carcinogenicity or allergenicity of the final products.[62] Prior iPSCs derived from food animals, including cattle, goats, sheep and pigs, formed teratomas in addition to having pluripotency.[63] The initial cells used for CM production must therefore undergo comprehensive analysis using different omics technologies as discussed earlier in this section, in order to monitor genetic stability and assess tumorigenicity.

---

[59] Smith-Uchotski and Wanjiru (2023).

[60] Cheli et al. (2024), Lanzoni et al. (2024), and Smith-Uchotski and Wanjiru (2023).

[61] Nunes et al. (2025).

[62] Singh et al. (2023).

[63] Verma et al. (2022).

3.2.2 Risks Related to Culture Medium Supplements

Culture medium growth supplements are conventionally derived from animal serum, sourced from fetal, newborn or adult cattle and horses. In particular, fetal bovine serum (FBS) is widely used as a universal growth supplement to promote cell division and differentiation. It is a by-product of the meat industry, obtained from the blood of calf fetuses after the slaughter of pregnant cows. Due to ethical concerns, batch-to-batch variability and high costs, FBS is considered unsuitable for CM production.[64] Commercial batches of FBS from different geographical locations have been screened for numerous known and emerging species of bovine viruses with single- or double-stranded DNA or RNA genomes.[65] Other contaminants, such as mycoplasma, endotoxins and prions have also been detected in FBS.[66] Efforts to replace FBS with natural (plant, insect, microbe) and recombinant derivatives are ongoing,[67] but require further extensive safety and efficacy testing before use in CM production. It is also essential to consider the safety limits of other nutritional additives, such as hormones, growth factors, vitamins and minerals, which are present in conventional meat but may be absent in CM and *vice versa*.[68] Identifying and mitigating the risks associated with nutritional hazards is crucial for ensuring the safety of final CM products.

3.2.3 Microbial Contaminants and Antimicrobial Usage

Although cell-based meat is produced in sterile environments, contamination by microorganisms such as bacteria (*Salmonella, Campylobacter, Escherichia Coli, Listeria*), fungi, viruses, mycoplasma and endotoxins, remains a significant concern. While the main objective is to ensure sterility, this is more easily achieved at lab-scale than in large-scale production. Contamination risks can arise in the biopsy stage (transport, storage), or during formulation of medium with animal serum, while non-sterile equipment or handling errors can occur at any stage of production. Even traces of contamination can impair overall product safety and quality.[69] To

---

[64] EFSA (2017), Van der Valk and Gstraunthaler (2017), and Chelladurai et al. (2021).

[65] Xia et al. (2011) and Paim et al. (2021).

[66] Xia et al. (2011), Paim et al. (2021), Gstraunthaler (2003), and Pecora et al. (2020).

[67] Chelladurai et al. (2021), Stout et al. (2022, 2023), and Dolgin et al. (2025).

[68] Cheli et al. (2024), Lanzoni et al. (2024), and Smith-Uchotski and Wanjiru (2023). The themes explored in the previous pages are based on a review and critical analysis of the academic literature as well as on the results of research activities as part of a Project funded under the National Recovery and Resilience Plan (NRRP), Mission 4 Component 2 Investment 1.3—Call for proposals No. 341 of 15 March 2022 of Italian Ministry of University and Research funded by the European Union—NextGenerationEU; Project code PE00000003, CUP D93C22000890001, Project title "ON Foods - Research and innovation network on food and nutrition Sustainability, Safety and Security – Working ON Foods".

[69] See FAO and WHO (2023), FAO (2009), and Smith-Uchotski and Wanjiru (2023).

mitigate microbial contamination, various antibiotics and fungicides, such as penicillin, streptomycin and gentamycin, are commonly used in cell cultures, particularly in preclinical research and development. Alternatively, culture media can be supplemented with chemical preservatives such as sodium benzoate to prevent yeast and fungal contamination.[70] While many CM industries are increasingly adopting antibiotic-free cell cultivation methods (filtration, heat sterilization, UV irradiation), some continue to use them, whereas others employ limited quantities during tissue biopsy of animals and cell isolation.[71] The presence of antibiotic residues in final CM products may contribute to antimicrobial resistance in consumers. FAO reports recommend diluting antibiotics to very low concentrations or washing the final products of large-scale production so that antibiotic levels are safer for consumers.[72] Strict quality control such as routine microbial testing, safety limit assessments for antimicrobial applications and use of pathogen-free materials, are also essential to ensure product safety.

3.2.4 Chemical Contaminants and Other Important Considerations

Cell-based meat production uses various non-biological components, including scaffolds, plasticware, bioreactor components, disinfectant and cleaning solutions, all of which may release chemical contaminants. Heavy metals such as lead, arsenic and cadmium may leach from bioreactor materials, culture media components and unfiltered water or air, potentially accumulating in the final CM product.[73] Plastic lab supplies (flasks, spray bottles, pipettes, single-use bioprocess containers), storage and packaging materials can release toxic leachables such as microplastics and bisphenols, raising concerns about endocrine-disrupting effects.[74] Other chemical contaminants may arise from cell culture additives, including pH indicators (phenol red), cryoprotective agents (dimethyl sulfoxide, formamide, sorbitol, inulin), media supplements (dexamethasone, indomethacin, xanthine, IBMX), stabilizers, artificial food colours and antifoaming agents, which could introduce unintended compounds that affect product safety. Materials used in scaffolds, whether synthetic or natural, may degrade and release harmful substances during cell maturation or processing.[75] To mitigate these risks, comprehensive contaminant testing, strict adherence to regulatory guidelines and good manufacturing practices are necessary before CM products enter the market. Implementation of Hazard Analysis and Critical Control Points (HACCP)-based food safety plans, established by the National Advisory

[70] See FAO and WHO (2023) and FAO (2009).

[71] See FAO and WHO (2023), FAO (2009), and McNamara and Bomkamp (2022).

[72] See FAO and WHO (2023) and FAO (2009).

[73] See FAO and WHO (2023) and FAO (2009).

[74] See FAO and WHO (2023), FAO (2009), Smith-Uchotski and Wanjiru (2023), and De Paula and Alves (2024).

[75] See Cheli et al. (2024), Lanzoni et al. (2024), and Zandonadi et al. (2025).

Committee on Microbiological Criteria for Foods (NACMCF) and approved by the FAO and EU Commission, can also help identify and mitigate risks associated with contaminants, toxins, allergens, chemicals and nutrient imbalances in CM products.[76]

3.2.5 Meat Quality and Nutritional Content Assessment

Preliminary data suggests that CM may not inherently match conventional meat in terms of nutrient composition. For example, comparisons between cultured cells and conventional meats from the same species have revealed significant differences in amino acid ratios.[77] A key objective of CM is therefore to match the nutritional value of conventional meat, while also offering the opportunity to enhance it. Multiomic approaches, including proteomics, metabolomics and lipidomics, are now applied to meat quality traits in many studies.[78] Proteomics can identify the quality and abundance of total proteins expressed in cell cultures, ensuring that they match that of conventional meat. Metabolomics and lipidomic profiling can determine the levels of essential nutrients like vitamins, minerals and fatty acids in CM. For example, mass spectrometry-based omics techniques have been used to investigate meat quality in poultry meat research.[79] Recent multiomic analysis and GEM of *Bos taurus* cattle have been used to optimize CM production.[80] Starting with bovine satellite cells, the researchers performed differential gene expression and gene set enrichment analyses using the RNA-seq data and identified the key regulators of muscle cell proliferation. Then, using human GEM as a template, they generated a bovine-specific model that enabled identification of muscle growth pathways. Using this information, they designed optimal culture media to support cell proliferation and differentiation.[81] Besides offering significant insights into understanding and modifying the nutritional content of CM products, multiomics can also provide important information on the molecules involved in fundamental properties of animal meat that should be present in CM. It is well known that protein profiles influence quality attributes such as colour, tenderness and flavour of meat, that can easily influence consumer meat purchase decisions and eating satisfaction.[82] Moreover, meat colour and tenderness are influenced by the post-mortem metabolic process, specifically the activity of proteolytic enzymes in the post-mortem period, which are extensively researched by proteomics approaches to unveil mechanisms that are not yet completely understood and can be of paramount importance for obtaining a palatable and satisfactory CM product. Flavour is another meat

---

[76] See FAO and WHO (2004, 2023) and FAO (2009).

[77] Stout (2023).

[78] Gagaoua et al. (2022), Zhou et al. (2024), Lee et al. (2024), and Power et al. (2023).

[79] Zhou et al. (2024).

[80] Lee et al. (2024).

[81] *Ibid.*

[82] Shackelford et al. (2001), Troy and Kerry (2010), and Ramanathan et al. (2022).

attribute that significantly impacts eating satisfaction and can be influenced by post mortem aging.[83] The molecules corresponding to specific flavour can be characterized by proteomic, peptidomic and metabolomic analyses. Although meat is mainly composed of skeletal muscle tissue, fat is also essential for texture and flavour. The characterization of lipid profile by lipidomics is often undertaken to identify the key regulators of adipogenesis[84] and to imitate the lipid composition of animal fat in CM.[85] Lipidomics can also be used to explore the fat distribution and marbling process that determines food palatability and taste. Based on multiomics research, CM can be tailored to meet specific taste, health and dietary needs, possibly even surpassing the nutritional value of conventional meat by incorporating functional or therapeutic food supplements.[86]

3.2.6 Meat Authentication, Safety and Regulatory Compliance

Food safety and authenticity rely on multiomics methods to implement qualitative discrimination and precise quantitative analysis. This includes targeted and non-target screening, as well as analyses ranging from multi-component to full component analysis to address these issues.[87] Omics sciences also play an important role in supporting regulatory compliance aspects for CM products by ensuring high standards of product quality, production chain safety and material/process traceability. Omics techniques can also ensure that CM products meet nutritional labelling requirements and standards set by food regulatory authorities. Providing accurate nutritional information is also essential to gain consumer trust and acceptance, especially for those with specific dietary needs or preferences.

3.2.7 Machine Learning Application

Although the application of machine learning to CM is still in its early stages, it has the potential to streamline experiments, predict optimal outcomes and significantly reduce experimentation time and resource consumption. These advances have important implications for environmental sustainability and cost reduction. In particular, machine learning can process vast amounts of multi-omics data, facilitating the identification and characterization of new cell lines, the design of optimized culture media and the enhancement of food processing techniques.[88] One of the most critical aspects of cell culture is the growth medium, which provides the essen-

---

[83] Garmyn et al. (2020).

[84] Martins et al. (2024).

[85] Wu et al. (2021).

[86] Stout (2023).

[87] Su et al. (2024).

[88] Todhunter et al. (2024) and Kamalapuram and Choudhury (2024).

tial nutrients for cell growth. Optimizing medium composition is crucial to improving yield, quality and process efficiency. It involves selecting the best concentrations of nutrients, amino acids, growth factors and other biochemical substances to support *in vitro* cell growth. The factors influencing the medium are complex and interdependent. Traditionally, an empirical approach has been used to optimize these parameters. Traditional experiments are time-consuming and resource-intensive. Machine learning methods offer a more efficient approach, allowing the creation of predictive models based on experimental data and thus reducing the number of physical experiments needed to find the optimal combination of medium components.[89]

Beyond medium optimization, machine learning also plays a pivotal role in scaffold design and production. Scaffolds provide the structural framework to support cell adhesion, proliferation and differentiation, mimicking the extracellular matrix of natural tissues.[90] Machine learning can assist in designing scaffolds with ideal porosity, mechanical properties and biocompatibility by analysing vast datasets of biomaterials and tissue growth dynamics. Machine learning-driven computational models can also predict how different scaffold architectures influence cell behaviour, leading to more efficient and scalable production methods. Advanced techniques such as generative design algorithms and reinforcement learning can further refine scaffold fabrication, ensuring optimal nutrient diffusion and tissue maturation while reducing trial-and-error iterations in the lab. By integrating machine learning in medium optimization and scaffold formation, the cultivated meat industry can accelerate research and development, enhance production efficiency and contribute to a more sustainable and scalable alternative to conventional meat production.

3.3 Cellular Agriculture: A Promising Approach to Environmental Sustainability Challenges

3.3.1 Ongoing Trends in Climate Change, Land Use, and Water Use in Global Livestock Sector

The World Meteorological Organization (WMO) reported that 2024 was the warmest year on record, with global temperatures rising 1.55 °C above pre-industrial levels.[91] This alarming increase is partially driven by GHG emissions from livestock supply chains, which contribute approximately 14.5% of total anthropogenic emissions as per 2023 FAO reports.[92] This report was generated by the FAO's latest

[89] *Ibid*; Escobar et al. (2021).

[90] Cinici et al. (2024).

[91] *Ibid*; World Meteorological Organization (WMO), https://wmo.int/news/media-centre/wmo-confirms-2024-warmest-year-record-about-155degc-above-pre-industrial-level. Accessed 10 August 2025.

[92] WMO (2025) and FAO (2023b).

Global Livestock Environmental Assessment Model (GLEAM), a geospatial framework that simulates the environmental impact of livestock systems to provide a comprehensive assessment of GHG emissions from livestock agrifood systems, comprising parameters as farm gate, land-use change (LUC) and supply chain processes.[93] As per the GLEAM assessment, methane (CH_4) accounts for approximately 50.3% of emissions (enteric fermentation, manure, rice feed production), followed by carbon dioxide (CO_2) 26.2% (feed production, pasture expansion, energy use, post farmgate) and nitrous oxide (N_2O) 23.3% (crop residues, manure and fertilizer use).[94] Notably, cattle alone contribute to over 60% of the global livestock emissions. It was also reported that among animal proteins such as meat, milk, and eggs, two-thirds of the emissions are associated with meat production across all species. Besides, one-third of emissions emerges from the production, processing, and transport of feed (see footnote 89). The livestock industry also consumes roughly 35% of total cropland and 20% of total blue water for fodder production.[95]

According to reports of the Organization for Economic Cooperation and Development (OECD), global GHG emissions are expected to increase by a further 50% and water demand by 55% by 2050 in order to sustain the growing population.[96] Sustainable animal production not only involves minimizing negative environmental impacts but also ensuring that animals are treated ethically throughout their lives. Practices that prioritize animal welfare, including provision of adequate living space, proper nutrition and reduction of pain and disease, can improve overall productivity and lower resource use.[97]

3.3.2 Addressing the Environmental Challenges of Livestock Food Production Through the EU Sustainable Development Goals

As per the European Commission's 2023 report "Sustainable development in the European Union",[98] the FAO and EU are increasingly addressing various environmental challenges including those associated with livestock food production systems by the introduction of 17 Sustainable Development Goals (SDGs) and 169 related targets to end poverty, protect the planet, and ensure prosperity and peace. The SDGs were first adopted by the UN General Assembly (UNGA) as part of the 2030 Agenda for Sustainable Development in September 2015,[99] which later integrated into the European policy framework in 2016 and they continue to guide the

---

[93] FAO (2023c); FAO (2017), p. 3.

[94] *Ibid.*

[95] FAO (2019) and Weindl et al. (2017).

[96] OECD (2012).

[97] Pulina and Bertoni (2023) and Pulina (2020). On this delicate topic, see also *Animal Welfare as a Constitutional Value: The Debate in Comparative Law and the Emerging Role of Cultivated Meat* by Piciocchi in this volume.

[98] European Commission (2023).

[99] UNGA (2015).

global efforts to attaining sustainability to this day.[100] Following these initiatives, the European Commission has adopted numerous other policies under the SDGs, such as the European Green Deal (adopted in December 2019),[101] EU Biodiversity Strategy for 2030 (adopted in May 2020),[102] 2030 Climate Target Plan (adopted in September 2020),[103] EU Zero Pollution Action Plan (adopted in May 2021),[104] and the New Circular Economy Action Plan (adopted in March 2020),[105] all of which emphasizes responsible resource use and sustainable food production practises, while becoming a climate-neutral continent by 2050.[106] Another important initiative is the EU Farm to Fork Strategy,[107] adopted in May 2020 as the heart of European Green Deal for redesigning the food system through new technologies and scientific discoveries to provide innovative solutions that ensure: Sustainable Food Production, Sustainable Food Processing and Distribution, Sustainable Food Consumption, and Food Loss and Waste Prevention.[108] Besides, the FAO Livestock Environmental Assessment and Performance (LEAP) partnership,[109] a multistakeholder initiative involving government organizations, private sectors, civil society organizations, and industries worldwide was established between the period of 2012 to 2021, and later extended through 2024. This partnership aims to accelerate the sustainable development of livestock sector through a science-based approach, focusing on research and development of methodological guidelines to enhance its productivity and

---

[100] European Commission (2020, 2021).

[101] Communication from the Commission to the European Parliament, the European Council, the Council, the European Economic and Social Committee and the Committee of the Regions—The European Green Deal, COM(2019)640 final.

[102] Communication from the Commission to the European Parliament, the European Council, the Council, the European Economic and Social Committee and the Committee of the Regions—EU Biodiversity Strategy for 2030 Bringing nature back into our lives, COM(2020)380 final.

[103] Communication from the Commission to the European Parliament, the European Council, the Council, the European Economic and Social Committee and the Committee of the Regions—Stepping up Europe's 2030 climate ambition Investing in a climate-neutral future for the benefit of our people, COM(2020)562 final.

[104] Communication from the Commission to the European Parliament, the European Council, the Council, the European Economic and Social Committee and the Committee of the Regions—Pathway to a Healthy Planet for All—EU Action Plan: 'Towards Zero Pollution for Air, Water and Soil', COM(2021)400 final.

[105] Communication from the Commission to the European Parliament, the European Council, the Council, the European Economic and Social Committee and the Committee of the Regions—A new Circular Economy Action Plan—For a cleaner and more competitive Europe, COM(2020) 98 final.

[106] European Commission (2020, 2021).

[107] Communication from the Commission to the European Parliament, the Council, the European Economic and Social Committee and the Committee of the Regions—A Farm to Fork Strategy for a fair, healthy and environmentally-friendly food system, COM(2020) 381 final, part of the European Green Deal, COM(2019) 640 final.

[108] European Commission (2020, 2021).

[109] https://www.fao.org/partnerships/leap/en. Accessed 10 August 2025.

sustainability while addressing key environmental impacts on GHG emissions, nutrient loss, resource depletion, water use efficiency, and biodiversity.[110]

3.3.3 Alternative Proteins Sources as Sustainable Foods

There is a growing global interest in alternative sources of dietary proteins such as plants, algae, insects, microbes, and cultivated foods as replacements or partial substitutes for conventional animal proteins due to their lower environmental impact and potential for high nutritional value.[111] Unlike conventional farming, production of these alternatives is often associated with lower water and land use, reduced GHG emissions, less waste generation, and the ability to be cultivated in a variety of environments, making them a promising solution for reducing the environmental footprint of food production.[112] However, the energy demands associated with the production of cultivated meat and microbial fermentations largely vary, with some reports indicated that they even exceed the energy requirements of conventional farming systems.[113] According to the 2024 report of the European Parliamentary Research Service (EPRS) on "Alternative protein sources for food and feed", about 13 million metric tonnes of total alternative proteins, including plant-based alternatives were consumed in 2020, accounted for approximately 2% of the animal protein market.[114] The exact contributions from individual sources as cultivated meat, algae, insects, and microbially-fermented products are unknown but are estimated to represent a smaller share of the total alternative protein market globally and in EU.[115] Another study projected that the global alternative protein market will continue to grow steadily between 2020 to 2030, with the current market value at about 20 billion USD (~2% of global meat market) in 2024 is expected to reach 150 billion USD (~10% of global meat market) by 2030.[116]

The shift towards alternative proteins for dietary requirements closely aligns with several SDGs, particularly the Farm to Fork strategy that aims to reduce the dependency on intensive animal agriculture, while supporting the transition to a more sustainable and resilient food system.[117] As per the EPRS report, alternative proteins are categorized into three groups, "(1) plant-based alternatives to conventional animal proteins, (2) protein sources that are relatively new to the EU but long

[110] FAO (2024).

[111] Medeiros et al. (2024) and D'Almeida and de Albuquerque (2025).

[112] Majumder et al. (2024) and D'Almeida and de Albuquerque (2025).

[113] EPRS (2024) and Sinke et al. (2023).

[114] *Ibid.*

[115] EPRS (2024).

[116] Medeiros et al. (2024).

[117] European Commission (2020, 2023).

used in other cultural contexts, and (3) entirely novel protein sources for both human and animal diets".[118]

4 Environmental and Sustainability Aspects of Cultivated Meat Production

4.1 Life Cycle Assessment of Land Use, Water Use, and Greenhouse Gas Emissions in Cultured Meat Production

Innovative approaches like cellular agriculture for livestock products such as meat and milk are emerging as a promising solution to meet the future demand for animal proteins. These initiatives focus on leveraging technologies and practices to sustainably enhance food production in order to sustain the growing population while preserving animal welfare.[119]

Although CM is not yet commercially available on a pilot-scale, several reports based on its ex-ante Life Cycle Assessment (LCA) provide valuable insights into its environmental footprint.[120] LCA is a method used to assess the environmental impacts of a product or service throughout its life cycle from production to waste management.[121] Accordingly, a recent 2023 LCA analysis by CE Delft, based on data collected from 15 companies involved in CM development and supply chains (chicken, pork, beef) has projected that cellular agriculture could convert crops into meat three times more efficiently than chickens, with significantly lower agricultural land use and GHG emissions. As CM production is devoid of manure, it also avoids associated nitrogen runoff and air pollution.[122] In support of this data, estimations of other studies suggest that transitioning to cellular agriculture by 2050 could potentially reduce annual GHG emissions by 52–96%, land use by 83–99% and water use by 82% compared to conventional livestock production.[123] However, these water consumption estimates seems to fluctuate as discussed in a 2025 report.[124]

Water used for CM production (culture medium production, cleaning, sterilization, evaporation losses) is 367–521 L/kg[125] compared to 313 L/kg for producing conventional chicken, 459 L/kg for pork and 550–700 L/kg for beef, which is not

---

[118] EPRS (2024).

[119] Eisner (2024) and Swartz and Bomkamp (2024); see also Post (2012); Post and Hocquette; Hocquette (2016).

[120] Tuomisto and de Mattos (2011), Tuomisto (2019), Sinke et al. (2023), and El Wali et al. (2024).

[121] Tuomisto et al. (2019).

[122] Tuomisto et al. (2011) and El Wali et al. (2024).

[123] *Ibid.*

[124] Hocquette et al. (2025).

[125] Tuomisto et al. (2011).

particularly promising in terms of water conservation.[126] The parameters used to assess water consumption in conventional meat production in the above study may have been limited. For example, the beef production data only considered water used for livestock housing, crop irrigation and animal consumption.[127] Incorporating additional factors, such as water use in slaughterhouses, processing and the supply chain, could significantly alter these estimates. On the other hand, the land use estimates are reasonable as bioreactors for CM production require less ground space when built with vertical stacking and they also eliminate land use for grazing. Building production plants near markets could further reduce transport costs and associated GHG emissions.[128] The LCA analysis also revealed variations in the profile of GHG emissions, with CO_2 predominating in CM production, whereas CH_4 and N_2O are the primary emissions in conventional livestock farming. The low CH_4 and N_2O emissions of CM are attributed to the absence of enteric fermentation and fertilizer use for fodder production, whereas the high CO_2 emissions are related to energy consumed (electricity, heat, or gas) for bioreactor operation (temperature regulation, cooling, aeration, agitation), production of culture medium ingredients (recombinant proteins, hydrolysates, amino acids synthesized microbially or chemically), as well as CO_2 emitted by aeration and pH control.[129] Total GHG emissions are lower for CM, but previous stimulations have indicated that cumulative CO_2 emissions may contribute more to global warming than conventional farming in the long run. This is because CH_4 emissions from livestock break down and do not persist in the atmosphere for as long as CO_2.[130] A recent study indicated that CO_2 has a lifetime of more than 100 years in the atmosphere compared to only about 12 years for CH_4.[131] Additionally, the high energy consumption of CM production remains a subject of debate. The energy demand of CM production is projected to surpass that of conventional livestock production, most of it attributed to production plant operation (>70%) and the chemical production of culture medium ingredients (~25%).[132] Adopting renewable energy sources and optimizing production processes, such as through omics technology,[133] could help mitigate these challenges.[134] The 2023 study of CE Delft as reported earlier has highlighted the critical role of renewable energy in determining the environmental sustainability of CM production. Accordingly, this study stimulated the use of renewable energy mixes, specifically 50% wind-based and 50% solar-based electricity, along with heat from geothermal sources for CM production and found that its CO_2 footprint is lower than that of beef

[126] Hocquette et al. (2025) and Doreau and Corson (2017).

[127] *Ibid.*

[128] Bhat et al. (2015).

[129] Sinke et al. (2023) and Tuomisto et al. (2022).

[130] Tuomisto and de Mattos (2011) and Sinke et al. (2023).

[131] Lynch and Pierrehumbert (2019).

[132] Sinke et al. (2023).

[133] Romero and Boyle (2023).

[134] Tuomisto et al. (2011), Sinke et al. (2023), and Tuomisto et al. (2014, 2022).

and, depending on the energy mixes, comparable to that of chicken and pork.[135] This analysis also noted that renewable energy integration is not only an environmental imperative but also a strategic consideration for scaling the CM industry. Since energy consumption is one of the largest contributors to CM's LCA impacts, decarbonizing the energy supply can substantially improve its long-term sustainability outcomes.

4.2 Modelling Feedstock Impacts on Environmental Sustainability of Cultured Meat Production

Another important aspect that regulates the environmental sustainability of CM production is the feedstock or otherwise the raw materials used in formulating cell culture scaffolds and culture media for supporting cell proliferation and differentiation.

In conventional livestock production, feedstock generally refers to crops such as corn, soy, feed crop byproducts, and other plant-based ingredients, which have significant impacts on land use, water consumption, and GHG emissions.[136] Similarly, in CM systems, feedstock for culture medium production can include sugars, amino acids, peptides, vitamins, salts, and growth factors, which are derived from industrial, microbial, or agricultural sources such as hydrolysates of rice, soy, yeast, cyanobacteria, and other plants.[137] The environmental profile of these feedstocks varies widely: for example, plant-derived hydrolysates may have higher land use impacts and GHG emissions, whereas nutrients produced via microbial fermentation can have comparatively low environmental impacts.[138] The first LCA study published in 2011, modelled the use of cyanobacteria hydrolysate (nutrient and energy source) as a sustainable feedstock for muscle cell cultivation in three different geographical regions (Spain, Thailand, California). In this analysis, diesel was used as the fuel for cyanobacteria operations and transportation of biomass, while electricity mixes (as per geographical region) were used for muscle cell cultivation and sterilization process. The result of this study showed that to produce 1000 kg of CM, approximately 26–33 GJ of energy, 367–521 m^3 of water, and 190–230 m^2 of land area are required, with 1900–2240 kg CO_2-eq of GHG emissions. When these estimates were compared with conventional European meat (beef, sheep, pork, and poultry), they represented approximately 7–45% lower energy use (except for poultry), 78–96% lower GHG emissions, 99% lower land use, and 82–96% lower water use depending on the meat product.[139]

[135] Sinke et al. (2023).
[136] Wang et al. (2024) and Sabia et al. (2025).
[137] Mattick et al. (2015) and Specht et al. (2018).
[138] Lynch and Pierrehumbert (2019).
[139] Tuomisto et al. (2011).

A following study by the same research group in 2014 assessed three alternative feedstocks such as cyanobacteria, wheat, and corn hydrolysate as sources of nutrient and energy for muscle cell cultivation in a hollow fiber-based bioreactor system.[140] Among the three feedstocks, cyanobacteria exhibited the lowest GHG emissions (611 kg CO_2-eq) and land use (0.046 ha), wheat had the lowest water use (71.8 m^3), and corn showed the lowest primary energy requirement (3.9 GJ) to produce 1000 kg of CM. Upon comparison with conventional livestock products, the energy requirement for CM was similar to that of beef, the GHG emissions and land use were lower (~95%) than all of the livestock products (beef, sheep, pork, and poultry), while, water use (~45–50%) was similar to that of poultry.[141]

Furthermore, a recent 2025 study established a linear programming model to evaluate the agricultural land use efficiency for different agricultural feedstocks for optimizing the environmental sustainability of cultivated pork production in southern Germany using solar energy. In the model, on a 100 ha plot, soybeans were cultivated on 50 ha and rapeseed on 20 ha, with varying propositions of winter wheat (7.59–13.33 ha), maize (0–9.63 ha), sunflower (0–13.33 ha), and potato (3.33–8.57 ha) in four different crop rotational scenarios. The results provided valuable insights into plant-based feedstocks and indicated that plants did not improve either land use efficiency or climate change impacts compared with conventional pork production. The study further highlighted the importance of exploring alternatives such as sourcing cell culture media from precision fermentation instead of relying solely on crops.

These findings suggest that, just as feed formulation strategies in livestock farming have been key to reducing environmental impacts, strategic feedstock management in CM production will play a pivotal role in determining its long-term environmental sustainability.[142]

4.3 *Waste Management in Cultured Meat Production*

Waste management in CM production is a crucial aspect for ensuring environmental sustainability and economic efficiency. Unlike conventional livestock farming, which generates large amounts of manure and slaughter by-products, CM production primarily generates waste in the form of used cell culture medium, cell debris and bioreactor materials.[143] It is reported that to produce 1 tonne of CM, approximately 20,000–140,000 L of proliferation and differentiation culture medium are required, depending on the operation size of the bioreactor plant. Cells need to be replenished with fresh culture medium every two days, further contributing to the

[140] Tuomisto et al. (2014).

[141] *Ibid.*

[142] Kossmann et al. (2025).

[143] Woiciechowski et al. (2024).

amount of waste.[144] GFI recently simulated a model to reduce the waste by recycling 80% of the culture medium in a long-term cell culture process. The method involved replacing 20% of the medium with fresh medium on the third day. On the sixth day, 100% of the medium is replaced with fresh medium. Thereafter, 20% of the medium is replaced with fresh medium every third day, retaining 80% of the original medium.[145] The GFI also reported that supplements like growth factor make up more than 95% of the total cost of animal-component-free media used in CM production.[146] Likewise, pH buffers account for about 38% of the total cost for medium.[147]

The expense of formulating animal-component-free culture medium and medium supplements is a major barrier for scaling CM productions. Reducing this cost through recycling could significantly advance research and pilot-scale production. Additionally, implementation of omics technology at all production steps, from cell selection to optimization of the finished products, can help reduce costs and set up a production pipeline that requires less water, energy and starting materials, while significantly increasing production yields.[148] For instance, systems biology-driven approaches have been successfully used to improve the biomass and productivity of Chinese hamster ovary cells by accelerating the development of cell-line-specific media and reducing research, development and production costs related to cell media and their optimization.[149] Another vital aspect of environmental sustainability is the effective management of waste effluent, such as spent culture medium that contains significant amounts of nutrients, salts, lactate, ammonia, nitrogen and many other biological and non-biological wastes and residues.[150] For each kilogram of CM produced, 1.57 kg of chemical oxygen demand (COD) is generated due to carbonaceous glucose and amino acid residues in the wastewater.[151] Several methods have been proposed for the efficient treatment of liquid waste, including bioremediation techniques that use microorganisms and plants to remove pollutants. The approach employs microalgae (*Chlorella vulgaris, Chlorococcum littorale, Scenedesmus rubescens*), bacteria (*Enterobacter cloacae*), or floating aquatic plants (*Pistia stratiotes, Lemna sp., Ipomoea aquatica, Salvinia natans*) to absorb and filter harmful pollutants from waste water. Chemical and electrochemical treatment processes are also being explored for efficient recovery and reuse of nitrogen and non-renewable resources such as phosphate from wastewater.[152] For example, electrochemical membrane reactors can establish a pH gradient, converting ammonium (NH_4^+) into ammonia (NH_3), which can be recovered with 99–100% efficiency.

[144] Woiciechowski et al. (2024) and Hubalek et al. (2022).

[145] Specht (2020).

[146] *Ibid.*

[147] *Ibid.*

[148] Romero et al. (2023).

[149] *Ibid.*

[150] Woiciechowski et al. (2024).

[151] Myers et al. (2023).

[152] Woiciechowski et al. (2024) and Sengupta et al. (2015).

Similarly, phosphate can be recovered by membrane filtration, although chemical precipitation is more efficient, removing 80–99% phosphate from wastewater streams.[153] Various other filtration techniques (microfiltration, ultrafiltration, osmosis, electrodialysis) can also be employed to recover valuable components and reduce overall waste from liquid effluent.[154]

5 Conclusions, Limitations and Future Perspectives

Cell-based meat may be a sustainable and ethical alternative to conventional meat products, addressing concerns related to food security, environmental impact and animal welfare. It provides high-quality dietary proteins with limited land and water resources, lower GHG emissions, while also eliminating the need for large-scale livestock farming. Despite these prospects, challenges remain in optimizing culture media, improving sensory attributes, scaling up production and reducing production costs. Advances in biotechnology, particularly omics-based and machine-learning approaches are crucial for refining these factors. While CM has the potential to significantly reduce the use of agricultural resources, energy efficiency remains a drawback. Although simulated LCA data provides positive insights into the use of renewable energy to meet the energy requirements of CM production, there is a substantial lack of real-world data on this and many other aspects, such as the environmental impact of the manufacture of culture medium, nutritional value, safety, quality, and the long-term effects of CM consumption on human and animal health. Without long-term comprehensive assessment of all these aspects, it is difficult to know whether CM can truly benefit the growing population as a replacement or supplement for conventional meat. This in turn can affect regulatory decisions and consumer acceptance, further complicating its broader implementation. Future research must focus on conducting long-term studies to evaluate the overall benefits and risks in order to ensure that CM can be a reliable and sustainable alternative in the global food system.

References

Alam AMMN, Kim CJ, Kim SH, Kumari S, Lee EY, Hwang YH, Joo ST (2024) Scaffolding fundamentals and recent advances in sustainable scaffolding techniques for cultured meat development. Food Res Int 189:114549

Alheib O, da Silva LP, Kwon IK, Reis RL, Correlo VM (2023) Preclinical research studies for treating severe muscular injuries: focus on tissue-engineered strategies. Trends Biotechnol 41(5):632–652

[153] Sengupta et al. (2015).

[154] Zielińska and Bułkowska (2025).

Allan SJ, De Bank PA, Ellis MJ (2019) Bioprocess design considerations for cultured meat production with a focus on the expansion bioreactor. Front Sustain Food Syst 3:44

Balkir P, Kemahlioglu K, Yucel U (2021) Foodomics: a new approach in food quality and safety. Trends Food Sci Technol 108:49–57

Ben-Arye T, Levenberg S (2019) Tissue engineering for clean meat production. Front Sustain Food Syst 3:46

Bezjak D, Orellana N, Valdivia G, Acevedo CA, Valdes JH (2024) Global transcriptome profiles provide insights into muscle cell development and differentiation on microstructured marine biopolymer scaffolds for cultured meat production. Sci Rep 14(1):10931

Bhat ZF, Kumar S, Fayaz H (2015) In vitro meat production: challenges and benefits over conventional meat production. J Integr Agric 14:241–248

Bonilla DA, Orozco CA, Forero DA, Odriozola A (2024) Techniques, procedures, and applications in host genetic analysis. Adv Genet 111:1–79

Chase P (2020) Billion dollar burger: inside big tech's race for the future of food. Penguin Random House, New York

Cheli F, Tedeschi G, Lanzoni D, Sundaram TS, Baldi A, Giromini C (2024) Carne a base cellulare: una nuova frontiera della ricerca. BioLaw J 4:151–164

Chelladurai KS, Selvan Christyraj JD, Rajagopalan K, Yesudhason BV, Venkatachalam S, Mohan M, Chellathurai Vasantha N, Selvan Christyraj JRS (2021) Alternative to FBS in animal cell culture – an overview and future perspective. Heliyon 7(8):e07686

Churchill W (1931) Fifty years hence. The Strand Magazine. https://www.nationalchurchillmuseum.org/fifty-years-hence.html. Accessed 31 Mar 2025

Cifuentes A (2017) Foodomics, foodome and modern food analysis. Trends Anal Chem 96:1

Cinici B, Yaba S, Kurt M, Yalcin HC, Duta L, Gunduz O (2024) Fabrication strategies for bioceramic scaffolds in bone tissue engineering with generative design applications. Biomimetics 9(7):409

Committee on Considerations for the Future of Animal Science Research; Science and Technology for Sustainability Program; Policy and Global Affairs; Board on Agriculture and Natural Resources; Division on Earth and Life Sciences; National Research Council (2015) Critical role of animal science research in food security and sustainability. National Academies Press (US), Washington (DC)

D'Almeida AP, de Albuquerque TL (2025) Is it possible to produce meat without animals? The potential of microorganisms as protein sources. Fermentation 11(1):24

Dagès BAS, Fabian JA, Polakova D, Rysova M, Topham PD, Souppez JRG, Hanga MP, Theodosiou E (2025) Edible electrospun materials for scalable cultivated beef production. Food Bioprod Process 149:118–129

Dai X, Shen L (2022) Advances and trends in omics technology development. Front Med 9:911861

Dayton WR, Hathaway MR (1990) Myogenic cell proliferation and differentiation. Poult Sci 70(8):1815–1822

De Carvalho JC, Karp SG, Mamani LDG, Biagini G, dos Santos Costa G, Herrmann LW, Carvalho CRS (2024) Downstream processes for cultivated meat. In: Soccol CL, Forte Maiolino Molento C, Glifke Reis G, Karp SG (eds) Cultivated meat. Springer, Cham, pp 145–166

De Paula LCP, Alves C (2024) Food packaging and endocrine disruptors. J Pediatr 100(Suppl 1):S40–S47

Dolgin J, Chakravarty D, Sullivan SF, Cai Y, Lim T, Yamaguchi P, Balkan JE, Xu L, Olawoyin AD, Lee K, Kaplan DL, Nair NU (2025) Microbial lysates as low-cost serum replacements in cellular agriculture media formulation. Food Res Int 201:115633

Dong H, Wang P, Yang Z, Xu X (2023) 3D printing based on meat materials: challenges and opportunities. Curr Res Food Sci 6:100423

Doreau M, Corson MS (2017) Production de viande et ressource en eau. La Revue Française de la Recherche en Viandes et Produits Carnés 33:1–8

EFSA Panel on Animal Health and Welfare (AHAW) (2017) Assessment of listing and categorisation of animal diseases within the framework of the Animal Health Law (Regulation (EU) 2016/429): Bluetongue. EFSA J 15(8):e04957
EFSA Panel on Nutrition, Novel Foods and Food Allergens (NDA) (2024) Guidance on the scientific requirements for an application for authorisation of a novel food in the context of Regulation (EU) 2015/2283. EFSA J 22(9):e8961
EFSA Scientific Colloquium (2024) Cell culture-derived foods and food ingredients. EFSA Supporting publication 27: EN-8664
Eisner MD (2024) Milk without animals – a dairy science perspective. Int Dairy J 156:105978
El Wali M, Golroudbary SR, Kraslawski A, Tuomisto HL (2024) Transition to cellular agriculture reduces agriculture land use and greenhouse gas emissions but increases demand for critical materials. Commun Earth Environ 5:61
EPRS (2024) Alternative protein sources for food and feed in the EU. Publications Office of the European Union. https://coilink.org/20.500.12592/pnvx6p3. Accessed 10 Aug 2025
Escobar R, Ignacia M, Cadena E, Nhu TT, Cooreman-Algoed M, De Smet S, Dewulf J (2021) Analysis of the cultured meat production system in function of its environmental footprint: current status, gaps and recommendations. Foods 10(12):2941
European Commission (2020) European Union's farm to fork strategy – for a fair, healthy and environmentally-friendly food system. https://food.ec.europa.eu/horizontal-topics/farm-fork-strategy_en. Accessed 10 Aug 2025
European Commission (2021) Report from the Commission to the European Parliament, the Council, the European economic and social committee and the committee of the regions. EU voluntary review on progress in the implementation of the 2030 Agenda for Sustainable Development. https://environment.ec.europa.eu/strategy/zero-pollution-action-plan_en. Accessed 31 Mar 2025
European Commission (2023) Sustainable development in the European Union – monitoring report on progress towards the SDGs in an EU context. Publications Office of the European Union, 2022. https://data.europa.eu/doi/10.2785/403194. Accessed 10 Aug 2025
European Institute of Innovation and Technology FOOD (EIT Food) (2024) GOURMEY files the first EU novel food submission for cultured meat. https://www.eitfood.eu/news/gourmey-files-the-first-eu-novel-food-submission-for-cultured-meat. Accessed 31 Mar 2025
FAO (2009) Global agriculture towards 2050: high-level expert forum on how to feed the world in 2050. https://www.fao.org/fileadmin/templates/wsfs/docs/Issues_papers/HLEF2050_Global_Agriculture.pdf. Accessed 31 Mar 2025
FAO (2017) Factsheet on livestock solutions for climate change. https://www.fao.org/family-farming/detail/en/c/1634679/. Accessed 31 Mar 2025
FAO (2019) Water use in livestock production systems and supply chains – guidelines for assessment (Version 1). Livestock Environmental Assessment and Performance (LEAP) Partnership. https://openknowledge.fao.org/items/34ad3136-0982-4129-bcf6-19f0460f43f6. Accessed 31 Mar 2025
FAO (2023a) Methane emissions in livestock and rice systems – sources, quantification, mitigation and metrics. https://doi.org/10.4060/cc7607en. Accessed 31 Mar 2025
FAO (2023b) World food and agriculture – statistical yearbook. https://openknowledge.fao.org/server/api/core/bitstreams/6e04f2b4-82fc-4740-8cd5-9b66f5335239/content. Accessed 31 Mar 2025
FAO (2023c) Pathways towards lower emissions – a global assessment of the greenhouse gas emissions and mitigation options from livestock agrifood systems. https://doi.org/10.4060/cc9029en. Accessed 10 Aug 2025
FAO (2024) Livestock Environmental Assessment and Performance Partnership. https://openknowledge.fao.org/handle/20.500.14283/cd1340en. Accessed 10 Aug 2025
FAO, WHO (2004) Second global forum for food safety regulations. European food safety control systems: new perspectives on a harmonized legal basis. Report by the Ministry of Agriculture,

Nature and Food Quality and the Food and Consumer Product Safety Authority of the Netherlands. https://www.fao.org/4/j2975e/j2975e.htm. Accessed 31 Mar 2025
FAO, WHO (2023) Food safety aspects of cell-based food. https://www.who.int/publications/i/item/9789240070943. Accessed 31 Mar 2025
Ford (2025) Dutch pioneer files EU's second lab-grown meat application. Politico. https://www.politico.eu/article/eu-mosa-meat-european-union-ceo-dutch-food-company-price/#:~:text=By%20Alessandro%20Ford-,Mosa%20Meat%20has%20lodged%20the%20European%20Union's%20second%20application%20for,food%20tech%20company%20announced%20Wednesday. Accessed 31 Mar 2025
Gagaoua M (2022) Recent advances in OMICs technologies and application for ensuring meat quality, safety and authenticity. Foods 11(16):2532
Garmyn A, Hardcastle N, Polkinghorne R, Lucherk L, Miller M (2020) Extending aging of beef longissimus lumborum from 21 to 84 days postmortem influences consumer eating quality. Foods 9(2):208
Godde CM, Mason-D'Croz D, Mayberry DE, Thornton PK, Herrero M (2021) Impacts of climate change on the livestock food supply chain; a review of the evidence. Glob Food Secur 28:100488
Godfray HCJ, Aveyard P, Garnett T, Hall JW, Key TJ, Lorimer J, Pierrehumbert RT, Scarborough P, Springmann M, Jebb SA (2018) Meat consumption, health and the environment. Science 361(6399):eaam5324
Gstraunthaler G (2003) Alternatives to the use of fetal bovine serum: serum-free cell culture. ALTEX 20(4):275–281
Hadi J, Brightwell G (2021) Safety of alternative proteins: technological, environmental and regulatory aspects of cultured meat, plant-based meat, insect protein and single-cell protein. Foods 10(6):1226
Hocquette JF (2016) Is in vitro meat the solution for the future? Meat Sci 12:167–176
Hocquette JF, Chriki S, Fournier D, Ellies-Oury MP (2025) Will "cultured meat" transform our food system towards more sustainability? Animal 19(1):101145
Hubalek S, Post MJ, Moutsatsou P (2022) Towards resource-efficient and cost-efficient cultured meat. Curr Opin Food Sci 47:100885
Jahir NR, Ramakrishna S, Abdullah AAA, Vigneswari S (2023) Cultured meat in cellular agriculture: advantages, applications and challenges. Food Biosci 53:102614
Kadim IT, Mahgoub O, Baqir S, Faye B, Purchas R (2015) Cultured meat from muscle stem cells: a review of challenges and prospects. J Integr Agric 14(2):222–233
Kamalapuram SK, Choudhury D (2024) Industry 4.0 technologies for cultivated meat manufacturing. Food Bioeng 3(1):14–28
Kaur L, Mao B, Beniwal AS, Kaur AR, Chian FM, Singh J (2022) Alternative proteins vs animal proteins: the influence of structure and processing on their gastro-small intestinal digestion. Trends Food Sci Technol 122:275–286
Kirsch M, Morales-Dalmau J, Lavrentieva A (2023) Cultivated meat manufacturing: technology, trends and challenges. Eng Life Sci 23(12):e2300227
Kossmann H, Moess T, Breunig P (2025) The climate impact and land use of cultivated meat: evaluating agricultural feedstock production. PLoS One 20(1):e0316480
Kumar P, Sharma N, Sharma S, Mehta N, Verma AK, Chemmalar S, Sazili AQ (2021) In-vitro meat: a promising solution for sustainability of meat sector. J Anim Sci Technol 63(4):693–724
Lanzoni D, Bracco F, Cheli F, Colosimo BM, Moscatelli D, Baldi A, Rebucci R, Giromini C (2022) Biotechnological and technical challenges related to cultured meat production. Appl Sci 12(13):6771
Lanzoni D, Rebucci R, Formici G, Cheli F, Ragone G, Baldi A, Violini L, Sundaram TS, Giromini C (2024) Cultured meat in the European Union: legislative context and food safety issues. Curr Res Food Sci 8:100722

Lee J, Kim J, Bae HW, Kim M, Jung BK, Kim J, Lee S, Kim HU (2024) Multi-omics analysis and genome-scale metabolic reconstruction of cattle bos taurus for optimal production of cultured meat. BioRxiv: 627468
Listrat A, Lebret B, Louveau I, Astruc T, Bonnet M, Lefaucheur L, Picard B, Bugeon J (2016) How muscle structure and composition influence meat and flesh quality. Sci World J 2016:3182746
Lynch J, Pierrehumbert R (2019) Climate impacts of cultured meat and beef cattle. Front Sustain Food Syst 3:5
Majumder R, Miatur S, Saha A, Hossain S (2024) Mycoprotein: production and nutritional aspects: a review. Sustain Food Technol 2:81–91
Martins B, Bister A, Dohmen RGJ, Gouveia MA, Hueber R, Melzener L, Messmer T, Papadopoulos J, Pimenta J, Raina D, Schaeken L, Shirley S, Bouchet BP, Flack JE (2024) Advances and challenges in cell biology for cultured meat. Annu Rev Anim Biosci 12:345–368
Mattick CS, Landis AE, Allenby BR, Genovese NJ (2015) Anticipatory life cycle analysis of in vitro biomass cultivation for cultured meat production in the United States. Environ Sci Technol 49(19):11941–11949
McNamara E, Bomkamp C (2022) Cultivated meat as a tool for fighting antimicrobial resistance. Nat Food 3:791–794
Medeiros F, Aleman RS, Gabríny L, You SW, Hoskin RT, Moncada M (2024) Current status and economic prospects of alternative protein sources for the food industry. Appl Sci 14(9):3733
Myers GM, Jaros KA, Andersen DS, Raman DR (2023) Nutrient recovery in cultured meat systems: impacts on cost and sustainability metrics. Front Nutr 10:1151801
Norton T (2021) From the lab to the supermarket: in vitro meat as a viable alternative to traditional meat production. J Food Law Policy 11(1):157–180
Nunes OB, Buranello TW, de Andrade Farias F, Rosero J, Recchia K, Bressan FF (2025) Can cell-cultured meat from stem cells pave the way for sustainable alternative protein? Curr Res Food Sci 10:100979
OECD (2012) OECD environmental outlook to 2050: the consequences of inaction. OECD Publishing, Paris. https://doi.org/10.1787/9789264122246-en. Accessed 31 March 2025
Oragui E, Nannaparaju M, Khan WS (2011) The role of bioreactors in tissue engineering for musculoskeletal applications. Open Orthop J 5(2):267–270
Paim WP, Maggioli MF, Falkenberg SM, Ramachandran A, Weber MN, Canal CW, Bauermann FV (2021) Virome characterization in commercial bovine serum batches - a potentially needed testing strategy for biological products. Viruses 13(12):2425
Pecora A, Pérez López J, Jordán MJ, Franco LN, Politzki R, Ruiz V, Alvarez I (2020) Analysis of irradiated Argentinean fetal bovine serum for adventitious agents. J Vet Diagn Invest 32(6):892–897
Post MJ (2012) Cultured meat from stem cells: challenges and prospects. Meat Sci 92(3):297–301
Post MJ (2014) Cultured beef: medical technology to produce food. J Sci Food Agric 94(6):1039–1041
Post MJ, Hocquette JF (2017) New sources of animal proteins: cultured meat. In: Purslow PR (ed) New aspects of meat quality – from genes to ethics. Elsevier, Amsterdam, pp 425–441
Post MJ, Levenberg S, Kaplan DL, Genovese N, Fu J, Bryant JC, Negowetti N, Verzijden K, Moutsatsouet P (2020) Scientific, sustainability and regulatory challenges of cultured meat. Nat Food 1:403–415
Power DM, Taoukis P, Houhoula D, Tsironi T, Flemetakis E (2023) Integrating omics technologies for improved quality and safety of seafood products. Aquacult Fish Fish 8(4):457–462
Pulina G (2020) Ethical meat: respect for farm animals. Anim Front 10(1):34–38
Pulina G, Bertoni G (2023) Ethics in meat production. In: Meiselman HL, Lorenzo JM (eds) Meat and meat replacements. Elsevier, Amsterdam, pp 197–224
Ramanathan R, Lambert LH, Nair MN, Morgan B, Feuz R, Mafi G, Pfeiffer M (2022) Economic loss, amount of beef discarded, natural resources wastage, and environmental impact due to beef discoloration. Meat Muscle Biol 6(1):13218

Reiss J, Robertson S, Suzuki M (2021) Cell sources for cultivated meat: applications and considerations throughout the production workflow. Int J Mol Sci 22(14):7513

Rocha LO, Silva NCC (2023) Application of omics-based technologies and the impact on food science. Front Microbiol 14:1155757

Romero SG, Boyle N (2023) Systems biology and metabolic modeling for cultivated meat: a promising approach for cell culture media optimization and cost reduction. Compr Rev Food Sci Food Saf 22(4):3422–3443

Rosser J, Thomas DJ (2018) Bioreactor processes for maturation of 3D bioprinted tissue. In: Thomas DJ, Jessop ZM, Whitaker IS (eds) 3D Bioprinting for reconstructive surgery. Elsevier, Amsterdam, pp 191–215

Sabia E, Braghieri A, Vignozzi L, Paolino R, Cosentino C, Di Trana A, Pacelli C (2025) Carbon footprint of by-product concentrate feed: a case study. Environments 12(2):42

Seah JSH, Singh S, Tan LP, Choudhury D (2022) Scaffolds for the manufacture of cultured meat. Crit Rev Biotechnol 42(2):311–323

Sengupta S, Nawaz T, Beaudry J (2015) Nitrogen and phosphorus recovery from wastewater. Curr Pollut Rep 1:155–166

Shackelford SD, Wheeler TL, Meade MK, Reagan JO, Byrnes BL, Koohmaraie M (2001) Consumer impressions of tender select beef. J Anim Sci 79(10):2605–2614

Siddiqui SA, Bahmid NA, Karim I, Mehany T, Gvozdenko AA, Blinov AV, Nagdalian AA, Arsyad M, Lorenzo JM (2022) Cultured meat: processing, packaging, shelf life and consumer acceptance. LWT J 172:114192

Singh A, Kumar V, Singh SK, Gupta J, Kumar M, Sarma DK, Verma V (2023) Recent advances in bioengineered scaffold for in vitro meat production. Cell Tissue Res 391(2):235–247

Sinke P, Swartz E, Sanctorum H, Van der Giesen C, Odegard I (2023) Ex-ante life cycle assessment of commercial-scale cultivated meat production in 2030. Int J Life Cycle Assess 28:234–254

Smith-Uchotski R, Wanjiru P (2023) Risk assessment unit science, evidence and research division. Risk Assessment Unit Science, Evidence and Research Division, FSA

Soleymani S, Naghib SM, Mozafari MR (2024) An overview of cultured meat and stem cell bioprinting: how to make it, challenges and prospects, environmental effects, society's culture and the influence of religions. J Agric Food Res 18:101307

Specht L (2020) An analysis of culture medium costs and production volumes for cultivated meat. The Good Food Institute. https://gfi.org/resource/analyzing-cell-culture-medium-costs/. Accessed 31 Mar 2025

Specht EA, Welch DR, Clayton EMR, Lagally CD (2018) Opportunities for applying biomedical production and manufacturing methods to the development of the clean meat industry. Biochem Eng J 132:161–168

Stout AJ (2023) Strategies for enhanced cultured meat products and processes. PhD thesis, Tufts University. https://dl.tufts.edu/pdfviewer/nc581234f/st74d564c. Accessed 31 Mar 2025

Stout AJ, Mirliani AB, Rittenberg ML, Shub M, White EC, Yuen JSK Jr, Kaplan DL (2022) Simple and effective serum-free medium for sustained expansion of bovine satellite cells for cell cultured meat. Commun Biol 5:466

Stout AJ, Rittenberg ML, Shub M, Saad MK, Mirliani AB, Dolgin J, Kaplan DL (2023) A beefy-r culture medium: replacing albumin with rapeseed protein isolates. Biomaterials 296:122092

Su G, Yu C, Liang S, Wang W, Wang H (2024) Multi-omics in food safety and authenticity in terms of food components. Food Chem 437(Pt 2):137943

Sundaram TS, Giromini C, Rebucci R, Lanzoni D, Petrosillo E, Baldi A, Cheli F (2025) Milk whey as a sustainable alternative growth supplement to fetal bovine serum in muscle cell culture. J Dairy Sci 108(5):4749–4760

Swartz E (2019) Meeting the needs of the cell-based meat industry. American Institute of Chemical Engineers. SBE Special Section: Industrial Biotechnology. https://www.aiche.org/resources/publications/cep/2019/october/sbe-special-section-industrial-biotechnology-complete-26-page-section. Accessed 31 Mar 2025

Swartz E, Bomkamp C (2024) The science of cultivated meat. https://gfi.org/science/the-science-of-cultivated-meat/. Accessed 31 Mar 2025
Tanaka RI, Sakaguchi K, Yoshida A, Takahashi H, Haraguchi Y, Shimizu T (2022) Production of scaffold-free cell-based meat using cell sheet technology. NPJ Sci Food 6(1):41
Todhunter ME, Jubair S, Verma R, Saqe R, Shen K, Duffy B (2024) Artificial intelligence and machine learning applications for cultured meat. Front Artif Intell 7:1424012
Tomiyama AJ, Kawecki NS, Rosenfeld DL, Jay JA, Rajagopal D, Rowat AC (2020) Bridging the gap between the science of cultured meat and public perceptions. Trends Food Sci Technol 104:144–152
Troy DJ, Kerry JP (2010) Consumer perception and the role of science in the meat industry. Meat Sci 86(1):214–226
Tuomisto HL (2019) The eco-friendly burger: could cultured meat improve the environmental sustainability of meat products? EMBO Rep 20(1):e47395
Tuomisto HL, de Mattos MJ (2011) Environmental impacts of cultured meat production. Environ Sci Technol 45(14):6117–6123
Tuomisto HL, Ellis M, Haastrup P (2014) Environmental impacts of cultured meat: alternative production scenarios. In: Schenck R, Huizenga D (eds) Proceedings of the 9th international conference on life cycle assessment in the agri-food sector. Vashon, ACLCA, pp 1360–1366
Tuomisto HL, Allan SJ, Ellis MJ (2022) Prospective life cycle assessment of a bioprocess design for cultured meat production in hollow fiber bioreactors. Sci Total Environ 851(Pt 1):158051
UNGA (2015) Transforming our world: the 2030 Agenda for Sustainable Development. A/RES/70/1. https://docs.un.org/en/A/RES/70/1. Accessed 10 Aug 2025
Van der Valk J, Gstraunthaler G (2017) Fetal bovine serum (FBS) – a pain in the dish? Altern Lab Anim 45(6):329–332
Van Eelen WF, Van Kooten WJ, Westerhof W (1998) Industrial production of meat from (in vitro) cell cultures. WIPO, Publ. of the Int. Appl. with Int. search report WO1998NL00721. https://patents.google.com/patent/WO1999031223A1/en. Accessed 10 Aug 2025
Vein JF (2004) Tissue engineered meat for consumption and a method for producing tissue engineered meat for consumption. European patent number EP1789063B1. https://patents.google.com/patent/EP1789063B1/en. Accessed 10 Aug 2025
Verma R, Lee Y, Salamone DF (2022) iPSC technology: an innovative tool for developing clean meat, livestock and frozen ark. Animals 12(22):3187
Wang K, Du C, Guo X, Xiong B, Yang L, Zhao X (2024) Crop byproducts supplemented in livestock feeds reduced greenhouse gas emissions. J Environ Manag 355:120469
Weindl I, Bodirsky BL, Rolinski S, Biewald A, Lotze-Campen H, Müller C, Dietrich J, Humpenöder F, Stevanović M, Schaphoff S, Popp A (2017) Livestock production and the water challenge of future food supply: implications of agricultural management and dietary choices. Global Biogeochem Cycles 47:121–132
Woiciechowski AL, Scapini T, Manzoki MC, Martínez-Burgos WJ, Molina-Aulestia DT, Karp SG, Soccol CR (2024) Waste management in cultivated meat production. In: Soccol CL, Forte Maiolino Molento C, Glifke Reis G, Karp SG (eds) Cultivated meat. Springer, Cham, pp 265–276
Wu B, Wei F, Xu S, Xie Y, Lv X, Chen H, Huang F (2021) Mass spectrometry-based lipidomics as a powerful platform in foodomics research. Trends Food Sci Technol 107:358–376
Xia H, Vijayaraghavan B, Belák S, Liu L (2011) Detection and identification of the atypical bovine pestiviruses in commercial foetal bovine serum batches. PLoS One 6(12):e28553
Zandonadi RP, Ramos MC, Elias FTS, Guimarães NS (2025) Global insights into cultured meat: Uncovering production processes, potential hazards, regulatory frameworks and key challenges – A scoping review. Foods 14(1):129
Zhou H, Quach A, Nair M, Abasht B, Kong B, Bowker B (2024) Omics based technology application in poultry meat research. Poult Sci 104(1):104643
Zielińska M, Bułkowska K (2025) Use of membrane techniques for removal and recovery of nutrients from liquid fraction of anaerobic digestate. Membranes 15(2):45

Safety Assessment of Cell-Culture-Derived Food: The Role of the European Food Safety Authority

Chiara Dall'Asta and Hilva Gjoni

Abstract Innovation is essential for driving the transition towards sustainable, healthier and more inclusive food systems, requiring progress across the entire spectrum from production to consumption. In recent decades, there have been noteworthy advances in food and feed production technologies, motivated by the need for safety, resource sustainability and responsiveness to societal needs. Among these innovations, cell-culture-derived products stand out as solutions to pressing global issues such as climate change, food security and animal welfare, attracting significant interest from various stakeholders. The European Food Safety Authority (EFSA) plays a vital role in evaluating these products under the Novel Food Regulation, ensuring the safety of food available on the European Union market. While plant-based cell-cultured foods are nearing market introduction, applications for cell-cultured animal products, such as Gourmey's foie gras alternative, are also emerging. This chapter examines EFSA's risk assessment frameworks for cell-cultured foods, emphasizing the importance of balancing innovation with public health in a complex environment. As new technologies like cell cultivation and tissue engineering develop, they raise important safety questions, particularly regarding allergenic potential and nutritional equivalence. Although cell-cultured foods offer promising sustainability benefits, several obstacles must be addressed before they can gain widespread acceptance. Transparency and thorough assessment will be a key as these preparations progress.

Keywords Innovation · Cell-culture derived food · Tissue engineering · EFSA · Risk assessment

---

C. Dall'Asta (✉) · H. Gjoni
University of Parma, Food and Drug Department, Parma, Italy
e-mail: chiara.dallasta@unipr.it; hilva.gjoni@unipr.it

G. Formici et al. (eds.), *Cell-Based Meat in the European Union and Beyond*,
https://doi.org/10.1007/978-3-032-14624-3_3

1 Introduction

Innovation is a crucial factor in speeding up the shift towards more sustainable, healthier and inclusive food systems, spanning from primary production to consumption. In recent decades, the food and feed production sector has seen significant advances in production technologies and the products they yield.[1] Key drivers for future development include changes in production technologies and global supply chains, the need to ensure food and feed safety, sustainable use of resources and addressing societal concerns. Innovative production technologies can create a new generation of foods and feed, requiring comprehensive pre-market evaluations to safeguard human and animal health as well as the environment.[2]

The rise of promising Novel Food innovations, such as cell-culture-derived food, addresses pressing global challenges including climate change, food security and animal welfare, sparking intense interest from industry stakeholders, policymakers and the public.[3] The successful integration of these products into the food supply relies on rigorous risk assessment and regulatory oversight to ensure public safety and trust. In the European Union, the European Food Safety Authority (EFSA) plays a central role in evaluating the safety and potential risks associated with cell-culture-derived food under the framework of the Novel Food Regulation.[4] However, the regulatory process surrounding cell-based food has become a focal point for lobbying and advocacy.[5] Industry players developing these products seek favorable regulatory outcomes to expedite market entry, while traditional animal-based product producers often advocate stricter safety standards to safeguard their market position. Concurrently, public advocacy groups emphasize environmental and ethical considerations, urging risk assessors and risk managers to adopt faster and more comprehensive approaches that extend beyond safety to address societal impacts.[6] This co-occurrence of pressures underscores the complex and highly politicized context in which the EFSA's scientific opinion must remain objective, while mediating between innovation, safety and competing interests. This chapter explores EFSA's role in the risk assessment of cell-based food, examining the scientific and procedural frameworks it employs and providing insights into how the EFSA can

[1] Aiking and De Boer (2020), p. 515.

[2] Rischer et al. (2020), p. 128.

[3] Santo et al. (2020), p. 2.

[4] Regulation (EU) 2015/2283 of the European Parliament and of the Council of 25 November 2015 on Novel Foods, amending Regulation (EU) no 1169/2011 of the European Parliament and of the Council and repealing Regulation (EC) no 258/97 of the European Parliament and of the Council and Commission Regulation (EC) no 1852/2001. For an in-depth analysis of the legislative framework on Novel Foods and the ongoing regulatory debate in the European Union, see *Cell-Based Meat in the European Union: A Regulatory Crossroads* by Corazza and Formici in this volume.

[5] See Greenpeace Unearthed (2024).

[6] As reported by Hall (2023). See also *Navigating Consumer Acceptance of Cultured Meat: Insights from Social Science Research* by Lin-Hi and Burdorf in this volume.

effectively navigate the delicate balance between promoting innovation and safeguarding public health in a rapidly evolving food landscape.

Urgent social issues, such as world hunger and climate change, pressure researchers to develop new technologies.[7] At the basis of food regulation stands the so-called General Food Law Regulation which sustains the idea that food is only food when it is safe to eat; thus no futuristic technology can hold if not safe for the consumer.[8] Regulatory frameworks in the food system are predominantly designed to mitigate known risks, focusing on issues that are well understood and grounded in established scientific knowledge. These frameworks prioritize safety, aiming to prevent harm from recognized hazards such as pathogens, contaminants and allergens. A clear demonstration of this is the precautionary principle which dictates that if there is potential risk—even in the absence of conclusive evidence—precautions may be taken. This reactive nature therefore often delays the swift integration of new technologies, such as cell-based food and genetically modified organisms, which may present uncertainties that regulators are not yet equipped to address.[9] The challenge arises from the difficulty of predicting emerging risks associated with new technologies, which often lack the clear, quantifiable risk profiles of traditional food products. Moreover, new discoveries may bring newer problems to the table while solving existing ones. As noted in the literature, current regulatory frameworks tend to be slow to adapt, delaying market introduction of beneficial innovations.[10]

2 The European Food Safety Authority

The European Food Safety Authority (EFSA) was established in response to growing concerns over food safety in the European Union (EU) and the need for a centralized, independent body that could assess and communicate risks associated with food and feed.[11] Prior to its establishment, the EU lacked a unified framework for evaluating food safety risks across Member States. Various national authorities were responsible for their own food safety assessments, but these were often fragmented, contradictory and not always transparent. This situation was further complicated by the increasing complexity of the global food system and the emergence of new

[7] See *Climate Change 2022: Impacts, Adaptation and Vulnerability – Chapter 18: Climate Resilient Development Pathways*, available at https://www.ipcc.ch/report/ar6/wg2/chapter/chapter-18/. Accessed 11 April 2025.

[8] Regulation (EC) no 178/2002, laying down the general principles and requirements of food law, establishing the European Food Safety Authority and procedures in matters of food safety. This regulation provides the foundation for food law in the European Union, ensuring a high level of protection for human life and health and safeguarding consumer interests.

[9] Le Bloch et al. (2025), p. 5.

[10] Dabic et al. (2022), p. 1777. See also Part "Legal and Policy Debate" of this volume on the legal and political debate.

[11] https://www.efsa.europa.eu/en/about/about-efsa-. Accessed 11 April 2025.

food-related risks, such as foodborne illnesses, chemical contamination and genetically modified organisms (GMOs). These challenges highlighted the need for an independent scientific authority that could evaluate food safety risks impartially, based on the best available scientific evidence, so as to ensure the protection of public health in the EU.

The EFSA was formally established in 2002 by Regulation (EC) no 178/2002 of the European Parliament and of the Council, which laid down the general principles and requirements of food law in the EU. This is especially important considering the number of food safety crises that broke out in the late 1990s, notably Bovine Spongiform Encephalopathy (BSE), or "mad cow disease", which revealed huge deficiencies in the EU's food safety infrastructure. The BSE crisis led to a lack of public trust in the ability of national and EU-level institutions to protect consumers from food-related risks. As a result, the European Commission recognized the need for a more robust and centralized approach to food safety, which culminated in the creation of EFSA as an independent scientific body to ensure that food and feed safety were rigorously evaluated and communicated across the EU.[12] This authority comprises a network of scientific experts and scientific panels, each with a particular area of food safety expertise. The panels are made up of independent scientists, specialized in fields such as toxicology, microbiology, food chemistry, nutrition and animal health, who conduct risk assessments based on the latest scientific evidence.

The scientific advice provided by the EFSA is independent, ensuring that risk assessments are based solely on the best available science. This independence is crucial for maintaining public trust in the food safety system and for ensuring that regulatory decisions are transparent and evidence based.

Indeed, EFSA's primary role is to provide scientific advice through risk assessments to support EU legislation and policymaking related to food and feed safety. One of its key functions is to assess and evaluate potential risks to human health posed by food and feed, including risks associated with chemical contaminants, foodborne diseases, food additives and the safety of Novel Foods and ingredients. The potential risks of new technologies in the food and agriculture sectors, such as GMOs, nanomaterials and cell-based food, are also evaluated. The EFSA may be commissioned by the EU to contribute to food safety legislation with scientific input during the legislative process. Indeed, whenever new legislation or regulations are proposed at EU level, the EFSA reviews the scientific evidence and checks the scientific basis of the laws. This is one way in which food safety legislation is kept current and addresses emerging risks. The EFSA also works closely with national food safety authorities, providing them with scientific support and fostering cooperation among EU Member States to ensure consistent implementation of food safety regulations across the EU. In addition to its risk assessment role, the EFSA plays a major role in risk communication. It communicates scientific findings to the public, policymakers and other stakeholders in a clear and accessible manner.

[12] Alemanno (2007), p. 2.

3 Authorisation Procedure for Approval of Cell-Cultured Foods in the EU

Cell-cultured foods, developed through cellular agriculture,[13] represent an innovative solution in food production. Using advanced tissue engineering techniques, plant or animal cells are cultivated to produce muscle and fatty tissue. Although such technologies have been used in the biomedical sector for decades, risk perception among consumers is still largely emotional and gaps in scientific knowledge persist.[14]

Cell-cultured products are classified as Novel Foods under EU law, as they utilize technologies and processes not commonly employed prior to May 1997.[15] Food Business Operators (FBOs) who want to market such products are required to submit detailed applications to the European Commission. While the EFSA conducts scientific risk assessment, risk management responsibilities lie with the European Commission and representatives of Member States, whereas risk communication pertains to both authorities.[16]

To initiate the approval process, FBOs must first ascertain with national authorities whether their products qualify as Novel Foods.[17] Cell-cultured products, such as those derived from animal cells grown in bioreactors and structured into edible tissues, fall in this category. FBOs are therefore required to prepare a comprehensive technical dossier according to EFSA guidelines. Although dedicated guidelines on cell-cultured food have yet to be published by the authority, the recently updated

[13] "Cellular agriculture" is a technical term referring to an emerging disciplinary field based on cell-cultivation technology. It focuses on the production of agricultural products from cell cultures using a combination of biotechnology, tissue engineering, molecular biology, and synthetic biology to create and design new methods of producing proteins, fats, and tissues that would otherwise come from traditional agriculture. The field has two primary branches: cultivated meat and precision fermentation. While the most well-known application is for animal-origin tissues (i.e. cultivated meat or cultivated eggs), the technology applies to plant tissues as well.

It should be noted that the term "cellular agriculture" is not yet defined by law in the EU.

[14] More information can be found in Stephens et al. (2018), p. 155; Bhat et al. (2019), p. 1192; Hadi and Brightwell (2021), p. 2; Post et al. (2020), p. 403. See also Part "Socio-Economic and Cultural Issues" of this volume, on the cultural and ethical issues linked to the consumption of cell-based meat and other Novel Foods. For a specific exposition of the scientific challenges related to cell-based meat, see *What Is Cell-Based Meat? Food Safety and Sustainability Aspects* by Cheli et al. in this volume.

[15] See *Cell-Based Meat in the European Union: A Regulatory Crossroads* by Corazza and Formici in this volume.

[16] Article 20 of Regulation (EC) no 178/2002 contains provisions about the Public Health Risk Assessment process in the EU.

[17] Commission Implementing Regulation (EU) 2018/456 of 19 March 2018 on the procedural steps of the consultation process for the determination of Novel Food status in accordance with Regulation (EU) 2015/2283 of the European Parliament and of the Council on Novel Foods.

general guidance presents a specific section for cell-based food.[18] Indeed, applicants are recommended to prepare the technical dossier documentation according to the requirements outlined in EFSA's "Guidance on the preparation and presentation of an application for authorization of a Novel Food in the context of Regulation (EU) 2015/2283".[19] In the technical dossier, FBOs must therefore provide detailed information on: (1) biological source (taxonomic information on family, genus, species, subspecies, variety) according to the international codes of nomenclature; (2) for plants, algae and fungi, determination of identity according to internationally recognized databases and methodology; (3) identity and characteristics of the cultured cells; (4) organ and tissue or part of the organism sourced; (5) production process, including the use of bioreactors and biomaterial scaffolds; (6) anticipated consumption levels; (7) nutritional and toxicological assessments.

Based on the authorization dossier submitted by the applicant to the European Commission, EFSA performs a comprehensive safety assessment. If necessary, further data can be requested of the applicant to permit robust evaluation. Once the EFSA's scientific opinion is released, the European Commission, together with Member State representatives to the Standing Committee on Plants, Animals, Food and Feed (PAFF), acts as risk manager to determine any necessary measures for mitigating potential risks. Within 7 months of EFSA's evaluation, the Commission drafts an implementing act authorizing the marketing of the product. The final authorization includes updating the Union's list of Novel Foods under Regulation (EU) 2017/2470,[20] ensuring that the product also complies with EU food safety standards and is marketed under clearly defined conditions.[21]

4 Risk Assessment of Cell-Cultured Foods: Open Challenges

Emerging technologies like cell culture, tissue engineering and precision fermentation are driving the development of Novel Foods, including lab-grown meat and dairy proteins produced by microorganisms. Cell and tissue engineering are longstanding technologies in the biomedical sector that enable the cultivation of cells and tissues separately from entire organisms. Precision fermentation, a technology that uses microorganisms to produce specific biomolecules, is well-established in

[18] The new updated guidance is available at https://efsa.onlinelibrary.wiley.com/doi/epdf/10.2903/j.efsa.2024.8961. Accessed 11 April 2025.

[19] EFSA NDA Panel (2021).

[20] https://food.ec.europa.eu/food-safety/novel-food/authorisations/union-list-novel-foods_en. Accessed 11 April 2025.

[21] For a detailed analysis of the specific provisions included in the Novel Foods Regulation, see *Cell-Based Meat in the European Union: A Regulatory Crossroads* by Corazza and Formici in this volume.

the biopharmaceutical sector, used for example to produce insulin.[22] These approaches are now mature for transfer to the food sector, for the production of food enzymes (e.g. for cheesemaking), functional ingredients (e.g. human-milk-identical oligosaccharides) and even foods (e.g. cell-cultured meat and cheese). Such products fall under the Novel Food Regulation and must therefore undergo thorough safety assessment for the general population and vulnerable groups.

4.1 Hazards Related to Cell-Cultured Food

The risk-assessment process involves four well-known steps. The Hazard Identification step considers potential biological, chemical and physical risks associated with the production process. The Hazard Characterization step aims to understand the nature and severity of any risks identified. The Exposure Assessment step estimates consumer exposure levels, considering anticipated consumption patterns and the nutritional profile of cell-cultured products. Finally, the Risk Characterization step unites the previous findings to assess the likelihood and impact of any risks identified in order to ascertain whether the product is safe and nutritionally comparable to conventional alternatives.[23]

Although the production processes enable identification of potential hazards related to cell-cultured food, the critical steps may be different from company to company due to the absence of processing standardization.

Recently the Food and Agriculture Organization (FAO) and the World Health Organization (WHO) reviewed safety aspects related to cell-cultured food.[24] Some common critical aspects may be found in the following stages:

Cell Selection and Storage The stages of cell sourcing, isolation and storage are critical for ensuring the safety and consistency of cell-cultured food production. In particular, the cell donor organism for animal-derived cultured food is a crucial starting point, a choice on which the whole process may depend.[25] Although cell

---

[22] Precision fermentation uses biotechnological modified microorganisms to produce specific proteins, enzymes, and other small molecules of biological interest. The term "precision" refers to the ability to precisely control the process, from the biotechnological modification of the microorganisms to the optimization of fermentation conditions. This technique allows us to produce high-value compounds with precision and consistency, with cost savings, less environmental impact and improved product quality. It has been applied for years in biopharmaceutical production (e.g. insulin, vaccines, gene therapies, etc.), and is currently used in food production to obtain specific "animal-free" components (e.g. collagen, lactoferrin, ovalbumin). More information can be found in Eastham and Leman (2024).

[23] For the EU risk assessment framework for feed and food, refer to the Regulation (EU) 2019/1381 on transparency and sustainability of the EU risk assessment in the food chain and to the General Food Law Regulation.

[24] See for discussion FAO and WHO (2023).

[25] Melzener et al. (2021), p. 7.

lines may be selected from biopsy samples from live or slaughtered animals, they must be able to proliferate and differentiate.[26] Often the selected cells are modified genetically to improve growth rates, nutrient absorption and overall product quality. However, these modifications carry risks such as the formation of novel proteins with unknown health implications, mainly allergic reactions. Zoonotic pathogens such as *Salmonella*, *E. coli*, *Listeria* and *Mycoplasma* spp., viruses and parasites are also concerns, though they occur at lower levels than in conventional livestock production. To preserve cells during storage and avoid the above hazards, antibiotics or cryoprotectants like dimethyl sulfoxide are commonly used.[27] These substances are generally diluted or removed during processing, to ensure consumer safety. However since they can leave trace residues in the final product, rigorous safety evaluations are necessary. Strict overall quality control and aseptic conditions are important and need to be described in detail to allow EFSA experts to properly assess the process outline and verify that critical points are specified and kept under control.

Biomass Production and Harvesting The cell cultivation stage requires strict sterile conditions to prevent microbial contamination. This phase centres on cultivating cells in bioreactors under controlled conditions, using cell culture medium to provide the nutrients and growth factors (e.g. glucose, inorganic salts, water-soluble vitamins) necessary for cell development, as well as other substances added to impart the organoleptic properties of the conventional counterpart.[28] Most cells used to produce cell-cultured food require scaffolds to replicate. These are made from natural or synthetic materials which must be safe for human consumption. Among potential microbiological hazards, *Mycoplasma* spp. is a critical concern, as it can impede cell growth and is challenging to eliminate. Additional risks include contamination by bacteria, fungi, viruses and prions derived from animal-based components of the medium. Mitigation strategies involve early detection of infections, good hygiene and medium formulations which are increasingly moving away from animal-derived components like fetal bovine serum (FBS) towards synthetic or serum-free alternatives.[29] Reducing antibiotic usage, essential to mitigate the risk of antimicrobial resistance, by means of chemical preservatives such as sodium benzoate, can be an alternative. Genetic drift in cell lines due to repeated sub-culturing poses another potential hazard. This risk can be controlled through the use of well-maintained, quality-controlled cell banks.[30]

The harvesting process, which involves removing residues from cell cultivation, may introduce contaminants such as enzymes or chemicals used for dissociation

[26] Ong et al. (2021), p. 5421.

[27] Best (2015), p. 422.

[28] Ong et al. (2021), p. 5421.

[29] Lee et al. (2022), p. 775.

[30] A comprehensive overview of the current landscape, limitations and opportunities offered by biotechnological tools for cultivated meat can be found in Gomez Romero et al. (2025) and in Piantino et al. (2025).

(e.g. proteases and EDTA). Growth factors and hormones, sourced from animal or non-animal origins, require meticulous handling to avoid adverse effects such as metabolic disruption and oncogenic risks. Proper harvesting techniques are vital for minimizing microbial contamination and ensuring product safety.[31]

Food Processing and Formulation The formulation of cell-based food products relies on additives to achieve the desired texture, flavour and preservation. However, ingredients such as microcarriers, scaffolds, flavour agents and preservatives may introduce allergenic risks, particularly when fish- or shellfish-derived cell lines or additives like gluten, soy or milk are involved. To safeguard consumers, all components must comply with safety and labelling regulations, including appropriate allergen labelling. Microbiological hazards, oxidation and enzymatic degradation during processing and storage must be controlled to minimize the formation of undesirable by-products.

The main hazards and potential concerns identified for cell-cultured food are reported in Fig. 1. Most are already well known as they also apply to conventionally produced food. For example, microbiological contamination can occur at any stage of any food production process, including that of cell-based food, and can be controlled by good manufacturing and hygiene practices, proper monitoring and appropriate Hazard Identification and Critical Control Points (HACCP). Food safety

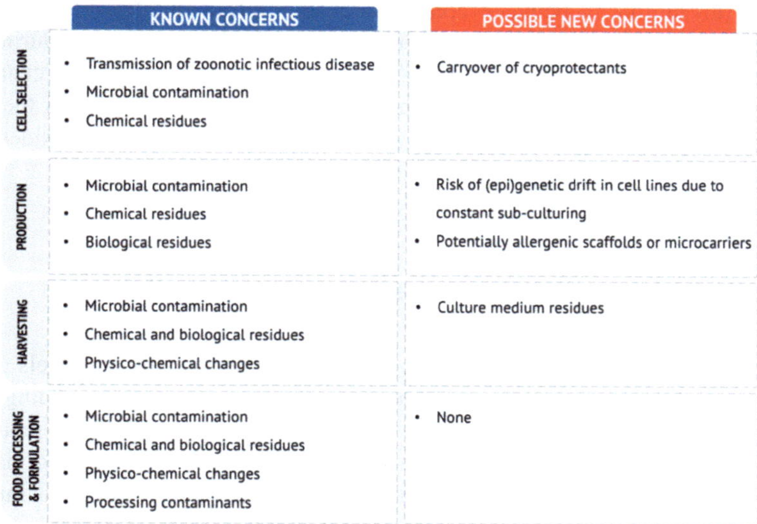

Fig. 1 Potential hazards of cell-cultured food [Adapted from FAO and WHO (2023)]

[31] For more information, Augustin et al. (2023).

plans should bear in mind that materials, ingredients and equipment used for cell growth can be different from those used in conventional food production.[32]

4.2 Nutritional Similarity of Cell-Cultured and Conventional Food

A core principle of the EU Novel Food regulation is that if a Novel Food is designed to replace an existing food, it must not differ in a manner that would make its consumption nutritionally disadvantageous for consumers.[33]

Achieving compositional similarity of a cell-cultured food with its conventional counterpart is still a challenge, as process plays a major role. This is also a condition for marketing cell-cultured products. Cultured meat, for instance, has been developed with the aim of replicating the nutritional composition of its conventional counterpart. However, studies suggest that while cell-based meat can be similar in macronutrient composition (proteins, fats and carbohydrates), its micronutrient content (e.g. vitamins, minerals and bioactive compounds) may differ depending on production methods.[34] The product may lack compounds naturally present in animal tissues, such as haem iron and specific lipids, unless these are intentionally added during production.[35]

This (dis)similarity between conventional and cell-cultured food may derive from the addition or supplementation of nutrients and/or compounds during cell cultivation. Such additions, while promoting growth, may change growth characteristics and the overall nutritional composition in unexpected ways. All the steps required for generating cell-based food should therefore be carefully considered for the safety and nutritional assessment of the final products.[36]

4.3 Toxicological Profile and Allergenic Potential

The compositional profiles, microbiological safety and contaminant levels of cultured meat products have been increasingly studied, yet knowledge regarding genotoxicity, subchronic toxicity and toxicokinetics remains limited.[37] This toxicological data is critical for the safety assessment of cultured meat as a Novel Food, particularly under EFSA guidelines. The safety assessment report must demonstrate that the Novel Food poses no toxicity risks. This includes evaluating systemic toxicity

[32] Powell et al. (2025).

[33] See for reference Article 7 of Regulation (EU) 2015/2283.

[34] *Ibid*, p. 2.

[35] Post (2022), p. 297.

[36] Lee et al. (2024).

[37] Ong et al. (2023), p. 2645.

(acute, sub-chronic and chronic), carcinogenicity, mutagenicity, reproductive and developmental toxicity, genotoxicity and other parameters. The toxicological safety evaluation of cultured meat should consider its unique but at the same time very familiar compositional profile, including any novel or engineered components.

It should be noticed that the toxicological evaluation of cell-cultured food, as any other Novel Food, should be assessed following the methodologies accepted by the Organisation for Economic Co-operation and Development (OECD) to ensure high quality standards and comparability.[38]

The potential risks associated with production processes, such as the use of animal-derived culture medium, synthetic scaffolds and growth factors, further emphasize the need for a tailored risk assessment strategy.[39] EFSA's tiered toxicity testing approach is well-suited for assessing cultured meat but requires adaptation due to the novel nature of its production. Cultured meat production introduces complexities that traditional toxicity methods may struggle to address. For instance, whole cultured-meat products cannot readily be tested using classic genotoxicity approaches. Instead, genotoxicity testing could focus on specific fractions of cultured meat, such as residual components of culture medium, additives and novel materials used in production.[40]

In vitro assays like the Ames and micronucleus tests can identify genetic damage, whereas *in vivo* studies can be considered only when uncertainties arise, to minimize ethical concerns.[41] Subchronic toxicity testing of cultured meat should examine potential adverse effects of its unique components, such as growth factors, genetically-modified cells and medium residues. For example, dose-dependent toxicity from residual growth factors or allergenic risk from plant-derived scaffold materials call for thorough evaluation.[42] Additionally, the undefined composition of some culture media presents challenges, because contaminants such as viruses, prions and bacteria may be present if the materials are not properly characterized. The production and processing of cultured meat also contribute to potential toxicological concerns. High temperatures and mechanical treatments during processing could create harmful by-products or alter the safety profile of additives. Scaffold materials, if not edible or degradable, may leave residues requiring removal and testing to ensure food safety. Regulatory standards for food-grade materials and bioreactor cleanliness are essential to mitigate these risks. In terms of post-market surveillance, consumer feedback and adverse

[38] The OECD is a forum and knowledge hub for data, analysis and best practices in public policy, see https://www.oecd.org/en.html. Accessed 11 April 2025.

[39] Gu et al. (2023), p. 491.

[40] Cobigo et al. (2024), p. 68.

[41] OECD guidelines are considered the benchmark reference for toxicological studies. In particular, OECD has released consensus documents on harmonisation of regulatory oversight in biotechnology, available at: https://www.oecd.org/en/search/publications.html?facetTags=oecd-content-types%3Apublications%2Freports%2Coecd-policy-issues%3Api55%2Coecd-serials%3Agl g3f7a8&path=%2Fcontent%2Foecd%2Fen%2Fpublications&orderBy=mostRelevant&page=0. Accessed 11 April 2025.

[42] For more information, Park et al. (2025), p. 692; Bomkamp et al. (2022), p. 3.

event monitoring are crucial for identifying unforeseen risks associated with cultured meat consumption. Nutritional vigilance systems, like those applied to other Novel Foods, could provide valuable data to refine risk assessments and ensure ongoing safety. Despite the advances in cultured meat technology, the lack of a history of safe use for many of its components highlights the importance of rigorous toxicological assessment. EFSA's recommendations state that risk assessments should address potential toxicological concerns arising from novel components, the influence of production processes and the absence of long-term data.[43]

Many food species, both of animal and plant origin, are known to cause allergic reactions because they express proteins with epitopes that can elicit an allergic response on ingestion. Cell-cultured food may therefore retain the allergenic potential of the cell source or undergo cross-contamination during production. In the former case, the allergen protein is seldom significantly eliminated or degraded during cell sourcing, production, harvesting or food processing. In the case of cross-contamination, the contaminating cell line may remain viable in culture and end up in the final product.[44] In such cases, allergens may be present in cell-based food, causing allergic reactions in some consumers. For species with limited consumption history, information on potential allergens is often scarce. Bioinformatic comparisons with known allergens in existing databases can help identify novel allergens or potential cross-reaction among proteins. However, since not all plant and animal sources have fully annotated genomes, additional data may be needed for comparisons. Cross-contamination between cell lines can be controlled by good manufacturing practices and record-keeping.

Among the main allergen-related safety challenges in cell-cultured food is the potential risk originating from genetic drift. During cultured meat production, cells replicate extensively in vitro, leading to potential cellular instability and genetic drift.[45] These factors are considered major contributors to phenotypic variations in cultured meat. Genome instability and genetic drift can be exacerbated by cell adaptation to in vitro selection pressure. Since the whole biomass is consumed, the cell output of cultured meat should be strictly controlled as it may contain undesirable proteins and/or metabolites that could exert adverse effects or be allergenic.[46] In order to create new or increased expression of allergens, cell modification must introduce or enhance substances capable of causing allergic reactions. The allergen must be expressed, remain in the cell culture, go undetected and be present in the final product at levels harmful for consumers.[47]

Proper allergenicity testing is therefore recommended, addressing new proteins when applicable or analysing expression levels of known ones. If a known allergen is detected or when cells are sourced from an unconventional species, the final product should be labelled accordingly.

[43] Hepburn et al. (2008), p. 12.

[44] Crevel et al. (2024), p. 2.

[45] See Li et al. (2019), p. 559; Soice and Johnston (2021), p. 2.

[46] Melzener et al. (2021), p. 7.

[47] Mariano et al. (2025), p. 10.

To address these challenges, EFSA has recently published a guidance document on the use of "read-across"[48] as a scientifically valid method for filling toxicological data gaps in the risk assessment of food and feed products.[49] This approach allows scientists to predict the toxicological properties of a novel substance (e.g., a new scaffold polymer or medium additive) by using data from structurally or functionally similar substances, known as source analogues. The read-across methodology involves a systematic comparison of chemical structure, biological activity, and metabolic pathways, along with uncertainty analysis and justification of similarity. This is especially relevant in the case of cultured meat, where many of the materials used have no history of safe use, and conducting full in vivo toxicity studies for every new component may not be practical, ethical, or scientifically necessary. By applying read-across in combination with New Approach Methodologies (NAMs) such as in vitro assays, in silico modelling, and omics-based tools, it is possible to build a robust safety assessment, while reducing reliance on animal testing.

This strategy ensures that risk assessments remain thorough and scientifically grounded, even within the context of emerging food technologies such as cultured meat. It also represents a shift toward evidence-based regulation that is responsive to advancements in both food innovation and toxicological science.

5 Cell-Cultured Food and EFSA: The Current Scenario

Despite the media hype and political debate, very little has yet been done to bring cell-cultured meat to market. Besides regulatory constraints, the main obstacles are still related to the process, mainly when market-scale production is required. In the meantime, EFSA has engaged with other international agencies to set up a common ground for evaluation.

The challenges in terms of safety assessment related to cell-culture-derived meat were addressed by EFSA in March 2023 during the Stakeholder Workshop "Future Food and Feed Lab",[50] together with other innovative technologies such as novel feed, nanotechnologies and edible insects. To delve deeper, EFSA organized a Scientific Colloquium on "Cell-culture-derived foods and food ingredients" in May

---

[48] Read-across is an approach used in chemical risk assessment for screening, classification, prioritisation and hazard assessment of substances. It is based on chemical/biological similarity. The process implies the use of known information from one or more (data-rich) source substances to predict the same property for a (data-poor) target substance. The full approach was outlined by OECD guidance documents, available at: https://www.oecd.org/en/publications/guidance-on-grouping-of-chemicals-second-edition_9789264274679-en.html. Accessed 11 April 2025.

[49] EFSA (2025). This is a very comprehensive document describing the use and the implementation of read-across methodologies in chemical risk assessment in food and feed, and it is of particular interest in the area of emerging technologies such as cell-cultivated food.

[50] More information at https://www.efsa.europa.eu/en/events/future-food-and-feed-lab-stakeholder-workshop-new-foodfeed-sources-and-technologies. Accessed 11 April 2025.

2023, bringing together 80 in-person and 550 online experts and stakeholders.[51] The main goal was to report ongoing activities in the field and discuss requirements for thorough risk assessment. Note that EFSA does not address cell-cultured meat *per se*, but it is always considered a part of cell culture-derived foods of animal or plant origin and food ingredients produced by precision fermentation. This is relevant, because while cell-cultured meat has increased the spotlight for public debate, plant or microbe cell-cultured food is closer to being ready for the market.

During the Colloquium, international case studies were described to showcase different national regulatory approaches. Representatives from the FAO and WHO discussed the status of emerging and advanced food production technologies, emphasizing sustainable practices and their support for Member States' policies. The FAO highlighted global developments in cell culture-derived food, including regulatory updates from Qatar and the USA, Israel's approval of cultured beef products, and Singapore's approval of cultured chicken ingredients.[52] The main safety issues related to production processes and methods were discussed, highlighting the critical points already discussed in the relevant section of the present chapter.

Among the major outcomes in terms of gaps and challenges, the experts underlined the need for methodological harmonization for assessment of the bioaccessibility/bioavailability of food in the gastrointestinal tract. Besides requirements set by the Organisation for Economic Co-operation and Development (OECD) consensus protocols for toxicological testing,[53] the experts also asked for New Approach Methodologies (NAMs)[54] to be included in risk assessment. Lastly, there was clear concern about the lack of consensus between risk assessors and risk managers on the acceptable level of identity/similarity between cell-cultured food and its conventional counterparts. Consensus is crucial for a comparative approach to risk assessment.

The EFSA positively assessed one Novel Food application for a plant-cell-culture-derived food ingredient and more specifically for the use of apple-cell-cultured biomass.[55] The latter was authorized in the European Union for use as a food supplement for an adult target-population in December 2023.[56] Interestingly,

[51] EFSA (2024).

[52] More information at https://www.sfa.gov.sg/food-safety-tips/food-risk-concerns/risk-at-a-glance/safety-of-alternative-protein. Accessed 11 April 2025.

[53] In order to ensure comparability and consensus, toxicological testing provided by FBOs to demonstrate the safety of Novel Foods must comply with OECD guidelines. The OECD is a forum and knowledge hub for data, analysis and best practices in public policy. Consensus documents for Novel Foods and feed are available at: https://www.oecd.org/en/topics/sub-issues/biosafety-novel-food-and-feed-safety/consensus-documents-on-the-safety-of-novel-foods-and-feeds.html. Accessed 11 April 2025.

[54] NAMs are considered to include any in vitro, in silico or chemistry-based method, as well as the strategies to implement them, that may provide information that could inform chemical safety assessment. For more information, see Westmoreland et al. (2022) and Sewell et al. (2024).

[55] EFSA (2023).

[56] Commission Implementing Regulation (EU) 2023/2847 of 20 December 2023 authorising the placing on the market of apple fruit cell-culture biomass as a Novel Food and amending Implementing Regulation (EU) 2017/2470.

the EFSA noted that the information provided on the cultivation and downstream process did not raise any safety concerns, whereas sharp differences in sodium, proteins and fibre have been reported between the composition of the apple-cell-culture biomass and fresh apples. In terms of potential risk, only allergenic potential was mentioned by the Panel.

In July 2024, the first application for cell-culture-derived food of animal origin was submitted to the European Commission. The request was submitted by the French food start-up Gourmey and concerned authorisation of "duck cells from cell culture".[57] The Novel Food, *foie gras*, is a traditional French product increasingly criticised for animal welfare reasons. Cell-cultured *foie gras* is an example of ethical use of cell-culture technologies, as the Novel Food would replace a highly controversial traditional preparation. *Foie gras* is made with livers of geese or ducks force-fed by gavage in preparation for slaughter. Although *foie gras* is a traditional delicacy in France, a number of EU governments have taken a stance against home production. The high market demand for the product makes cell-cultured *foie gras* an ethical and socially sustainable alternative. In the case of a positive assessment, Gourmey's duck product would be available to high-end European restaurants by 2026, while retail production would need time to scale up and reduce production costs. At the time of writing, the outcome of the risk assessment is not yet known.

While cell-cultured foods have transformative potential, particularly in terms of sustainability and public health benefits, including reducing greenhouse gas emissions and eliminating the need for antibiotics, their production faces several challenges. Key obstacles include optimizing cell culture media, scaling up bioreactors for mass production and achieving regulatory acceptance of novel biomaterials used in scaffolding. These challenges need to be overcome before cell-cultured foods can be accepted in the EU food system, and at the same time ensure safety, nutritional value and consumer trust. While there is still no urgent need for safety assessments of cell-based food products, preparedness is the key.[58] The competent authorities should therefore open dialogues with stakeholders, including consumers, the private sector, civil society, partner agencies and policy makers, not only to define the regulatory framework but also all the monitoring steps required to bring safe and nutritious cell-cultured food to the market in full transparency.

References

Aiking H, De Boer J (2020) The next protein transition. Trends Food Sci Technol 105:515–522

Alemanno A (2007) The evolution of European food regulation: why the European Food Safety Authority is not a EU-style FDA. In: Ansell C, Vogel D (eds) What's the beef? The contested

[57] The public summary of the application submitted by Suprême SAS (Gourmey) is available at: https://food.ec.europa.eu/document/download/45e22e48-b72b-4cbf-9ab6-5aac3debdbcb_en?filename=novel-food_sum_ongoing-not_2024-22222.pdf. Accessed 11 April 2025.

[58] More information can be retrieved in FAO (2022).

governance of European food safety. MIT Press, Bocconi Legal Studies Research Paper No. 1007451
Augustin MA, Hartley CJ, Malone G, Tyndall S (2023) Innovation in precision fermentation for food ingredients. Crit Rev Food Sci Nutr 64(18):6218–6238
Best BP (2015) Cryoprotectant toxicity: facts, issues and questions. Rejuvenation Res 18:422–436
Bhat ZF, Morton JD, Mason SL, Bekhit AEA, Bhat HF (2019) Technological, regulatory and ethical aspects of in vitro meat: a future slaughter-free harvest. Compr Rev Food Sci Food Saf 18:1192–1208
Bomkamp C, Skaalure SC, Fernando GF, Ben-Arye T, Swartz EW, Specht EA (2022) Scaffolding biomaterials for 3D cultivated meat: prospects and challenges. Adv Sci (Weinh) 9(3):e2102908
Cobigo LM, Habauzit D, Mehennaoui K, Gutleb AC, Serchi T, Fessard V, Hogeveen K (2024) Identifying in vitro toxicity testing approaches for (novel) proteins in the context of food and feed risk assessment. EFSA Supporting Publication 21(11): EN-9099
Crevel RWR, Verhoeckx K, Bøgh KL, Buck N, Chentouf A, Flanagan S, Galano M, Garthoff JA, Hazebrouck S, Yarham R, Borja G, Houben G (2024) Allergenicity assessment of new or modified protein-containing food sources and ingredients. Food Chem Toxicol 189:114766
Dabic M, Hjortsø CN, Marzi G, Vlačić B (2022) Guest editorial. Open innovation in the food industry: what we know, what we don't know, what we need to know. Br Food J 24:1777–1785
Eastham JL, Leman AR (2024) Precision fermentation for food proteins: Ingredient innovations, bioprocess considerations, and outlook – a mini-review. Curr Opin Food Sci 58:101–194
EFSA (European Food Safety Authority) (2024) EFSA Scientific colloquium 27: cell culture-derived foods and food ingredients. EFSA Supporting publication 2024: 8664–8710. https://www.efsa.europa.eu/en/events/efsas-scientific-colloquium-27-cell-culture-derived-foods-and-food-ingredients. Accessed 11 Apr 2025
EFSA NDA Panel (EFSA Panel on Dietetic Products, Nutrition and Allergies), Turck D, Bresson JL, Burlingame B, Dean T, Fairweather-Tait S, Heinonen M, Hirsch-Ernst KI, Mangelsdorf I, McArdle HJ, Naska A, Neuhäuser-Berthold M, Nowicka G, Pentieva K, Sanz Y, Siani A, Sjödin A, Stern M, Tomé D, Vinceti M, Willatts P, Engel KH, Marchelli R, Pöting A, Poulsen M, Salminen S, Schlatter J, Arcella D, Gelbmann W, de Sesmaisons-Lecarré A, Verhagen H, van Loveren H (2021) Guidance on the preparation and submission of an application for authorisation of a novel food in the context of Regulation (EU) 2015/2283 (Revision 1). EFSA J 19(3):6555
EFSA NDA Panel (EFSA Panel on Nutrition, Novel Foods and Food Allergens), Turck D, Castenmiller J, De Henauw S, Hirsch-Ernst KI, Maciuk A, Mangelsdorf I, McArdle HJ, Naska A, Pelaez C, Pentieva K, Siani A, Thies F, Tsabouri S, Vinceti M, Cubadda F, Frenzel T, Heinonen M, Marchelli R, Neuhauser-Berthold M, Poulsen M, Sanz Y, Schlatter JR, van Loveren H, Gelbmann W, Knutsen HK (2023) Scientific opinion on the safety of apple fruit cell culture biomass as a novel food pursuant to Regulation (EU) 2015/2283. EFSA J 21(7):8065
EFSA Scientific Committee, Bennekou SH, Allende A, Bearth A, Casacuberta J, Castle L, Coja T, Crépet A, Halldorsson T, Hoogenboom LR, Jokelainen P, Knutsen H, Koutsoumanis K, Lambré C, Nielsen S, Turck D, Civera AV, Villa RE, Zorn H, Serafimova R (2025) Guidance on the use of read-across for chemical safety assessment in food and feed. EFSA J 23(7):e9586
FAO (2022) Thinking about the future of food safety – a foresight report. https://doi.org/10.4060/cc4855en. Accessed 11 Apr 2025
FAO and WHO (2023) Food safety aspects of cell-based food. https://www.who.int/publications/i/item/9789240070943. Accessed 11 Apr 2025
Gomez Romero S, Spielmann I, Boyle N (2025) Genome-scale metabolic models in cultivated meat: advances, challenges, and future directions. Curr Opin Biotechnol 94:103313
Greenpeace Unearthed (2024). https://unearthed.greenpeace.org/2024/07/30/cultivated-backlash-livestock-industry-lobbying-europe-lab-grown-meat/. Accessed 11 Apr 2025
Gu Y, Li X, Chan ECY (2023) Risk assessment of cultured meat. Trends Food Sci Technol 138:491–499
Hadi J, Brightwell G (2021) Safety of alternative proteins: technological, environmental and regulatory aspects of cultured meat, plant-based meat, insect protein and single-cell protein. Foods 10(6):1226

Hall R (2023) Social movements, technological solutions to problems and political conflict: the case of alternative proteins. Purdue University Graduate School Thesis. https://doi.org/10.25394/PGS.24719775.v1. Accessed 11 Apr 2025

Hepburn P, Howlett J, Boeing H, Cockburn A, Constable A, Davi A, de Jong N, Moseley B, Oberdörfer R, Robertson C, Wal JM, Samuels F (2008) The application of post-market monitoring to novel foods. Food Chem Toxicol 46(1):9–33

Le Bloch J, Rouault M, Langhi C, Hignard M, Iriantsoa V, Michelet O (2025) The novel food evaluation process delays access to food innovation in the European Union. NPJ Sci Food 9:117

Lee DY, Lee SY, Yun SH, Jeong JW, Kim JH, Kim HW, Choi JS, Kim GD, Joo ST, Choi I, Hur SJ (2022) Review of the current research on fetal bovine serum and the development of cultured meat. Food Sci Anim Resour 42:775–799

Lee M, Kim H, Kang GH, Park S, Lee JH, Kim HJ (2024) Cultured meat with enriched organoleptic properties by regulating the biochemical compositions of fat. Nat Commun 15:44359

Li J, Settivari RS, LeBaron MJ (2019) Genetic instability of in vitro cell lines: implications for genetic toxicity testing. Environ Mol Mutagen 60:559–562

Melzener L, Verzijden KE, Buijs AJ, Post MJ, Flack JE (2021) Cultured beef: from small biopsy to substantial quantity. J Sci Food Agric 101:7–14

Ong KJ, Johnston J, Datar I, Sewalt V, Holmes D, Shatkin JA (2021) Food safety considerations and research priorities for the cultured meat and seafood industry. Compr Rev Food Sci Food Saf 20:5421–5448

Ong KJ, Tejeda-Saldana Y, Duffy B, Holmes D, Kukk K, Shatkin JA (2023) Cultured meat safety research priorities: regulatory and governmental perspectives. Foods 12(14):2645

Park SM, Ryoo JH, Kwon HC, Han SG (2025) Scaffold biomaterials in the development of cultured meat: a review. Food Sci Anim Resour 45(3):688–710

Piantino M, Muller Q, Nakadozono C, Yamada A, Matsusaki M (2025) Towards more realistic cultivated meat by rethinking bioengineering approaches. Trends Biotechnol 43(2):364–382

Post MJ (2022) Cultured meat from stem cells: challenges and prospects. Meat Sci 92:297–301

Post MJ, Levenberg S, Kaplan DL, Genovese N, Fu J, Bryant CJ, Negowetti N (2020) Scientific, sustainability and regulatory challenges of cultured meat. Nat Food 1(7):403–415

Powell DJ, Li D, Smith B, Chen WN (2025) Cultivated meat microbiological safety considerations and practices. Compr Rev Food Sci Food Saf 24(1):e70077

Rischer H, Szilvay GR, Oksman-Caldentey KM (2020) Cellular agriculture – industrial biotechnology for food and materials. Curr Opin Biotechnol 61:128–134

Santo RE, Kim BF, Goldman SE, Dutkiewicz J, Biehl EMB, Bloem MW, Neff RA, Nachman KE (2020) Considering plant-based meat substitutes and cell-based meats: a public health and food systems perspective. Front Sustain Food Syst 4:134

Sewell F, Alexander-White C, Brescia S, Currie RA, Roberts R, Roper C, Vickers C, Westmoreland C, Kimber I (2024) New approach methodologies (NAMs): identifying and overcoming hurdles to accelerated adoption. Toxicol Res 13(2):tfae044

Soice E, Johnston J (2021) Immortalizing cells for human consumption. Int J Mol Sci 22(21):11660

Stephens N, Di Silvio L, Dunsford I, Ellis M, Glencross A, Sexton A (2018) Bringing cultured meat to market: technical, socio-political and regulatory challenges in cellular agriculture. Trends Food Sci Technol 78:155–166

Westmoreland C, Bender HJ, Doe JE, Jacobs MN, Kass GEN, Madia F, Mahony C, Manou I, Maxwell G, Prieto P, Roggeband R, Sobanski T, Schütte K, Worth AP, Zvonar Z, Cronin MTD (2022) Use of new approach methodologies (NAMs) in regulatory decisions for chemical safety: report from an EPAA Deep Dive Workshop. Regul Toxicol Pharmacol 135:105261

Legal and Policy Debate

Cell-Based Meat in the European Union: A Regulatory Crossroads

Maria Giulia Corazza and Giulia Formici

Abstract In recent years, we have witnessed the emergence of cell-based meat as a transformative innovation, bearing significant implications for the ongoing debate on the European Union (EU) regulatory framework. The inherently novel and ethically debated nature of cultivated meat calls for a comprehensive and forward-looking regulatory approach—one that can address not only scientific and safety concerns, but also societal, economic, ethical and environmental dimensions. As the discussion around the legitimacy of national provisions limiting the marketing of this product continues to evolve, it remains to be seen whether the current Novel Foods Regulation will prove sufficiently robust, or if further legislative reforms will be required to effectively integrate the emerging sector of cellular agriculture into the EU's regulatory landscape. This chapter explores the recent debate and political initiatives put forward by various Member States, as well as the responses of different EU institutions, offering an in-depth analysis of future scenarios and challenges.

Keywords Regulation of cell-based meat · Novel Foods · Regulation (EU) 2015/2283 · Harmonization · Precautionary principle

While the chapter has been elaborated jointly by the co-Authors, Sects. 1 and 2 were written by M G Corazza, and Sects. 3 and 4 by G Formici.

M. G. Corazza · G. Formici (✉)
Department of Law, Politics and International Studies, Department of Excellence 2023–2027 "Food for Future", University of Parma, Parma, Italy
e-mail: mariagiulia.corazza@unipr.it; giulia.formici@unipr.it

G. Formici et al. (eds.), *Cell-Based Meat in the European Union and Beyond*,
https://doi.org/10.1007/978-3-032-14624-3_4

1 Alternative Protein Sources and the European Union Agri-Food Political Agenda: An Ongoing Discussion

Finding new alternative protein sources is a recently debated objective on the European Union (EU) agri-food political agenda.[1] In fact, EU Institutions themselves have progressively recognized this necessity,[2] recently acknowledging the importance of innovative protein sources as "additional mean to preserve food security and sustainability".[3] Non-traditional proteins encompass a wide range of sources and types. These include plant-based, fungal, microbial, insect-based[4] and algal (macro- and microalgae) options, as well as emerging biotechnologies[5] and cultured meat.[6] Such diversification of protein sources could revolutionize the agri-food market, addressing major global challenges such as climate change, the environmental impact of food production and food insecurity, reducing the reliance on traditional animal-based proteins.

At the same time, advances in this field and cutting-edge scientific innovations have enabled the development of protein alternatives that closely mimic the sensory and nutritional characteristics of conventional proteins from animal sources.[7] Nonetheless, the potential negative externalities of such technologies should not be

---

[1] See Lähteenmäki-Uutela et al. (2021).

[2] Among the recent frequently cited documents adopted by the EU, one traces increasingly explicit recognition of the need to support sustainable protein production, from the Communication from the Commission to the European Parliament, the Council, the European Economic and Social Committee and the Committee of the Regions—A Farm to Fork Strategy for a fair, healthy and environmentally-friendly food system, COM(2020) 381 final, part of the European Green Deal, COM(2019) 640 final, to the European Parliament resolution of 19 October 2023 on a European Protein Strategy (2023/2015(INI)).

[3] See *Research and Innovation*, updated version December 2024, p. 12, https://agriculture.ec.europa.eu/farming/crop-productions-and-plant-based-products/cereals/reducing-plan-protein-deficit-eu_en (European Commission 2024). Accessed 31 March 2025.

[4] On this vast topic, see Formici and Scaffardi (2022) and Montanari et al. (2021), and for a strict legal analysis, Paganizza (2019).

[5] In this regard, see the Communication from the Commission to the European Parliament, the Council, the European Economic and Social Committee and the Committee of the Regions—Building the future with nature: Boosting Biotechnology and Biomanufacturing in the EU, COM(2024)137 final.

[6] For an overview of alternative protein source policies, see *Cell-Based Meat and Food Policy* by Mancini in this volume.

[7] In this context, comprehensive innovation is also acknowledged as being of fundamental importance by the EU institutions. See *Research and Innovation*, p. 13: "R&I is further needed to develop solutions to improve taste/texture, reduce production costs and increase the market uptake of alternative proteins." https://agriculture.ec.europa.eu/farming/crop-productions-and-plant-based-products/cereals/reducing-plan-protein-deficit-eu_en. Accessed 31 March 2025. See also the very recent Communication from the Commission to the European Parliament, the Council, the European Economic and Social Committee and the Committee of the Regions—A Vision for Agriculture and Food: Shaping together an attractive farming and agri-food sector for future generations, COM(2025) 75 final.

overlooked: to cite two well-known examples, the food safety and energy sustainability of these products have often been questioned by the public opinion and further studies and research have been invoked by the scientific community.[8] Moreover, while several of these advancements promise to reduce the environmental and ethical concerns associated with livestock farming, they also raise unprecedented and delicate questions regarding animal welfare.[9] In this dynamic and rapidly evolving landscape, cultivated meat has emerged as a particularly prominent and promising area of innovation, gaining increasing momentum in the political and legislative spheres.

The present chapter therefore provides an overview of the regulatory issues and hurdles related to the marketing of cell-based meat in the EU. The starting point is the so-called Novel Food Regulation (NFR),[10] the current legal framework pertaining to cultured foods.[11] Section 2 delves into the established authorisation process, paying specific attention to the objectives and evolution of the abovementioned regulation (Sect. 2.1), as well as to the roles of different institutional actors and the procedural solutions to guarantee comprehensive risk management (Sects. 2.2 and 2.3).

Notwithstanding the solid Novel Foods legislative framework, the topic of cell-based foods and cultivated meat has recently sparked unprecedented debate, particularly since 2023, when Member State initiatives and calls for reform were first actioned. In fact, that year inaugurated what have been defined as "centrifugal forces"[12] or national regulatory solutions departing from the existing supranational Novel Foods discipline, either by promoting or banning cell-based meat in their territory. Section 3.1 examines such interventions and the complex and controversial discussion that has arisen, especially with regard to their compliance with EU law.

The push towards legislative fragmentation has also been accompanied by the promotion of different "centripetal forces", advocating a reformative response at

[8] See *Who Hates Cultured Meat? Examining Public Opposition to Cellular Agriculture* by Broad and Glenna and *What Is Cell-Based Meat? Food Safety and Sustainability Aspects* by Cheli et al. in this volume.

[9] Considering cell-based meat, its production has been promoted as a more humane alternative to conventional meat, as it avoids the slaughter of animals. However, as noted in *Animal Welfare as a Constitutional Value: The Debate in Comparative Law and the Emerging Role of Cultivated Meat* by Piciocchi in this volume, there are still concerns about aspects such as the use of Fetal Bovine Serum in the cultivation process and its implications for animal welfare. In general, the evolving legal and ethical discussions surrounding cultivated meat but also insects cannot be therefore separated from broader debates on animal rights and the role of animals in food production, discussing how existing frameworks for animal welfare intersect with emerging food technologies.

[10] Regulation (EU) 2015/2283 of the European Parliament and of the Council of 25 November 2015 on novel foods, amending Regulation (EU) no 1169/2011 of the European Parliament and of the Council and repealing Regulation (EC) no 258/97 of the European Parliament and of the Council and Commission Regulation (EC) no 1852/2001.

[11] This chapter focuses on the marketing authorisation procedure for these products. For the rules concerning labelling, see *The Regulatory Landscape of Cultured Meat and Fish: Terminology and Consumer Information Provisions* by Sirsi and Rolandi in this volume.

[12] Formici (2024) and Formici and Scaffardi (2025).

supranational level. Several Member States claim that the current EU regulatory scenario is inadequate for dealing with the specific challenges of cell-based foods,[13] and urge modification of the Novel Foods Regulation, expressing a position that seems to be backed by the European Parliament. Analysing this multi-level dialogue, in Sect. 3.2 we also consider the—at least, by now—clear standpoint of the European Commission, underlining the importance of a harmonised approach and the effectiveness of the current regulatory framework.

Moving from this examination, Sect. 4 promotes some reflections on the possible consequences of the different yet interrelated centripetal and centrifugal forces. Setting aside merely political evaluations, based on ideological and nationalist reasoning, the regulatory debate on cell-based meat merits thorough examination: by seriously challenging—for the first time—the Novel Foods Regulation, the discussion opens vaster questions concerning the relationship between EU institutions and national legislators in matters of risk management, the role of independent agencies, and the interplay between scientific considerations and political decisions. While the conclusions cannot provide definitive answers, they offer elements for reflection on the potentially disruptive implications of the current confusing regulatory discussion in the EU.

2 The EU Discipline Applicable to Cell-Based Meat: The Novel Foods Regulation

Evidence of the pioneering potential of cultivated meat is apparent in the increasing number of authorisation requests, as well as the growing breadth of policy and industry discussions surrounding the topic. The submission of applications to market this new product in the European Union has just begun. In July 2024, the French company Gourmey submitted the first request for the approval of cultivated *foie gras*,[14] marking a milestone in the history of the authorisation process in the EU market. This was followed in January 2025 by Mosa Meat's application[15] for the authorisation of cultivated beef fat, further highlighting the sector's scientific progress, while underscoring the urgent need for a thorough assessment of the

[13] See Rossi and Cazzini (2025).

[14] See EIT Food (2024), https://www.eitfood.eu/news/gourmey-files-the-first-eu-novel-food-submission-for-cultured-meat. Accessed 31 March 2025; Good Food Institute—GFI (2024), https://gfieurope.org/blog/first-ever-regulatory-submission-made-for-cultivated-meat-in-the-eu/. Accessed 31 March 2025. The application is available in the *Summary of applications and notifications* of the European Commission website, https://food.ec.europa.eu/document/download/45e22e48-b72b-4cbf-9ab6-5aac3debdbcb_en?filename=novel-food_sum_ongoing-not_2024-22222.pdf. Accessed 31 March 2025. See also Reinhardt et al. (2024).

[15] See Mosa Meat's declaration, https://mosameat.com/blog/submitting-our-first-eu-market-authorisation-request. Accessed 31 March 2025 (Mosa Meat 2025).

robustness of the regulatory framework governing the authorisation and commercialisation of cell-based foods.

However, discussion on the regulatory challenges posed by the advent of these innovations predates these formal applications, as explored in the following sections. A pivotal starting point for understanding this evolving debate is the current NFR, with particular emphasis on the authorisation procedure for new foods.[16]

2.1 *The Evolution and Objectives of the Novel Foods Regulation in Europe*

The commercialisation of Novel Foods (NF)[17] provides an illustration of the regulatory challenges brought about by food innovation and food sustainability. Besides traditional foods coming from Third Countries, this category includes cutting-edge food products, ranging from those with modified molecular structures to those developed by non-traditional production methods, such as innovative foods derived from cell cultures.[18] At European level, Novel Foods have been disciplined since 1997, initially under Regulation (EC) no 258/1997[19] and subsequently under Regulation (EU) 2015/2283. The overarching ambition of this regulatory framework is twofold: firstly, to ensure the safety of foodstuffs through the establishment of rigorous hygiene and health standards, as delineated by an extensive European legislative framework; secondly, to support the functioning of the Single Market while fostering innovation, sustainability and food security, as emphasised in Recital

[16] The authorisation is not the only procedure established by the NFR: the notification procedure under Article 14 *et seq.* of Regulation (EU) 2015/2283 is also provided. It is not addressed here, as it pertains to a different category of Novel Foods, namely those coming from Third Countries. For a definition of "traditional foods", see Article 3 (c) of Regulation (EU) 2015/2283 "traditional food from a third country means novel food as defined in point (a) of this paragraph, other than novel food as referred to in points (a) (i), (iii), (vii), (viii), (ix) and (x) thereof which is derived from primary production as defined in point 17 of Article 3 of Regulation (EC) no 178/2002 with a history of safe food use in a Third Country". With reference to this particular category of 'new' foods, the applicant is asked to demonstrate that the product has been consumed continuatively "in at least one third country for at least 25 years as a part of the customary diet of a significant number of people. The history of safe food use should not include non-food uses or uses not related to normal diets" (see Recital 15 of Regulation (EU) 2015/2283). On this specific category, see Scaffardi (2022) and Formici (2020). For a detailed analysis of the removal of the notification procedure for substantially equivalent foods, see Paganizza (2024), p. 802. For a comprehensive overview of the notification procedure, see Rusconi and Cesana (2024), specifically pp. 593–594.

[17] Note that Novel Foods do not include genetically modified organisms (GMOs). Food products obtained through the use of GMOs therefore do not fall within the scope of the NFR, as they are governed by the specific regulatory framework applicable to GMOs, in particular Regulation (EC) no 1829/2003 on genetically modified food and feed. See Seehafer and Bartels (2019).

[18] Scaffardi (2020a, b).

[19] Regulation (EC) no 258/1997 of the European Parliament and of the Council of 27 January 1997 concerning novel foods and novel food ingredients.

29 of Regulation (EU) 2015/2283. It also aspires to ensure the protection of consumer interests.[20]

Under both Regulations, a food product is considered "novel" if it meets two criteria: (1) it was not used for human consumption to a significant degree[21] in the EU before 15 May 1997 (temporal criterion); (2) it falls in one of the ten subcategories defined in Article 3(2) of the current Regulation (substantive criterion).[22] Cultivated meat clearly conforms to the definition of a Novel Food as outlined in this context. Indeed, it was not available for human consumption in the EU before 15 May 1997, and it falls in the category outlined in Article 3(2)(a)(VI) of Regulation (UE) 2015/2283, which explicitly includes "food consisting of, isolated from or produced from cell culture or tissue culture derived from animals".[23] Cultivated meat is therefore subject to the regulatory procedures established by the NFR for market approval.[24]

The structured authorisation procedure has evolved over time. The previous regulatory framework required pre-market authorisation, based on the assumption that new food products could pose risks to consumer health.[25] This procedure was partly decentralized: the initial evaluation of authorisation applications was handled at national level, Member States being responsible for assessing the scientific documentation on potential risks submitted by applicants. If no objections were raised, the authorisation process concluded at national level; otherwise, the procedure was taken to EU level for further assessment.[26]

Although innovative for its time, the system was slow, costly and inefficient, particularly in view of the increasing number of Novel Food applications.[27] Regulation (EU) 2015/2283 was then introduced. It centralised the authorisation procedure with a view to streamlining it and unified the authorisation iter in order to

---

[20] See Article 1(2) Regulation (EU) 2015/2283. For a detailed analysis of the objectives of the NFR, see also Paganizza (2024), p. 814, where she explains that it is a "regulation ensuring a reinforced risk assessment process to guarantee the marketing of safe foods that do not mislead consumers in any way, while also safeguarding the proper functioning of the internal market" (translation).

[21] See Information and Guidance Document "Human Consumption to a Significant Degree" in https://food.ec.europa.eu/document/download/7d4bba03-4e60-4a2a-be97-76db96d47974_en?filename=novel-food_guidance_human-consumption_en.pdf. Accessed 31 March 2025.

[22] Formici and Scaffardi (2022), in particular Volpato (2022).

[23] Albisinni (2025), p. 107; Paganizza (2024), p. 814.

[24] Monaco (2025).

[25] In this respect, see also the reference to the precautionary principle in Recital 20 of the current Regulation (EU) 2015/2283, which states: "Novel foods should be authorised and used only if they fulfil the criteria laid down in this Regulation. Novel foods should be safe and if their safety cannot be assessed and scientific uncertainty persists, the precautionary principle may be applied. Their use should not mislead the consumer. Therefore, where a novel food is intended to replace another food, it should not differ from that food in a way that would be nutritionally less advantageous for the consumer." See also Article 7 (a) of Regulation (EU) 2015/2283.

[26] Scaffardi (2020b); Volpato (2022), p. 20; Paganizza (2024), p. 806. See also Schebesta and Purnhagen (2024) and Rizzioli (2025).

[27] Volpato (2022), p. 21. See also Scarpa and Dalfra (2008) and Hyde et al. (2017).

avoid the creation of an inequitable environment, that might otherwise result from the disparate positions of the various Member States involved.[28] The revised framework requires applications to be submitted directly to the Commission, rather than to individual Member States. It establishes clear timelines and designates a single authority, the European Food Safety Authority (EFSA),[29] to conduct rigorous risk assessments.

2.2 *The Authorisation Procedure Between Risk Assessment and Risk Management: Key Actors and Phases*

Since 1 January 2018, when Regulation (EU) 2015/2283 came into force, the European Commission has been responsible for the initial evaluation of Novel Food applications. The competent authority, namely the Commission itself, may initiate the procedure either *ex officio* or on request from a Member State, a non-Member State, a legal entity or even a natural person.[30] The central role of the Commission in this process is unmistakable: the ultimate decision on how to manage the risks associated with Novel Foods rests with this Institution, which retains the decision-making authority over the risk management phase.[31] In fact, in order to reach a well-founded decision and appropriately balance the various interests at stake,[32] the Commission conducts a thorough multifaceted analysis of the potential consequences of allowing a Novel Food into the Single Market—an evaluation that ultimately leads to a decision of a political nature, that should also be approved through the vote of the Standing Committee on Plants, Animals, Food and Feed (PAFF). This intricate process is reflected in the authorisation procedure that Novel Foods have to undergo before they can be marketed in EU.

[28] See Brookes (2007), Volpato (2022) and Paganizza (2024).

[29] On EFSA's role, see Martini et al. (2020) and Alemanno and Gabbi (2014).

[30] The dossier submitted by the original applicant must include all the elements required under Article 10(2) of Regulation (EU) 2015/2283. Also, refer to the guidelines provided by EFSA (2024a) https://efsa.onlinelibrary.wiley.com/doi/10.2903/sp.efsa.2024.EN-9041. Accessed 31 March 2025.

[31] Regarding the concept of risk management and the responsibility of the Commission, see Volpato (2022), p. 28, where "tensions... related to risk regulation, where scientific complexities, value judgments, and consumer sensibilities need to be accommodated in the design of the decision-making procedure" are highlighted, to be coordinated with the progressive creation of a European space of integrated administration, in the context of a strongly harmonised and non-derogable procedure.

[32] "Such as innovation and safety in the agri-food sector, sustainability concerns and ethical values, as well as cultural sensibilities and consumer perceptions", see Volpato (2022), p. 32.

In particular, once an application passes a formal preliminary check,[33] it is published on the Commission's website, notifying Member States.[34] Within 1 month, at the Commission's request,[35] the dossier is forwarded to EFSA, which conducts a scientific evaluation within a period of 9 months[36] to assess the product's safety and identify potential risks.[37] This crucial stage of risk assessment culminates in a scientific opinion, which forms the basis for the evaluations conducted by the Commission and other actors involved in the authorisation procedure.[38] The scientific risk assessment carried out by the independent authority based in Parma is preliminary to all other evaluations (economic, social, political) for a rather intuitive reason: although EFSA's opinion is not legally binding, if the independent authority identifies a potential risk to human health posed by a Novel Food, the procedure is presumably halted.[39] While equally important in the subsequent risk management phase, a positive scientific evaluation is therefore, *de facto*, foundational for the later evaluations of a different nature.[40]

---

[33] The scientific and administrative requirements for the submission of a valid application are set out in Commission Implementing Regulation (EU) 2017/2469 of 20 December 2017 laying down administrative and scientific requirements for applications referred to in Article 10 of Regulation (EU) 2015/2283 of the European Parliament and of the Council on novel foods. The application must be accompanied by a cover letter, a technical dossier containing administrative and scientific data sufficient to enable a comprehensive risk assessment, and an executive summary of the dossier — see Rusconi and Cesana (2024), pp. 586–587.

[34] The publication of authorisation proposals and accompanying scientific dossiers (see also note 1, *Summary of applications and notifications*) has raised important considerations regarding the protection of data submitted as part of such applications. On this point, see Article 23 of Regulation (EU) 2015/2283, which allows applicants, with appropriate justification, to request the confidentiality of specific parts of their dossier. Regarding transparency and the sustainability of risk analysis in the EU food chain, see also Rusconi and Cesana (2024), p. 587, particularly their commentary on Regulation (EU) 2019/1381 of the European Parliament and of the Council of 20 June 2019 on the transparency and sustainability of the EU risk assessment in the food chain and amending Regulations (EC) no 178/2002, (EC) no 1829/2003, (EC) no 1831/2003, (EC) no 2065/2003, (EC) no 1935/2004, (EC) no 1331/2008, (EC) no 1107/2009, (EU) 2015/2283 and Directive 2001/18/EC. Further provisions on data protection are set out in Article 26 of Regulation (EU) 2015/2283: see Volpato (2022), p. 27; Holle (2014); Simpson (2016).

[35] The Commission is not, in fact, obliged to request an opinion from EFSA where the information provided in the original dossier is deemed sufficient to ensure a high level of scientific certainty regarding the safety of the food in question. See Article 10(3) of Regulation (EU) 2015/2283.

[36] This period may, however, be extended in cases where EFSA requires additional information from the applicant. See Article 11(4) of Regulation (EU) 2015/2283.

[37] The structure and content of EFSA's scientific opinion are defined by Implementing Regulation (EU) 2017/2469. See Rusconi and Cesana (2024), p. 590; see also Dall'Asta (2022). See also EFSA (2024b).

[38] See *Safety Assessment of Cell-Culture-Derived Food: The Role of the European Food Safety Authority* by Dall'Asta and Gjoni in this volume.

[39] As stated in Article 12(1)(c) Regulation (EU) 2015/2283.

[40] The risk assessment procedure carried out by EFSA is defined as "a scientifically-based process consisting of four steps: hazard identification, hazard characterisation, exposure assessment and risk characterisation", https://www.efsa.europa.eu/en/glossary/steps-risk-assessment. Accessed 31

Following EFSA's risk assessment,[41] the Commission drafts an implementing act,[42] act, approving or not the product's market entry. The draft should be prepared by the Commission based not only on EFSA's scientific evaluations, as previously discussed, but also taking into account the fundamental principles of the European General Food Law (GFL),[43] particularly the precautionary principle, among others.[44] The Commission shall also consider the so-called "other legitimate factors relevant to the application",[45] the importance of which cannot be ignored, especially in the case of a controversial Novel Food like cell-based meat. This provision broadens the scope of the risk management, preventing the Commission from relying solely on scientific evidence and urging it to evaluate additional factors deemed equally important, "including societal, economic, traditional, ethical, and environmental considerations, as well as the feasibility of controls."[46]

The draft, in the form of a proposal for an implementing act, including conditions of use and specific labelling requirements, is then reviewed by the PAFF Committee, which comprises representatives from Member States. The PAFF Committee was established pursuant to Article 58(1) of Regulation (EC) no 178/2002, with the specific mandate of assisting the European Commission and deliberating on matters within the scope of its functions as defined by EU legislation. In this context, the Committee forms part of the EU Comitology system: "a

---

March 2025 (EFSA 2024c). On the relationship between the scientific risk assessment and the subsequent risk management phase entrusted to the Commission, see Gabbi (2007).

[41] On the elements to be taken into account by EFSA in this phase, see Article 11 of Regulation (EU) 2015/2283.

[42] The draft implementing act must be submitted within 7 months of the date of publication of the scientific opinion. See Article 12(1) of Regulation (EU) 2015/2283.

[43] Regulation (EC) no 178/2002 of the European Parliament and of the Council of 28 January 2002 laying down the general principles and requirements of food law, establishing the European Food Safety Authority and setting out procedures in matters of food safety.

[44] See Article 12(1)(b) of Regulation (EU) 2015/2283, which refers to Article 7 of Regulation (EC) no 178/2002.

[45] See Article 12(1)(d) of Regulation (EU) 2015/2283. For insightful legal considerations on the topic, see Volpato (2022), Bellenghi and Knuth (2024) and Formici (2024). For considerations regarding the specific category of "other legitimate factors" to be taken into account, see Part "Socio-Economic and Cultural Issues" of this volume.

[46] Regulation (EC) no 178/2002, Recital 19. For a critical analysis of the political role of the Commission's decisions concerning the consideration of "other legitimate factors", see Bellenghi and Knuth (2024), pp. 58–61. The reference to "other legitimate factors" has already been addressed in the White Paper on Food Safety, COM(1999) 719 final, Recital 15: "In the decision-making process in the EU, other legitimate factors relevant for the health protection of consumers and for the promotion of fair practices in food trade can also be taken into account. The definition of the scope of such legitimate factors is presently being studied at the international level, particularly in Codex Alimentarius. Examples of such other legitimate factors are environmental considerations, animal welfare, sustainable agriculture, consumers' expectations regarding product quality, fair information, and the definition of the essential characteristics of products and their process and production methods." Further references to such evaluations can be found in Formici (2024) and Volpato (2022). A previous (and ultimately predictive) evaluation is that of Szajkowska (2011).

framework of procedures, including committee meetings with Member State representatives, that enables Member States to influence the adoption of implementing acts".[47] In fact, the adoption of the act proposed by the Commission must be supported by the PAFF Committee, which evaluates the proposal through an examination procedure.[48] The Committee discusses the proposal and votes by qualified majority, the vote outcome determining the subsequent steps in the process: if the draft is approved by a qualified majority, the Commission is obliged to adopt it, although exceptions may be made if contrary new circumstances arise. In the case of a negative opinion, however, the Commission is blocked from adopting the act, effectively resulting in a veto. The Commission can then choose between two alternative courses of action: amend the act or refer it to the Appeal Committee.[49]

If, after the PAFF Committee phase, the final authorisation is granted—now in a general rather than product-specific manner, as was the case under the previous framework[50]—the Novel Food is added to the Union List established by Regulation (EU) 2017/2470,[51] thereby becoming legally available for sale and consumption across the EU. The implementing act (typically in the form of a regulation to ensure general applicability) concludes the procedure, leading to updating of the Union List of Novel Foods.[52]

---

[47] As stated in https://commission.europa.eu/law/law-making-process/adopting-eu-law/implementing-and-delegated-acts/comitology_en. Accessed 31 March 2025.

[48] Article 12(1) of Regulation (EU) 2015/2283, which refers to Article 30 of Regulation (EU) 2015/2283.

[49] For this procedure, see Volpato (2022), p. 25; on the role of comitology, also, Bellenghi and Knuth (2024).

[50] See Volpato (2022), p. 27: "the authorisation is not in the form of a decision with an individual addressee, namely the applicant, but it has general effects".

[51] Commission Implementing Regulation (EU) 2017/2470 of 20 December 2017 establishing the Union list of novel foods in accordance with Regulation (EU) 2015/2283 of the European Parliament and of the Council on novel foods.

[52] See Article 9 of the Regulation (EU) 2015/2283. It should be emphasised, however, that the authorisation applies solely and specifically to the food product for which the application was originally submitted—as a consequence, with regard to cell-based meat, the hypothetical authorisations resulting from the aforementioned submissions would apply only to cultured *foie gras* and beef fat, and not to cell-based meat in general. If the authorisation is denied, the decision is binding solely for the applicant, without extending *erga omnes*. It is the responsibility of the relevant authorities in the Member States to assess, on a case-by-case basis, whether a product marketed within their jurisdiction corresponds to the Novel Food for which authorisation has been denied. It must be noted that this interpretation originates from case law relating to the previous Regulation (EC) no 258/1997, as demonstrated in Court of Justice of the European Union (CJEU) Judgement (14 April 2011) Case C-327/09, *Mensch und Natur AG v Freistaat Bayern*, para 36: "A Commission decision taken on the basis of Article 7 of Regulation (EC) no 258/97 concerning novel foods and novel food ingredients, refusing authorisation to place a food or food ingredient on the European Union market, is not binding on any persons other than the person or persons whom that decision specifies as its addressees. By contrast, the competent authorities of a Member State must determine whether a product marketed in the territory of that Member State, the characteristics of which appear to match those of the product subject to that Commission decision, is a novel food or novel

2.3 Regulatory Outcomes and Systemic Implications

The Commission currently employs two legal instruments to classify all potential Novel Foods that are required to undergo the authorisation procedure. The first of these is the Union List, a binding instrument that defines the constitutive effects of authorisation for trade within the EU. It also imposes mandatory requirements[53] related to the specification, conditions of use, additional specific labelling requirements,[54] and post-market monitoring obligations associated with the authorisation.[55] Importantly, inclusion of a Novel Food in the Union List is the result of an implementing act adopted by the Commission, which is legally binding in its entirety and directly applicable in all Member States.

The second instrument is the Catalogue of Novel Foods, a non-binding reference tool indicating the legal status of certain foods by categorizing them accordingly. Its effects are purely declaratory, serving as a useful reference for stakeholders, particularly in the preliminary stages of an application for authorisation or notification.

It classifies products into the following categories:

- NOT NOVEL IN FOOD – According to the information available to the Member States' competent authorities, this product was used for human consumption to a significant degree within the Union before 15 May 1997. Thus, it is not considered to be "novel" according to the provisions of the Novel Food Regulation (EU) 2015/2283 and its access to the market is not subject to the pre-market authorisation in accordance with Regulation (EU) 2015/2283.[56]
- NOT NOVEL IN FOOD SUPPLEMENTS – According to the information available to Member States' competent authorities, this product was used in food supplements in the EU before 15 May 1997. Therefore, its use in food supplements is not considered to be

food ingredient within the meaning of Article 1(2) of that regulation, and, where necessary, must require the person concerned to comply with the provisions of that regulation."

[53] See Article 9(3) of the Regulation (EU) 2015/2283.

[54] The general conditions for inclusion of novel foods in the Union List, as outlined in Article 7 of Regulation (EU) 2015/2283, are mirrored and aligned with the objectives of the authorisation procedure, as defined in Article 1 of the same Regulation. Specifically, these conditions stipulate that "The Commission shall only authorise and include a novel food in the Union list if it complies with the following conditions: (a) the food does not, based on the scientific evidence available, pose a safety risk to human health; (b) the food's intended use does not mislead the consumer, especially when the food is intended to replace another food and there is a significant change in its nutritional value; (c) where the food is intended to replace another food, it does not differ from that food in such a way that its normal consumption would be nutritionally disadvantageous for the consumer." See also *The Regulatory Landscape of Cultured Meat and Fish: Terminology and Consumer Information Provisions* by Sirsi and Rolandi in this volume.

[55] See Article 24 of the Regulation (EU) 2015/2283.

[56] The European Commission's website also states as follows: "However, other legislation may restrict the placing on the market of this product as a food in the EU or in some Member States. It is therefore recommended to check with the competent authority(ies) of the Member State(s)." See https://ec.europa.eu/food/food-feed-portal/screen/novel-food-catalogue/search. Accessed 31 March 2025.

novel and is not subject to the pre-market authorisation in accordance with Regulation (EU) 2015/2283.[57]

- NOVEL FOOD – According to the information available to Member States' competent authorities, this product was not consumed in the EU to a significant degree as a food before 15 May 1997. Therefore, a pre-market authorisation in accordance with Regulation (EU) 2015/2283 is required before it can be placed as food on the EU market.
- AUTHORISED NOVEL FOOD – This is an authorised novel food and it has been included in the Union list established by Commission Implementing Regulation (EU) 2017/2470.
- SUBJECT TO A CONSULTATION REQUEST – This product is currently the subject of a consultation request to a Member State submitted pursuant to the requirements of Implementing Regulation (EU) 2018/456 on the procedural steps of the consultation process for determination of novel food status. The outcome of the request will be made publicly available on this website (https://ec.europa.eu/food/safety/novel_food/consultation-process_en).[58]

This analysis of the NFR suggests several provisional conclusions. Despite enduring criticalities,[59] the Novel Foods Regulation has proven to be an effective instrument in safeguarding food safety, fostering innovation and preserving the integrity of the harmonised Single Market in the context of emerging food technologies.[60] This can be attributed to convergence of the key actors in the procedure (the Commission, EFSA and PAFF Committee), who worked together to prevent inconsistencies and ensure that the authorisation system operates smoothly and efficiently.[61]

[57] *Ibid*: "Any other food uses of this product, other than as, or in, food supplements, might be considered novel and may consequently need to be authorised pursuant to the requirements of the Novel Food Regulation (EU) 2015/2283 before they can be marketed as food in the EU. In addition, other legislation may restrict the marketing of this product as food in the EU or in some member states. It is therefore recommended to check with the competent authority(ies) of the member state(s)."

[58] The Catalogue of Novel Foods is available for consultation on the Commission's website; the categories are outlined at the following link https://ec.europa.eu/food/food-feed-portal/screen/novel-food-catalogue/search. Accessed 31 March 2025.

[59] For instance, some issues identified in practice include: frequent failure to comply with the timeframes for completion of the various authorization phases; certain ambiguities in definitions; implementation challenges concerning the regulation of traditional foods originating from Third Countries; and complexities related to data protection and industrial property rights. Whether the NFR slows down innovation is increasingly questioned, the length and cost of the authorization procedure being cited as major obstacles. See Monaco and Reinhardt (2025), Paganizza (2024), Monaco and Purnhagen (2022), Monaco (2023), Lähteenmäki-Uutela et al. (2021) and Holle (2018).

[60] For specific numerical data on the authorizations granted, see the official website of the European Commission, https://food.ec.europa.eu/food-safety/novel-food/authorisations_en. Accessed 31 March 2025. See also Monaco and Reinhardt (2025); Formici (2024), pp. 176–177.

[61] For a critical perspective, see Bellenghi and Knuth (2024), pp. 60–61: "In this respect, the law remains confined to formulating procedural conditions and norms of conflict in the 'triangular relationship between politics, science, and law'. Yet, administrative law and its judicial application can contribute to fostering deliberative quality of EU decision-making by strengthening procedural duties. In the past, refinement of procedural duties has often focused on the scientific quality of the information base. In the absence of a political safety valve within the current system of EU execu-

Notably, up to now, there have been no particularly contentious cases in which the Commission has deviated from EFSA's scientific opinion or where Member States have opposed the Commission's draft implementing decision.[62] This absence of major conflicts has contributed to the stability and predictability of the system. However, the advent of cultivated meat poses unique challenges that may strain this equilibrium, potentially introducing new tensions among institutional actors and testing the robustness of the current regulatory framework.

3 Centrifugal and Centripetal Forces: Member States' Initiatives and Reform Calls

3.1 National Legislative Initiatives: Different Regulatory Approaches Between Promotion and Bans

Although authorisation procedures continue under the NFR, as already described, this harmonised discipline has recently been subject to centrifugal forces, the full impact of which remains to be clearly assessed. In fact, prior to the applications submitted by Gourmey and Mosa Meat, several Member States adopted regulatory solutions for cell-based meat that reflected different—if not outright opposite—approaches. On one hand, some national initiatives have aimed to facilitate research and development of innovative cultured foods. On the other hand, legislative efforts have more frequently sought to block the expansion of cellular agriculture industries and restrict the circulation of such products. However, this restrictive stance has given rise to significant concerns regarding its compatibility with existing EU law.

In the Netherlands, the Parliament approved a motion[63] allowing the tasting of cultivated meat, albeit under controlled conditions, on the basis of a code of conduct established by the government in July 2023.[64] At the same time, substantial investments have been directed toward promoting and advancing research and

tive rulemaking, controversial cases should lead to the acknowledgement, not only in the law but also in practice, that societal and ethical implications are relevant aspects requiring due attention by the responsible risk manager."

[62] Volpato (2022).

[63] See *Kamerstuk, Motie Van De Leden Tjeerd De Groot En Valstar over onder gecontroleerde omstandigheden proeverjen van kweekvlees mogelijk maken*, 27428/383, 8 March 2022. https://www.tweedekamer.nl/kamerstukken/moties/detail?id=2022Z04324&did=2022D08835. Accessed 31 March 2025.

[64] *Uitvoering van de motie van de leden Tjeerd de Groot en Valstar over onder gecontroleerde omstandigheden proeverijen van kweekvlees mogelijk maken (Kamerstuk 27428-383)*, 5 July 2023. https://www.tweedekamer.nl/kamerstukken/brieven_regering/detail?id=2023D30899&did=2023D30899. Accessed 31 March 2025.

development in the field of alternative proteins.[65] Without conflicting with the supranational legislative framework, this centrifugal force highlighted the need to provide more concrete support for private and public entities financing the production of cell-based foods.

The Italian case moves in precisely the opposite direction. Italian Law No. 172 of 1 December 2023 (Official Gazette No. 281, 1 December 2023) is paradigmatically entitled "Provisions on the prohibition of the production and placing on the market of food and feed composed of, isolated from, or produced from cell cultures or tissues derived from vertebrate animals, as well as on the prohibition of the designation of meat for processed products containing vegetable proteins."[66] Leaving aside the latter, albeit interesting, provisions on the labelling of plant-based foods,[67] the focus of the present analysis is primarily on the content of Article 2. This provision bans food and feed business operators from "using in the preparation of food, beverages and feed, selling, holding for sale, importing, producing for export, serving or distributing for human consumption, or promoting for such purposes, food or feed consisting of, isolated from or produced from cell culture or tissues derived from vertebrate animals".[68]

This measure, along with the corresponding sanctions,[69] is justified—as explicitly stated in Article 1—by invoking the precautionary principle[70] as well as the need to "ensure the protection of human health and citizens' interests, as well as to preserve the agri-food heritage as a set of products that reflect Italy's socio-economic and cultural evolution" (Article 1, para 1).[71]

The introduction of this ban sparked broad and heated political and social debate, involving consumers, farmers and breeders associations, the media, politicians and

---

[65] See *Cell-Based Meat and Food Policy* by Mancini in this volume.

[66] See Law No. 172 of 1 December 2023, *Disposizioni in materia di divieto di produzione e di immissione sul mercato di alimenti e mangimi costituiti, isolati o prodotti a partire da colture cellulari o di tessuti derivanti da animali vertebrati nonché di divieto della denominazione di carne per prodotti trasformati contenenti proteine vegetali.* (23G00188) (Official Gazette General Series No. 281 of 1 December 2023), https://www.gazzettaufficiale.it/eli/id/2023/12/01/23G00188/sg. Accessed 31 March 2025.

[67] See on this topic *The Regulatory Landscape of Cultured Meat and Fish: Terminology and Consumer Information Provisions* by Sirsi and Rolandi in this volume. See also Di Lauro (2025).

[68] Translation. The decision to limit to vertebrate animals is questionable. For comments on the Italian law, see Panebianco (2023), Spoto (2024), Rubino (2024), Planchenstainer (2024), and Laus (2024).

[69] See Article 5 of Law No. 172 of 1 December 2023.

[70] See Article 7 of Regulation (EC) no 178/2002.

[71] Translation. In particular, as stated by the Minister of Agriculture, Food Sovereignty and Forestry in an interview published on the Ministry's website (https://www.masaf.gov.it/ddl_cibo_sintetico_via_libera_camera_legge#, accessed 31 March 2025), the main reasons behind the legislative choice include the need to protect cultural identity and traditional production methods, as well as to safeguard farmers' incomes and the Italian food system more broadly. The cultural and socio-economic dimensions of this legislation are explored in greater depth in *Who Hates Cultured Meat? Examining Public Opposition to Cellular Agriculture* by Broad and Glenna in this volume.

the scientific community.[72] Some recent developments, occurring more than a year after the Italian law came into force, highlight the concerns surrounding this controversial food. For instance, Coldiretti[73] organized a protest in the city of Parma, where EFSA is based, calling for greater attention and protection for consumers and producers against so-called "artificial" foods. The organization also advocated reform of the EU law governing cell-based products, suggesting a regulatory approach more akin to that applied to pharmaceuticals. This position, echoing views expressed by several Member States (see Sect. 3.2), rests on the idea that cell-based foods are closer to medicinal products than to conventional food.[74]

At the same time, several Italian researchers published a petition calling not only for protection of freedom of scientific research, which has been hindered in its ability to secure funding, but also for preservation of the decision-making process established by current EU legislation. In particular, they emphasized the importance of observing the proper sequence of this process, which first requires scientific assessment by EFSA, followed only then by political evaluation. The need to promote cross-sectoral research and to uphold a principle of "regulatory diligence" have been considered fundamental for advancement of the field.[75]

While these examples demonstrate the polarization of the discourse on cell-based meat in Italy,[76] the stricter legal implications of the national ban cannot be overlooked. In this regard, two closely interrelated aspects merit thorough examination: (1) from a substantive perspective, compliance of national law with the existing EU law regulating the marketing of Novel Foods and the precautionary principle, as defined by hard law instruments and interpreted in the case-law of the Court of Justice of the EU (CJEU); (2) from a "formal" or "procedural" perspective, conformity of the Italian provisions with the so-called Technical Regulation Information System (TRIS) Directive.[77]

---

[72] See Corazza (2023) and Formici (2024).

[73] Coldiretti (2025) https://www.coldiretti.it/notizie/parma-20mila-agricoltori-in-piazza-per-difendere-la-salute-di-tutti. Accessed 31 March 2025.

[74] Coldiretti (2022) https://www.coldiretti.it/economia/una-firma-contro-il-cibo-sintetico-scatta-la-mobilitazione-coldiretti. Accessed 31 March 2025.

[75] Fino et al. (2024) https://iris.unito.it/retrieve/43a08a42-3b62-4bd3-9b36-074b30707335/Nota_critica_One_Earth__ITA.pdf. Accessed 31 March 2025.

[76] As reported in a Parliamentary Interrogation of 21 March 2025 (https://aic.camera.it/aic/scheda.html?core=aic&numero=4/04673&ramo=CAMERA&leg=19, accessed 31 March 2025), and similarly recalled by Coldiretti (see https://www.coldiretti.it/notizie/parma-20mila-agricoltori-in-piazza-per-difendere-la-salute-di-tutti, accessed 31 March 2025), the Italian Minister of Health and Minister for Agriculture, Food Sovereignty and Forestry established an interministerial technical working group by Interministerial Decree dated 25 January 2024, to study the potential overall impact of the marketing of cell agriculture foods derived from vertebrate animals. On 7 February 2025, two documents attributed to this working group were published on the Ministry of Health's website. The first contains technical comments on the updated EFSA guidelines on Novel Foods of 27 June 2024, while the second presents additional considerations, concluding that clinical and preclinical tests are necessary to assess the food safety of so-called cultivated meat.

[77] Directive (EU) 2015/1535 of the European Parliament and of the Council of 9 September 2015 laying down a procedure for the provision of information in the field of technical regulations and

In assessing the compatibility of the Italian provisions with EU law, the first regulation to consider is the NFR. Contrary to the position—at least initially—affirmed by the Italian legislator,[78] cell-based meat clearly falls within the scope of application of the EU Novel Foods discipline. As repeatedly emphasized in Sect. 2 of the present chapter, this legislation mandates the centralization of risk assessment and risk management at EU level. Notably, the 2015 NFR has not reaffirmed the "safeguard clause" previously established under Article 12 of Regulation (EC) no 258/1997.[79] As a result, the current framework ensures full harmonization and attributes general effect to authorisations, valid across the entire EU.[80] In other words, "By enacting the Novel Foods Regulation, the EU has exercised its internal market shared competence, and in principle, prevents Member States from undermining its harmonization efforts by adopting diverging measures on the same subject."[81]

These considerations also hold for the precautionary principle, invoked in this context. In fact, as interpreted by extensive CJEU case-law,[82] the applicability of this fundamental principle, also enshrined in Article 7 of Regulation (EU) no 178/2002, is excluded in matters governed by fully harmonized legislation.[83] Moreover, precautionary measures must be provisional (thus imposing a review obligation), proportionate, non-discriminatory and limited to what is necessary to achieve a high level of health protection. Generally speaking, looking at EU law in fields outside food regulation, the preconditions for legitimately applying this principle lie in the existence of: (a) the possibility of harmful effects on environmental, human, animal or plant health; (b) a persistent situation of scientific uncertainty that prevents the risks from being assessed with sufficient, conclusive or reliable scientific evidence; (c) a scientific evaluation based on "empirical input and/or modelling

---

of rules on Information Society services.

[78] Formici (2024), Rubino (2024) and Cerini (2025).

[79] Article 12 of Regulation (EC) no 258/1997 stated: "1. Where a Member State, as a result of new information or a reassessment of existing information, has detailed grounds for considering that the use of a food or a food ingredient complying with this Regulation endangers human health or the environment, that Member State may either temporarily restrict or suspend the trade in and use of the food or food ingredient in question in its territory. It shall immediately inform the other Member States and the Commission thereof, giving the grounds for its decision. 2. The Commission shall examine the grounds referred to in paragraph 1 as soon as possible within the Standing Committee for Foodstuffs; it shall take the appropriate measures in accordance with the procedure laid down in Article 13. The Member State which took the decision referred to in paragraph 1 may maintain it until the measures have entered into force."

[80] Rossi Dal Pozzo and Rubino (2024), p. 322.

[81] Bellenghi and Knuth (2024), pp. 48–49. Here the reference is to Article 2, para 2, Treaty on the Functioning of the European Union (TFEU), establishing a sort of pre-emption, preventing Member State intervention in areas of shared competence in which the EU has already adopted legislative provisions.

[82] See Bellenghi and Knuth (2024), and more generally on this principle, Borghi (2011), Tallacchini (2012), Costato (2016) and Canfora (2024), but also Weimer (2019) and De Smedt and Vos (2022).

[83] Rubino (2024), Cerini (2025) and Lanni and Ragone (2025).

outputs that lead to the plausible scientific hypothesis that serious harm appears possible".[84] These conditions cannot therefore be considered met by adopting a "purely hypothetical approach to risk, founded on mere suppositions which are not yet scientifically verified."[85]

Looking at these criteria, at least some doubts can be cast on the applicability of the precautionary principle in the specific case of cell-based meat, given the full harmonization established by the current NFR as well as the national legislator's failure to propose provisional measures or to present specific and detailed scientific evidence demonstrating the potential harm of cultured meat to consumer health (and/or to animal welfare and the environment).[86]

Based on the above evaluations related to substantial aspects, the Italian ban raises significant concerns regarding its compliance with EU law. Under the NFR, in the absence of prior supranational authorisation, the marketing of cell-based meat is currently prohibited in the EU. Consequently, the national prohibition adopted in Italy mirrors the restrictions already so far established by EU legislation. However, the emerging contrast at least with Regulation (EU) 2015/2283 suggests a potential breach of EU law, not only in the present—should the ban be enacted—but also, hypothetically, in the future. In particular, although more specific remarks on potential forthcoming scenarios will be offered in the next section, it is worth underlining already the complex situation that could arise in the event of a positive outcome of cell-based meat authorisation procedure(s). In such a case, the Italian discipline on cultured products derived from vertebrate animals would be incompatible with the supranational approval, thereby creating an unprecedented—at least in the context of Novel Foods—contrast between national and EU provisions, and contributing to a fragmented regulatory landscape.[87]

In addition to these crucial and sensitive considerations concerning the substantive aspects of Italy's "headlong rush", the formal and procedural dimensions of the national ban also warrant thorough examination. Given its potential to introduce a technical regulation that may affect the free circulation of industrial, agricultural and fishery products, the Italian legislative proposal on cell-based meat was subject to the obligation of notification to the European Commission. In fact, according to the TRIS Directive, national legislative drafts—prior to coming into force—must be communicated to the Commission. This requirement enables both the Commission and other Member States to evaluate the draft's compatibility with EU law, and if necessary, to issue reasoned opinions. The ultimate objective of this procedure is to prevent the emergence of internal market barriers before they take effect, thereby anticipating and mitigating negative consequences for private business operators.

---

[84] De Smedt and Vos (2022), p. 168.

[85] Decision CJEU Judgement (9 September 2003) Case C-236/01 *Monsanto Agricoltura Italia SpA and Others* v *Presidenza del Consiglio dei Ministri and Others,* para 106.

[86] Paganizza (2025) and Cerini (2025).

[87] Rubino (2024), Bellenghi and Knuth (2025) and Monaco (2025).

To this end, Article 6 of the TRIS Directive establishes a *status quo* period of 3 months (extendable), during which the notified technical rule cannot be adopted definitively by the national legislator. However, although the Italian law banning cell-based meat was notified on 1 December 2023, when it was also published in the Official Gazette, it came into force on 16 December 2023, during the *standstill* period.[88] In light of this procedural irregularity, the European Commission published a "premature closure of the notification procedure" on 29 January 2024, noting failure to comply with the deferral requirement under Article 6 of the TRIS Directive. The Commission also invited "the Member State in question to inform it of the actions taken, also in light of the relevant case law of the Court of Justice."[89] The latter reference appears to point to the many well-established decisions by Luxembourg judges on violation of the TRIS Directive, which have consistently held that national technical rules adopted before the conclusion of the standstill period are inapplicable.[90] As a result, from a procedural standpoint, the Italian law should be regarded as inapplicable and unenforceable due to its non-compliance with the requirements set out in the TRIS Directive.[91]

As has clearly emerged, in the Italian case the Commission—differently from the opinion submitted, for instance, by Lithuania[92]—did not address the technical rule's compatibility with EU law. Rather, it confined its intervention to a formal/procedural review, an approach that some have deemed timid.[93]

Nonetheless, the Commission had the opportunity to take a substantive stance on the compliance of national bans on cell-based meat in the framework of the TRIS notification procedure triggered by Hungary. Following Italy's example, in fact, other Member States also attempted to adopt similar legislation aimed at prohibiting the production and marketing of cultivated meat. While in France two *Propositions de loi* have been submitted to the *Assemblée Nationale* and remain under debate,[94]

---

[88] For details on the notification procedure that instigated a complex debate even between the legislator and the President of the Republic, see Panebianco (2023).

[89] Communication from the Commission—TRIS/(2023) 0244, https://technical-regulation-information-system.ec.europa.eu/it/notification/25152. Accessed 31 March 2025.

[90] On this point, see the case-law listed in the document prepared by the European Commission titled *Case law relating to Directive (EU) 2015/1535*, https://technical-regulation-information-system.ec.europa.eu/en/about-the-20151535/case-law. Accessed 31 March 2025.

[91] Rubino (2024), Bellenghi and Knuth (2024), Monaco (2025), Paganizza (2025) and Cerini (2025).

[92] Detailed Opinion of the Republic of Lithuania on the technical regulation prepared by the Italian Republic (Note 2023/0675/it), Message 115 to the Notification: 2023/0675/IT, https://technical-regulation-information-system.ec.europa.eu/it/notification/25152/message/106844/EN. Accessed 31 March 2025.

[93] Rubino (2024), p. 52.

[94] See *Proposition de loi* No. 1965 of 5 December 2023 and No. 2172 of 13 February 2024. It is worth noting that Article L. 230-5 of the *Code rural et de la pêche maritime*, as amended by *Loi* No. 2024-317 of 8 April 2024, Article 30, also establishes a ban on "*denrées alimentaires qui se composent de cultures cellulaires ou tissulaires dérivées d'animaux ou qui sont isolées ou produites à partir de cultures cellulaires ou tissulaires dérivées d'animaux*", with specific reference to "*services de restauration scolaire et universitaire ainsi que des services de restauration des établisse-*

the Hungarian Parliament adopted draft legislation "prohibiting the production and placing on the market of laboratory-grown meat",[95] promptly notified under the TRIS procedure. In the absence of formal irregularities, the Commission proceeded to assess the Hungarian ban from a substantive perspective.

In its opinion dated 9 October 2024, the Commission first stated that considering the existing Novel Foods Regulation, "a ban is unnecessary, as currently the prohibition to market results from Union law and applies to all the Union territory". Furthermore, given EFSA's central role in the authorisation procedure—ensuring "that foods to be placed on the EU market are safe and do not present risk for human health", a national ban is deemed "unjustified, since it could pre-empt the harmonised authorisation procedure for novel foods at EU level."[96] With this statement, the Commission appears to affirm the adequacy of the supranational NFR in ensuring food safety, also with regard to cell-based meat. Contrary to the claims made by the Italian government in its previous notification, this position thereby highlights the existence of a proper regulatory framework and a solid scientific risk assessment mechanism.

As a consequence and in line with the concerns expressed by several other Member States in their observations,[97] the Commission confirmed that the draft legislation notified "does not comply with the provisions laid down in Chapters II to VI of Regulation (EU) 2015/2283 concerning the authorisation procedure for novel foods".

In the face of such an unequivocal position, pursuant to Article 6, para 2 of the TRIS Directive, Hungary is now required "to report to the Commission on the action that it proposes to take on this detailed opinion". Should the Member State fail to comply, the Commission retains the right to begin an infringement procedure under Article 258 of the Treaty on the Functioning of the European Union (TFEU).[98]

---

ments d'accueil des enfants de moins de six ans, des établissements de santé, des établissements sociaux et médico-sociaux et des établissements pénitentiaires." A similar debate is also happening in the United States. See *The Legal Debate on Cultured Meat Around the World: Comparative Insights from the United States, Singapore and Israel* by Pitto in this volume. Similarly, in Romania, a legislative proposal to ban cultured meat was submitted for parliamentary scrutiny in 2023.

[95] See Notification 2024/0394/HU—Draft act prohibiting the production and placing on the market of laboratory-grown meat ("a laboratóriumi hús"), https://technical-regulation-information-system.ec.europa.eu/en/notification/26066. Accessed 31 March 2025.

[96] Notification 2024/0394/HU—Draft act prohibiting the production and placing on the market of laboratory-grown meat ("a laboratóriumi hús")—Delivery of a detail opinion pursuant to Article 6(2) of Directive (EU)2015/1535, https://technical-regulation-information-system.ec.europa.eu/en/notification/26066. Accessed 31 March 2025.

[97] Specifically Lithuania, Sweden, The Netherlands, Denmark, Czechia (see https://technical-regulation-information-system.ec.europa.eu/en/notification/26066. Accessed 31 March 2025).

[98] Reportedly, according to some news websites, the legislative proposal banning cultured meat has recently been re-submitted for Parliament scrutiny, notwithstanding the opinion expressed by the Commission within the TRIS procedure. https://english.atlatszo.hu/2025/03/28/hungarys-law-proposal-to-ban-laboratory-meat-is-against-the-eu-law/. 31 March 2025.

In conclusion, the stand taken with regard to the Hungarian draft legislation appears to confirm the doubts and criticalities previously raised in relation to the Italian ban, thereby opening the door to a number of important considerations concerning possible future scenarios, albeit still largely unclear, at supranational and national levels. In particular, the following aspects require careful observation: (a) Hungary's reaction, i.e. whether or not the draft legislation will ultimately be abandoned; (b) whether the Italian legislator or government will reconsider the national ban in light of the Commission's recent opinion on the very similar Hungarian law; (c) whether the Commission will initiate infringement proceedings for violation of the TRIS Directive or for non-compliance with EU law; (d) whether the firm stance adopted in the Hungarian case will deter other Member States from pursuing similar centrifugal approaches.

While these aspects remain uncertain and future developments are difficult to predict, two key conclusions emerge from our analysis in this section: first, the unprecedented "advance move" by Italy is not an isolated approach, but rather the most apparent and visible example of a widespread trend; second, although the regulatory debate unfolding at national level raises complex questions in terms of compliance with the EU law, it reflects a call for a profound reassessment of the existing EU regulatory framework. Indeed, the centrifugal forces here examined should be considered alongside the centripetal forces seeking to foster supranational debate through the EU's institutional and legislative channel.

3.2 "Much Ado About Nothing" or a Necessary Reform? The Positions Expressed by the European Commission, the European Parliament and Several Member States

At a meeting of the Agriculture and Fisheries Council on 23 January 2024, Austria, France and Italy, with the broad support of other Member States,[99] presented a note[100] regarding the concerns and scepticism about what were called "lab-grown artificial cell-based foods". According to the signatories, these controversial foods were "a threat to primary farm-based approaches and genuine food production methods" and therefore called for a thorough evaluation not only involving Member States and the Commission, but also stakeholders and citizens. Besides the ethical, economic, environmental, social and food health concerns expressed,[101] the note

[99] Specifically, the Czech, Cypriot, Greek, Hungarian, Luxembourg, Lithuanian, Maltese, Romanian, Slovak and Spanish delegations.

[100] See *CAP's role in safeguarding high-quality and primary farm-based food production—Note from the Austrian, French and Italian Delegations*, 5469/2024-REV1, https://data.consilium.europa.eu/doc/document/ST-5469-2024-REV-1/en/pdf. Accessed 31 March 2025.

[101] See *ibid*, p. 2; it is particularly interesting to underline the socio-economic concerns expressed in the analysed Note, where is clearly stated that "Lab-grown cell-based meat production requires economies of scale to operate cost effectively. Given the high fixed costs, there are high risks that

questioned the adequacy of the existing regulatory framework and in particular the capacity of the NFR to "provide a suitable and comprehensive framework to assess the potential risks associated with these products while fully considering the precautionary principle".[102] It urged the EFSA to adopt revised and dedicated guidelines. The signatory States called for these guidelines to include "pre-clinical and clinical studies", akin to those required by pharmaceutical regulations.

A "moratorium" was therefore proposed so that the Commission could conduct a comprehensive impact assessment "before taking any market authorization decisions".[103] These evaluations were to be grounded in accurate, science-based data and information, and were to avoid what was described as "deceptive greenwashing campaigns" that misleadingly overstate unproven positive impacts on the environment and animal welfare, which have been challenged by certain referenced studies.

During the meeting, the Commissioner for Health and Food Safety, Stella Kyriakides, responded to the note by first reaffirming the effectiveness of the existing NFR. She emphasized that this legislative framework already imposes proper consideration of the precautionary principle; moreover, under Article 12, the authorisation process for Novel Foods grants the Commission the possibility of evaluating other relevant "factors", as detailed in Sect. 2 of the present chapter. According to the Commissioner, these aspects are "legitimate issues" that are necessarily part of the authorisation process.[104]

By acknowledging the possibility of weighing other factors apart from the scientific food safety evaluation, the position adopted by the Commission thus far essentially confirmed the strength and comprehensiveness of the Novel Foods discipline, recognizing its capacity to address the challenges posed by cell-based meat. On one hand, this stance appears to dampen the calls for *ad hoc* regulatory intervention, and on the other to counteract centrifugal moves by national legislators. By emphasizing that "on food safety, we are on very solid ground", the Commissioner stressed the importance of the risk assessment entrusted to EFSA: although no authorisation requests had yet been submitted at that time, the Commission had already invited the independent authority to review and update its guidelines and the documentation required for risk assessment, with the aim of encouraging the "introduction of state-of-the-art provisions for the safety and nutritional assessment of lab-grown and

---

artificial meat production would consequently be monopolized amongst a few large-scale industrial producers. This could lead to crisis-prone dependencies along the value chain and create additional pressure on small-scaled family owned farms." For further considerations on these delicate aspects, see *Who Hates Cultured Meat? Examining Public Opposition to Cellular Agriculture* by Broad and Glenna, as well as *Navigating Consumer Acceptance of Cultured Meat: Insights from Social Science Research* by Lin-Hi and Burdorf, both in this volume.

[102] *Ibid*, p. 3.

[103] *Ibid.*

[104] See Opening Remarks by Commissioner Stella Kyriakides at the Agrifish Council Meeting—High-quality and primary farm-based food production of 23 January 2024, https://ec.europa.eu/commission/presscorner/detail/en/SPEECH_24_384. Accessed 31 March 2025.

cell-based foods".[105] At the same time, Kyriakides underlined the complexity and multiple challenges of the marketing of cell-based foods, as well as the limited scientific data currently available. She therefore underscored the importance of promoting and financing research and development projects in this field.[106]

Despite the clear position adopted by the Commission, the "centripetal" debate remains far from settled. Increasingly differentiated and potentially conflicting positions continue to emerge. While several Member States support and align with the Commission's view and even advocate strategies and investments aimed at studying and producing alternative proteins, including cell-based meat, others maintain a more sceptical perspective, calling for specific regulatory intervention, in the absence of which they view a ban to be preferable.

These divergent approaches and views are not only observed between Member States but also between EU Institutions themselves and in civil society. Contrary to the position broadly expressed by the Commission, the European Parliament's "European Protein Strategy" clearly suggests that the NFR "is not fit for purpose" in the case of cell-based foods.[107] While the Parliament's outlook on other alternative proteins, such as insects and fermentation-derived products, is cautious but more positive, its stance on cell-based meat is much more critical. This criticism extends beyond ethical, social, environmental and economic issues to include perceived inadequacies of the current EU legislative framework.

Lastly, looking at civil society initiatives at EU level, a petition has been launched for a European Citizens' Initiative called "Stop cruelty, stop slaughter." The aim of this initiative is to urge the Commission to provide more significant incentives and investment for a real transition to plant-based proteins and cell-based meat, not only for human consumption, but also for pet food; the petition also urges research and development studies in this area with the final goal of ensuring ever "cleaner" and "more ethical" cell-based foods.[108]

Taken together, the centripetal forces outlined here reveal a deeply fragmented social and political landscape, as well as a regulatory and legal framework under mounting pressure from different directions, such that its future course is still highly uncertain.

---

[105] *Ibid.* The reference here is to the subsequently adopted EFSA Panel on Nutrition, Novel Foods and Food Allergens (2024) https://efsa.onlinelibrary.wiley.com/doi/10.2903/j.efsa.2024.8961. Accessed 31 March 2025. See also *Safety Assessment of Cell-Culture-Derived Food: The Role of the European Food Safety Authority* by Dall'Asta and Gjoni in this volume.

[106] Opening Remarks by Commissioner Stella Kyriakides. A similar position was expressed by Executive Vice-President Vestager on behalf of the European Commission, answering the Parliamentary question E-001778/2023 on 2 August 2023. https://www.europarl.europa.eu/doceo/document/E-9-2023-001778-ASW_EN.html. Accessed 31 March 2025.

[107] European Parliament resolution of 19 October 2023 on a European protein strategy (2023/2015(INI)), para 44.

[108] European Citizens' Initiative "Stop Cruelty Stop Slaughter", Commission registration number: ECI(2024)000008. https://citizens-initiative.europa.eu/initiatives/details/2024/000008_en. Accessed 31 March 2025.

4 Conclusive Remarks: An Unprecedented Debate on the EU Table

The NFR appears to be at a crossroads and undergoing what could be described as a stress-test. While several Member States have on various occasions in recent years attempted to deviate from the full harmonisation established by the supranational legislation, especially with regard to insect-based food products,[109] the case of cell-based meat is at the basis of a unique potential legal disruption that seriously calls into question the effectiveness and resilience of the current authorisation procedure. Although the existing mechanism has functioned well for a wide range of Novel Foods, many of which are already authorised, marketed and consumed in the EU, the controversy[110] surrounding a few of them has significantly strained the current regulatory framework.[111]

Though future developments remain difficult to predict, some reflections from at least three analytical perspectives are possible: (a) the ongoing authorisation procedures under the current NFR, promoted by Gourmey and Mosa Meat; (b) the potential institutional and legal reactions to gathering centrifugal forces; (c) the evolution of reform calls, shaped by the dynamic interplay between centripetal and centrifugal forces.

(a) Regarding the first strand, close and thorough observation of the upcoming steps in the authorisation path will be crucial. The possible outcomes and hypothetical scenarios are numerous and highly uncertain, particularly in relation to the risk management phase. Assuming a positive opinion from EFSA, affirming the food safety of cultivated *foie gras* and/or cultivated beef fat,[112] and therefore an absence of risks for human health, the European Commission will face two primary alternatives when drafting the implementing act. On one hand, it may choose to rely solely on EFSA's scientific risk assessment; on the other, it could adopt a "bolder" approach,[113] carefully integrating the broader set of "other legitimate factors" that have become increasingly central to the political, social and academic debate on cell-based meat. Such an approach would involve moving beyond the more "comfortable" and seemingly safer reliance on scientific food safety evaluation, and instead embracing the complexity inherent in politi-

[109] Paganizza (2024), p. 814.

[110] See *Who Hates Cultured Meat? Examining Public Opposition to Cellular Agriculture* by Broad and Glenna in this volume.

[111] Reinhardt et al. (2024).

[112] As already underlined *supra* footnote n. 52, the authorization procedure applies to the specific cell-based product for which approval has been requested; to date, in the EU, this has concerned only cultivated foie gras and cultivated beef fat.

[113] See Bellenghi and Knuth (2024), p. 59: "Within pre-marketing authorisation procedures, the Commission's role has been metaphorically described as a blind driving a car whilst receiving directions from EFSA as its back seat passenger" citing Vos and Wendler (2006). The authors also expand this reasoning by affirming the need to "re-conceptualise" comitology functioning (p. 61).

cal risk management. However, this more challenging option should not simply end up in decisions driven solely by ideological or "emotional"[114] reasoning but must be grounded in the evaluation of data, studies and evidence on environmental, ethical and socio-economic factors, all of which can be a foundation for a well-informed and responsible decision-making process. When dealing with controversial products and production methods, the biphasic process established by the EU legislators back in 2015 is of even greater importance: it equips the European Commission and the Member States' representative in the PAFF Committee with the necessary tools and powers to tackle their full political responsibilities, enabling broader and more articulated risk management that extends well beyond the simple food safety risk assessment phase.

In the authors' view, it is precisely these broader evaluations that will determine the future of cell-based meat in the EU. An in-depth consideration of the "other legitimate factors", which have played a pivotal role in shaping the abovementioned national legislative interventions and proposals, may prove crucial for the future acceptance (or denial) of centralized authorisation. In other words, the European Commission's ability to meaningfully incorporate sustainability, environmental, socio-economic and ethical considerations into the political discourse during the risk management phase could profoundly affect future dynamics, especially in terms of the persistence of centrifugal and centripetal forces.

(b) Concerning initiatives promoted by Member States, the responses of national governments/legislators and EU Institutions will require careful monitoring. The firm stance recently taken by the Commission in the context of the TRIS Procedure could halt or restrain centrifugal tendencies by discouraging Member States from pursuing *ad hoc* disciplines in contrast with the NFR. The Commission could even threaten activation of infringement proceedings. Whether this will happen is hard to say. It can however be inferred from the strong position expressed—or at least discussed—by several Member States that any hypothetical authorisation of cell-based meat may encounter strong opposition[115] and could lead to a clash between different levels of governance. This could ultimately result in a fragmented scenario, with national actors seeking to counter a supranational decision for example by invoking derogatory measures such as those found in Articles 53 and 54 GFL.[116] Should this occur,

---

[114] Volpato (2022).

[115] Planchenstainer (2024) refers to a Council vote simulation, based on the abovementioned Note, elaborated by Francesco Montanari: "As observed by some food law scholar, following the rules of comitology, if the same Member States opposed a Commission Implementing Act approving a cell-based novel food, such qualified majority could successfully curtail the ability of placing such product on the EU market", p. 70. This could ultimately lead to an unprecedented "clash" between the European Commission and the PAFF Committee, as well as between the scientific food safety assessment and the political risk management evaluation. On this point, see also Volpato (2022) and Bellenghi and Knuth (2024).

[116] Formici (2023).

conflicting interpretations of EU food law could spark litigation in national and supranational courts, echoing the contentious landscape previously seen with reference to the discipline of GMOs.[117] Such an articulated and unclear situation could, in turn, undermine legal certainty and deter food business operators from investing in cell-based meat research and development.

(c) The previous two scenarios and the multitude of unsolved questions surrounding them are closely intertwined with the broader reform process increasingly advocated by the European Parliament and several Member States. Although at the time of writing, the Commission has confirmed the suitability and adequacy of the current NFR, without envisaging the need to revise or amend it, the centripetal forces—together with several individual national initiatives—should not be ignored. Indeed, they ought to serve as a foundation for critical reflection on the existing framework. Such reflection could lead to evaluation of possible targeted reforms aimed not at radically altering the centralized procedure, but rather at enhancing its efficacy. For instance, one might explore the opportunity of introducing multiple and concurrent risk assessments and "proceduralized" scientific evaluations that go beyond food safety alone. These assessments could also encompass the environmental and animal-welfare implications of cell-based meat production, as well as any socio-economic risks and consequences.[118]

This form of procedural integration would remain perfectly consistent with the biphasic authorisation procedure established by the NFR. At the same time, it would respond to some of the concerns raised by various institutional actors by providing the political decision-makers with scientifically grounded, evidence-based data and insights from independent authorities. While not binding, these additional assessments could offer a valuable basis for the risk management phase, and in particular for the evaluation of "other legitimate factors"; at the same time, these modifications would not jeopardise full harmonization, thereby avoiding fragmentation.[119]

Conversely, the advocated reclassification of cell-based products from foods to pharmaceutical products would require far more transformative legislative intervention; such a significant change in the regulatory approach does not yet appear to have been seriously considered or thoroughly investigated by EU institutions.

A further question arises in the event of a hypothetical authorisation: could the fragmented reactions of Member States, as currently expressed by various centrifugal

---

[117] See Lee (2008), Ragone (2015), Costato (2016), Sirsi (2017) and Pitto (2022).

[118] It may also be worth discussing whether the "other legitimate factors" should be more clearly determined, defined and how they should be interpreted, as well as if and how to make the political risk management phase more transparent.

[119] Other regulatory interventions could also be inspired by the pioneering solutions adopted in the United Kingdom, where the Food Standard Agency announced in March 2025 the launch of a sandbox programme, funded by the Department of Science, Innovation and Technology's Engineering Biology Sandbox Fund. https://www.food.gov.uk/news-alerts/news/fsa-launches-pioneering-regulatory-programme-for-cell-cultivated-products. Accessed 31 March 2025. Another noteworthy experience to be observed is the one concerning Switzerland; on this legal system, see Delerm and Levy (2024).

positions, eventually lead to broader discussion on regulatory reforms? Such a debate might explore the possibility of granting Member States exceptional discretionary powers, a regulatory path not unlike that chosen for GMOs. For the sake of legal certainty and to avoid a confused and conflict-ridden multi-level governance scenario, the Commission could be asked to consider departing from full harmonization, instead granting exemptions to national legislators. Should such a shift occur, it would be essential to establish clear conditions for the exercise of these discretionary powers, in order to avoid arbitrariness. While these projections remain speculative, they provide food for thought in relation to the present and future trajectory of cell-based food in the EU.

In conclusion, we have described a confused and still evolving debate marked by many unresolved questions and shaped by a wide range of actors. The debate, which first emerged prominently in Italy but has since expanded across other Member States and EU Institutions, cannot be overlooked. Rather, it must serve as the foundation for an ongoing, constructive and inclusive dialogue among all relevant stakeholders.

Far from being a result of ideological campaigns, the regulatory hype surrounding this innovative source of alternative proteins is emblematic of the broader challenges inherent in governing techno-scientific advances in the agri-food sector. The doubts and concerns facing the regulation of cell-based meat touch deeper systemic issues: the fraught relationship between science and law, the balance between the precautionary principle and the so-called, still debated "principle of innovation",[120] the tension between the "scientification" of political decisions and the politicisation of highly scientific matters,[121] and ultimately the complex task of ensuring the correct functioning of the internal market, while guaranteeing the proper role attributed to different institutional actors in the multi-level governance. More generally, the cultured meat case clearly underscores the delicate and essential mission of determining an appropriate balance between innovation, food safety and food security (and sustainability).

Promoting a regulatory discourse capable of engaging with, and possibly untangling, all these multifaceted challenges is not only vital for the destiny of cell-based meat but also for the attractiveness of the EU market, the resilience of its agri-food sector, the protection of consumer freedom of choice and the achievement of long-term sustainability goals.

References

Albisinni F (2025) Quale carne? Rivista di Diritto Alimentare 1:104–112

Alemanno A, Gabbi S (eds) (2014) Foundations of EU food law and policy: ten years of the European Food Safety Authority, 1st edn. Routledge, London

Bellenghi G, Knuth L (2024) EU food law and the politics of the internal market: the challenge of cultivated meat. Rev Eur Adm Law 17(3/4):39–62

Borghi P (2011) Il rischio alimentare e il principio di precauzione. In: Costato L, Rook Basile E, Germanò A (eds) Trattato di diritto agrario. Utet, Milano, pp 53–72

[120] Critical of this "principle", see Reins and Ala-Lahti (2024); see also Ducuing (2022).

[121] Bellenghi and Knuth (2024), p. 59.

Brookes G (2007) Economic impact assessment of the way in which the European Union novel food regulatory approval procedures affect the European food sector. Briefing paper. http://www.fooddrinkeurope.eu/documents/news_events/finalreport2july2007.pdf. Accessed 31 Mar 2025

Canfora I (2024) I principi: Principio di precauzione, analisi del rischio, trasparenza. In: Borghi P, Canfora I, Di Lauro A, Russo L (eds) Trattato di diritto alimentare. Giuffrè, Milano, pp 112–120

Cerini D (2025) 'From farm to fork' vs. 'from factory to lab': Riflessioni su sostenibilità sociale e benessere animale a partire dalla l. 172/2023 in materia di alimenti e mangimi da colture cellulari. Rivista di Diritto Alimentare 1:76–103

Coldiretti (2022). https://www.coldiretti.it/economia/una-firma-contro-il-cibo-sintetico-scatta-la-mobilitazione-coldiretti. Accessed 31 Mar 2025

Coldiretti (2025). https://www.coldiretti.it/notizie/parma-20mila-agricoltori-in-piazza-per-difendere-la-salute-di-tutti. Accessed 31 Mar 2025

Corazza MG (2023) Alla frontiera della produzione alimentare: La carne c.d. coltivata nell'Unione europea fra divieti e mancanza di pareri scientifici. Forum di Quaderni Costituzionali 4:212–221

Costato L (2016) Principles and rules of European and global food law. In: Costato L, Albisinni F (eds) European and global food law. Cedam, Milano, pp 1–15

Dall'Asta C (2022) Why 'new' foods are safe and how they can be assessed. In: Formici G, Scaffardi L (eds) Novel foods and edible insects in the European Union – an interdisciplinary analysis. Springer, Cham, pp 81–95

De Smedt K, Vos E (2022) The application of the precautionary principle in the EU. In: Mieg HA (ed) The responsibility of science, Studies in history and philosophy of science. Springer, Cham, pp 163–186

Delerm F, Levy M (2024) Beyond the butcher: navigating the legal landscape of cultured meat. Sui Generis, pp 197–206

Di Lauro A (2025) Carne 'artificiale' e meatsounding: Un orizzonte cannibale possibile e la retorica del 'senza'. Rivista di Diritto Alimentare 1:4–16

Ducuing C (2022) A legal principle of innovation? Need for an assessment against the principle of democracy. Law Innov Technol 14:237–266

EFSA (2024a) Administrative guidance for the preparation of novel food applications in the context of Article 10 of Regulation (EU) 2015/2283. EFSA Supporting Publications 21(10):9041E. https://efsa.onlinelibrary.wiley.com/doi/10.2903/sp.efsa.2024.EN-9041. Accessed 31 Mar 2025

EFSA (2024b) EFSA's activities on emerging risks in 2023. EFSA J 21(2):9198E. https://efsa.onlinelibrary.wiley.com/doi/10.2903/sp.efsa.2024.EN-9198. Accessed 31 Mar 2025

EFSA (2024c). https://www.efsa.europa.eu/en/glossary/steps-risk-assessment. Accessed 31 Mar 2025

EFSA Panel on Nutrition, Novel Foods and Food Allergens (2024) Guidance on the scientific requirements for an application for authorisation of a novel food in the context of Regulation (EU) 2015/2283. EFSA J 22(9): e8961. https://efsa.onlinelibrary.wiley.com/doi/10.2903/j.efsa.2024.8961. Accessed 31 Mar 2025

EIT Food (2024) Gourmey files the first EU novel food submission for cultured meat. https://gfieurope.org/blog/first-ever-regulatory-submission-made-for-cultivated-meat-in-the-eu/. Accessed 31 Mar 2025

European Commission (2024) Research and innovation for reducing the EU dependency on imported plant protein. https://agriculture.ec.europa.eu/farming/crop-productions-and-plant-based-products/cereals/reducing-plan-protein-deficit-eu_en. Accessed 31 Mar 2025

Fino MA, Anzà B, Bairati L, Bertini I, Biolatti B, Biressi S, Cannizzo FT, Cavallarin L, Conti L, Deriu M, Gargioli C, Loera B, Lo Sapio L, Marchisio D, Pallante L, Stano S, Torri L, Bertero A, Massai D (2024) Cultivated meat beyond bans: ten remarks from the Italian case toward a reasoned decision-making process. https://iris.unito.it/retrieve/43a08a42-3b62-4bd3-9b36-074b30707335/Nota_critica_One_Earth__ITA.pdf. Accessed 31 Mar 2025

Formici G (2020) Novel food tra esigenze di mercato, sicurezza alimentare e sviluppo sostenibile: La complessa disciplina degli alimenti tradizionali provenienti da paesi terzi. BioLaw J 2:67–87
Formici G (2023) "Meating the future": alcune riflessioni sulla necessità di promuovere un attento dibattito regolatorio in materia di c.d. carne sintetica. Forum di Quaderni Costituzionali 2: 15–19.
Formici G (2024) Un assaggio di futuro? La carne a base cellulare e il vivace dibattito regolatorio dentro e fuori i confini dell'Unione europea. BioLaw J 4:165–197
Formici G, Scaffardi L (eds) (2022) Novel foods and edible insects in the European Union – an interdisciplinary analysis. Springer, Cham
Formici G, Scaffardi L (2025) Cell-based meat. In: Costato L, Albisinni F, Georgopoulos T (eds) European and global food law, 3rd edn. Wolters Kluwer, Milano, pp 714–727
Gabbi S (2007) The interaction between risk assessors and risk managers. Eur Food Feed Law Rev 2(3):126–135
GFI (2024) First-ever regulatory submission made for cultivated meat in the EU. https://gfieurope.org/blog/first-ever-regulatory-submission-made-for-cultivated-meat-in-the-eu/. Accessed 31 Mar 2025
Holle M (2014) The protection of proprietary data in novel foods – how to make it work. Eur Food Feed Law Rev 9(5):280–284
Holle M (2018) Pre-market approval and its impact on food innovation: the novel foods example. In: Bremmers H, Purnhagen K (eds) Regulating and managing food safety in the EU. Springer, Cham, pp 291–330
Hyde R, Hartley S, Millar K (2017) European novel foods policy at a critical juncture: drawing lessons for future novel food governance through a retrospective examination of Regulation 258/97. Food Drug Law J 72:472–505
Lähteenmäki-Uutela A, Rahikainen M, Lonkila A, Yang B (2021) Alternative proteins and EU food law. Food Control 130:108336. https://www.sciencedirect.com/science/article/pii/S0956713521004746. Accessed 31 Mar 2025
Lanni S, Ragone G (2025) Il rapporto fra novel foods e carne coltivata. Rivista di Diritto Alimentare 1:17–32
Laus F (2024) L'approccio precauzionale al rischio nella sicurezza alimentare: Riflessioni intorno alla legge sullo stop alla carne coltivata. Federalismi.it 5:61–93
Lee M (2008) EU regulation of GMOs: law and decision making for a new technology – biotechnology regulation series. Edward Elgar, Cheltenham
Martini D, Del Bo C, Riso P (2020) Legislazione europea e ruolo di EFSA nella valutazione della sicurezza d'uso dei novel foods: Principi e prospettive. BioLaw J 2:9–23
Monaco A (2023) Data protection under the Novel Food Regulation: valuable instrument or barrier to innovation? Insights from the insect sector. Eur Food Feed Law Rev 18(3):172–178
Monaco A (2025) A perspective on the regulation of cultivated meat in the European Union. NPJ Sci Food 9:21
Monaco A, Purnhagen KP (2022) Risk triggers as innovation triggers? Risk analysis and innovation's promotion under the Novel Food Regulation. Eur Food Feed Law Rev 17:219–227
Monaco A, Reinhardt T (2025) How innovation friendly is the EU Novel Food Regulation? The case of cellular agriculture. Fut Foods 11:100574
Montanari F, Pinto de Moura A, Cunha LM (2021) Production and commercialization of insects as food and feed. Identification of the main constraints in the European Union. Springer, Cham
Mosa Meat (2025) Submitting our first EU market authorisation request. https://mosameat.com/blog/submitting-our-first-eu-market-authorisation-request. Accessed 31 Mar 2025
Opening Remarks by Commissioner Stella Kyriakides at the Agrifish Council Meeting - High-quality and primary farm-based food production of 23 January 2024. https://ec.europa.eu/commission/presscorner/detail/en/SPEECH_24_384. Accessed 31 Mar 2025
Paganizza V (2019) Bugs in law. Insetti e regole: dai campi alla tavola. Cedam, Padova

Paganizza V (2024) I nuovi alimenti ("novel foods"). In: Borghi P, Canfora I, Di Lauro A, Russo L (eds) Trattato di diritto alimentare italiano e dell'Unione europea, 2nd edn. Giuffrè, Milano, pp 802–816

Paganizza V (2025) Carne coltivata: una comparazione tra Australia e Unione europea, passando da Singapore. Rivista di Diritto Alimentare 1:62–75

Panebianco M (2023) Brevi considerazioni sul comunicato della Presidenza della Repubblica del 1° dicembre 2023. Consultaonline 3:1072–1075

Pitto S (2022) Genetically modified organisms and coexistence measures: the role of domestic courts in the evaluation of national limitations. Diritto Pubblico Comparato ed Europeo Online 55(4):2339–2346

Planchenstainer F (2024) Meat me in Italy: the Italian ban on sounding names and cell-cultured meat. Eur Food Feed Law Rev 19(2):66–71

Ragone G (2015) La disciplina degli OGM tra Unione europea e stati nazionali: A chi spetta il diritto all'ultima parola su questioni scientifiche controverse? BioLaw J 1:115–130

Reinhardt T, Monaco A, Purnhagen K (2024) Cultivated foie gras files into Europe – prepare for legal disruption. European Law Blog. https://www.europeanlawblog.eu/pub/ys66nyqh. Accessed 31 Mar 2025

Reins L, Ala-Lahti T (2024) Innovation, precaution and sustainable development in EU environmental law – a false trichotomy? https://papers.ssrn.com/sol3/papers.cfm?abstract_id=4697948. Accessed 31 Mar 2025

Rizzioli S (2025) Novel foods. In: Costato L, Albisinni F, Georgopoulos T (eds) European and global food law, 3rd edn. Wolters Kluwer, Milano, pp 703–713

Rossi EA, Cazzini F (2025) The future of food safety: international policies and domestic legislations on cultivated meat. In: Campanella P (ed) Towards safe food – rights, rules, values in the meat value chain. Editoriale Scientifica, Napoli, pp 73–84

Rossi Dal Pozzo F, Rubino V (2024) Il quadro regolatorio per l'autorizzazione all'immissione in commercio della carne coltivata nella UE. Eurojus 2:317–325

Rubino V (2024) La battaglia della carne coltivata dalle aule parlamentari a quelle di giustizia? Considerazioni a margine della legge 172/2023 fra armonizzazione, leale cooperazione e margini di autonomia degli stati membri. Rivista di Diritto Alimentare 1:34–55

Rusconi G, Cesana O (2024) Il diritto alimentare. Wolters Kluwer, Milano

Scaffardi L (2020a) Novel food, una sfida ancora aperta tra sicurezza alimentare, innovazione e sviluppo sostenibile. In: Scaffardi L, Zeno-Zencovich V (eds) Cibo e diritto. Una prospettiva comparata, vol 2. Roma TrE-Press, Roma, pp 735–773

Scaffardi L (2020b) I novel food, un futuro ancora da definire. BioLaw J 2:43–66

Scaffardi L (2022) A peculiar category of novel foods: traditional foods coming from third countries and the regulatory issues involving sustainability, food security, food safety, and the free circulation of goods. In: Formici G, Scaffardi L (eds) Novel foods and edible insects in the European Union – an interdisciplinary analysis. Springer, Cham, pp 37–58

Scarpa B, Dalfra S (2008) Regulating the novel foods sector: moving forward. Eur Food Feed Law Rev 3(5):292–299

Schebesta H, Purnhagen K (2024) Pre-market authorization of novel products: Novel Food Regulation. In: Schebesta H, Purnhagen K (eds) EU food law. Oxford University Press, Oxford, pp 159–173

Seehafer A, Bartels M (2019) Meat 2.0. The regulatory environment of plant-based and cultured meat. Eur Food Feed Law Rev 14(4):323–331

Simpson C (2016) Data protection under food law post: in the aftermath of the Novel Food Regulation. Eur Food Feed Law Rev 4:309–314

Sirsi E (2017) OGM e agricoltura. Evoluzione del quadro normativo, strategie di comunicazione, prospettive dell'innovazione. Editoriale Scientifica, Napoli

Spoto G (2024) Libertà di ricerca e paternalismo giuridico in Italia: La carne coltivata in laboratorio. Rivista di Diritto Alimentare 1:56–61

Szajkowska A (2011) Different actors, different factors. Science and other legitimate factors in the EU and national food safety regulation. Eur J Risk Regul 2:523–539
Tallacchini M (2012) Sicurezze e responsabilità in tempi di crisi. Rivista di Diritto Alimentare 1:1–19
Volpato A (2022) Novel food in the EU integrated administrative space: an institutional perspective. In: Formici G, Scaffardi L (eds) Novel foods and edible insects in the European Union – an interdisciplinary analysis. Springer, Cham, pp 15–36
Vos E, Wendler F (2006) Food safety regulation at the EU level. In: Vos E, Wendler F (eds) Food safety regulation in Europe: a comparative institutional analysis. Intersentia, Antwerp, pp 65–138
Weimer M (2019) Risk regulation in the internal market. Oxford University Press, Oxford

The Legal Debate on Cultured Meat Around the World: Comparative Insights from the United States, Singapore and Israel

Simone Pitto

Abstract This chapter offers a comparative legal analysis of the regulation of cultured meat in the United States, Singapore and Israel, jurisdictions that have authorised the commercialization of cell-based meat products. In particular, it addresses regulatory definitions, labelling standards and the role of public authorities in licensing and oversight. While all three share common features, significant differences in institutional competences, legal classifications and political dynamics emerge. In the United States, a dual-agency model involving the Food and Drug Administration (FDA) and United States Department of Agriculture (USDA) has been established, although the field remains marked by legal uncertainty and growing political opposition at state and federal levels. Singapore and Israel have adopted supportive frameworks aligned with broader sustainability and food security goals, including dedicated innovation policies. Notably, Singapore recently enacted the 2025 Food Safety and Security Act, creating a comprehensive framework for novel and cell-based foods. In 2024, Israel became the first country to authorise the sale of cultured beef, under a Novel Food framework involving multiple actors and emphasizing safety, while supporting the industry with dedicated institutions and investment. The chapter explores these developments with the aim of identifying common trends, divergences and broader implications for food governance.

Keywords Cultured meat regulation · Comparative food law · Novel Foods · Cell-based meat and alternative proteins · Food safety and security

---

S. Pitto (✉)
University of Genoa, Faculty of Law, Genoa, Italy
e-mail: simone.pitto@edu.unige.it

G. Formici et al. (eds.), *Cell-Based Meat in the European Union and Beyond*,
https://doi.org/10.1007/978-3-032-14624-3_5

1 Introduction

The growing debate on cell-based or cultured meat is driven by the need to address major contemporary challenges in food systems,[1] such as global population growth and the consequent increased demand for protein sources, as well as the environmental impact of food production and its contribution to climate change.[2] This is coupled with rising public interest in animal welfare and a growing stigma against unnecessary animal suffering.

Biotechnology and cellular agriculture are offering possible alternatives to traditional protein sources. According to a significant part of the scientific community, cultured meat is a promising technology, albeit in its initial phases, as it may provide alternative sources of protein and reduce environmental impacts, land use and consumption of water.[3]

Yet questions persist regarding the regulation, safety and long-term sustainability of cultured meat.[4] In this regard, existing legal frameworks on cell-based meat around the world reveal certain common elements but also significant divergence. Common features include pre-market procedures, substantial investment by public bodies, but also widespread criticism, political opposition and in some cases, limitations that often accompany the introduction (or the proposal of introduction) of the first cultured meat products. At the same time, differences in the competent authorities responsible for the relevant processes, the design and types of the procedures (authorisation, approval or voluntary consultation), the types of cultured meat released, as well as the market characteristics and underlying political motivations that shaped the legislation can be observed in the three countries.

The present chapter compares the different regulatory and legal frameworks adopted by countries engaged in the development of cultured meat, with particular attention to the case studies of the United States, Singapore and Israel, where cultured meat has reached a significant degree of development.[5]

Section 2 offers a brief general introduction and outlines the methodological framework for the comparative analysis. Section 3 discusses the legal framework of the United States, highlighting recent developments and controversial debates. Section 4 explores Singapore's regulatory landscape. Singapore has adopted a generally favourable stance towards food tech innovation and a comprehensive food

[1] According to a widely accepted definition of the Food and Agriculture Organization (FAO), food systems cover "the entire range of actors and their interlinked value-adding activities involved in the production, aggregation, processing, distribution, consumption and disposal of food products that originate from agriculture, forestry or fisheries, and parts of the broader economic, societal and natural environments in which they are embedded". FAO (2018), p. 1.

[2] Cheli et al. (2024), p. 152.

[3] Ferri et al. (2024), p. 22.

[4] See *What Is Cell-Based Meat? Food Safety and Sustainability Aspects* by Cheli et al. in this volume.

[5] Other countries such as Brazil, Qatar and Australia, however, are dealing with this technology. Olenic and Thorrez (2023), p. 751; Choudhury et al. (2020), pp. 574–576.

safety and security reform came into force in 2025. Section 5 turns to Israel, where a strong biotech sector and substantial government support have positioned the country as a global leader of alternative proteins. It is in this context that in 2024, Israel became the first country to authorise the commercialisation of cell-based beef. The Israeli case also highlights the cultural dimension of food regulation, which may be appreciated, *inter alia*, in the debate on the possibility of considering cultured meat *kosher*.[6]

Following the analysis of individual countries based on the methodology described in Sect. 2, the last section offers a comparative overview, identifying shared elements, divergences and emerging trends in the global regulatory *avant-garde* of cultured meat beyond Europe, and conclusive remarks.

2 Cultured Meat and Legal Research: Brief Methodological Notes for a Legal Comparison

In their work on comparative food law, Van Der Meulen and Wernaart proposed a fundamental methodological framework based on functional methods of comparative law[7] for legal comparative analysis in the food sector.[8] Recalling the classical identification of the purposes of comparative methods highlighted by Zweigert and Kötz,[9] they advanced a systematic approach to comparative food-law research. This methodological framework suggests an approach to comparative analysis based on identifying the "who, what and how in food law".[10] The "who-question" includes recognizing the relevant actors in a particular jurisdiction, including the institutions and public authorities involved in food policy, as well as its addressees (e.g. individuals, food businesses and stakeholders). The "what-question" seeks to define the concepts, principles and objectives underlying food law provisions, while the "how-question" addresses the specific regulation of individual segments of the food law frameworks. In this regard, it is worth recalling another relevant distinction of general comparative studies, which is the distinction between micro and macro comparative analysis. While the former focuses on specific rules or legal issues in a particular jurisdiction, macro-comparison casts a larger glance over entire legal systems, comparing broader legal components with those of other countries.[11]

The advantages of a structured method of this kind in comparative legal analysis include the reduction of common biases (such as the "optical aberrations" resulting from a scholar's own legal culture), the ability to focus on key elements of the

[6] See *Food Justice in Religious Laws: Hypotheses on Cultivated Meat* by Anello in this volume.

[7] Michaels (2019), p. 680.

[8] Van Der Meulen and Wernaart (2023), p. 201.

[9] See Zweigert and Kötz (1998).

[10] Van Der Meulen and Wernaart (2023), p. 201.

[11] Monateri (2022).

regulatory frameworks under study, and a better comparability of results. This also ensures that the analysis is based on homogeneous (and thus comparable) areas across the selected jurisdictions.

Building on the framework proposed by Van Der Meulen and Wernaart, the following pages adopt a micro-comparative approach, focusing on the legal treatment and implications of cultured meat in selected jurisdictions and analysing the "who", "what" and "how" of the legal discipline.

In this regard, it also seems opportune to address some definitory issues. This chapter uses the expression "cultured meat" to refer to various production techniques, also referred to in legal literature as "lab-grown meat" or "cell-based meat", involving the cultivation of animal cells in controlled artificial environments known as "bioreactors" to produce edible muscle fibres and tissues, often combined with plant-based ingredients and adapted for human consumption.[12] Though commercial interest is only recent, the technology dates back at least some decades,[13] with early achievements including National Aeronautic Space Administration (NASA)'s space food research in the 1990s and Willem Van Eelen's 1997 patent.[14] The technology may be suitable for extension to other nonanimal-sourced food products such as fish and even milk.[15]

Notably, the very reference to the products of this technology as "meat" is neither neutral nor to be taken for granted. Indeed in some cases, the inclusion and labelling of products not derived directly from raised and slaughtered animals as "meat" has been firmly contested.[16] Furthermore, given the plurality of techniques and production methods currently in use, there is still no clear or universally accepted definition of this category of food.[17] As such, a further criterion for comparative analysis will be the definitory dimension of the phenomenon, which will allow us to clarify the scope of relevant legislation, assess whether cultured meat products are assimilated to conventional meat products, and determine whether such assimilation is consistent or divergent across selected jurisdictions.

As previously noted, the case studies of this chapter are limited to the United States, Singapore and Israel. Indeed, at the time this chapter was written, the United

---

[12] Olenic and Thorrez (2023), p. 749. See *What Is Cell-Based Meat? Food Safety and Sustainability Aspects* by Cheli et al. in this volume.

[13] Olenic and Thorrez (2023), p. 749. While the idea has been traced back to the 1897 scientific novel Auf Zwei Planetem by Treich (2021), p. 38, the first cultured burger is credited to the physiologist Mark Post in 2013. See Hall (2023), p. 5.

[14] Ferri et al. (2024), p. 3.

[15] See Saija (2025), pp. 42–45.

[16] See the Petition proposed by the United States Cattlemen's Association (USCA) before the Food Safety Inspection Service (FSIS) in 2018 for the imposition of beef and labelling requirements deemed to exclude products such as cultured meat not derived directly from slaughtered animals. See https://www.fsis.usda.gov/sites/default/files/media_file/2020-07/18-01-Petition-US-Cattlement-Association020918.pdf. Accessed 18 February 2025.

[17] On this aspect see *The Regulatory Landscape of Cultured Meat and Fish: Terminology and Consumer Information Provisions* by Sirsi and Rolandi in this Volume, who recall the recent work of the Codex Alimentarius Commission and its definitory relevance.

States, Singapore and Israel were the only countries where cultured meat had been authorized in some form, and had been available for commercialization, albeit subject to certain limitations, for a sufficiently long period to allow a meaningful assessment. This makes these countries suitable case studies for examining the "who", "what" and "how" of current cultured meat regulation from a comparative perspective.

Nevertheless, it should be noted that these jurisdictions are not the only ones where regulatory developments on cultured can be traced. For instance, in 2023 the Australian Vow Group Pty Ltd ('Vow') submitted an application to amend the Australia New Zealand Food Standard Code in order to use cell-cultured quail as an ingredient.[18] The application was assessed by the Food Standards Australia New Zealand (FSANZ) with two rounds of public consultation and sought approval by FSANZ in 2025.[19] Notably, in June 2025, the Australia New Zealand Food Standards Code was changed accordingly with new standards to allow the use of cell-cultured quail as a novel food ingredient.[20]

3 The United States

The United States (US), one of the world's leading consumers of meat, in recent years has experienced a significant increase in sales of plant-based food and alternative proteins sources.[21] This brought a wave of investments in alternative protein technologies, including cell-based meat. According to a 2022 survey of the Good Food Institute (GFI), almost one-third (43) of the 156 companies active in cell-based meat are based in the US, making the country the one with the largest concentration of businesses in the sector.[22]

Against this backdrop, the regulatory debate entered a central stage in 2018–2019. Driven by Congress, the United States Department of Agriculture (USDA) and the Food and Drug Administration (FDA), entered an agreement to establish mutual responsibilities regarding the regulation of cultured meat, formalized in 2019.[23] A

---

[18] FSANZ—Food Standards Australia New Zealand (2023). See Paganizza (2025), pp. 66–67.

[19] FSANZ (2025a).

[20] Namely, amendments to the Australia New Zealand Food Standards Code include, *inter alia*: Standard 1.5.4., setting out labelling requirements with the use of the expression 'cell cultured' or cell-cultivated; Standard 3.4.1., outlining food safety requirements; a new schedule (Schedule 25A), listing the cell-cultured food which are permitted, Schedule 27, which has been modified to include relevant microbiological limits. FSANZ (2025b).

[21] This was estimated to be a 29% increase in sales since 2017, compared to only 4% for general retail sales of food products. Bernstein (2021), p. 2.

[22] Good Food Institute (2023); Congressional Research Service (CRS) (2023), p. 5.

[23] Formal Agreement Between FDA and USDA Regarding Oversight of Human Food Produced Using Animal Cell Technology Derived from Cell Lines of USDA-amenable Species, at https://www.fda.gov/food/human-food-made-cultured-animal-cells/formal-agreement-between-fda-

voluntary pre-market consultation process was also established, leading to approval of products by companies such as Upside Foods and Good Meat for the commercial sale of cultured chicken.[24] However, recent legal developments have proven contentious, as traditional meat-industry concerns and lobbying have led to proposed federal bans in Congress and adoption of state-level restrictions on the sale of cultured meat in States like Alabama[25] and Florida.[26]

Regarding the "who" question outlined in Sect. 2, it is worth recalling briefly that in the US food law system, different federal agencies share competence in food regulation. The Department of Health and Human Services (HHS) is responsible for public health protection, while the FDA, under its authority, oversees food safety and medical products.[27] The USDA,[28] through its Food Safety and Inspection Service (USDA-FSIS), regulates agriculture and breeding, also being responsible for meat and poultry production.[29]

Since cultured meat lies at the intersection of the competencies of these agencies, a notable initial regulatory challenge in the US concerned identifying the authority responsible for overseeing its market authorisation.[30] To resolve this uncertainty, as anticipated, the FDA and USDA formalized a joint agreement, establishing the division of responsibilities in overseeing operations related to cultured meat products.

Based on the Agreement, the FDA oversees the premarket consultation process, as well as cell collection, differentiation operations and the maintenance of cell banks in the first phases of the culture process. This voluntary premarket procedure is intended to assess food safety in the early stages of production rather than an approval process.[31] It primarily focuses on ensuring that producers observe the relevant manufacturing practices and all necessary controls to verify that cultured meat products are safe to eat.[32] At the same time, a comprehensive review is conducted by

---

and-usda-regarding-oversight-human-food-produced-using-animal-cell. Accessed 23 February 2025.

[24] FDA (2022). See Formici (2024), pp. 183–185.

[25] Alabama S.B. 23, signed into law on 7 May 2024.

[26] Florida S.B. 1084, passed on 1 May 2024. For more, see National Agricultural Law Center (2024).

[27] The Federal Food, Drug and Cosmetic Act (FFDCA; 21 USC para 301 *et seq.*) is the primary legal framework governing the FDA's action on food safety. Additionally, the Agency's authority stems from the Public Health Service Act, 42 USC para 201 *et seq.* Cf. Bruno (2017), pp. 155–156.

[28] The Federal Meat Inspection Act (FMIA), codified at 21 USC para 601 *et seq.*, may be regarded as the legal basis of the USDA's competences over meat products and sale, jointly with the Poultry Products Inspection Act, 21 USC para 451 *et seq.*

[29] Sanchez (2018), p. 25. See in particular Section 5, Poultry Products Inspection Act, 21 U.S.C. para 454, regulating Federal and State cooperation.

[30] Formici (2024), pp. 184–185.

[31] FDA (2022) available at https://www.fda.gov/food/hfp-constituent-updates/fda-completes-first-pre-market-consultation-human-food-made-using-animal-cell-culture-technology. Accessed 25 April 2025.

[32] *Verbatim*: "The voluntary pre-market consultation is not an approval process. Instead, it means that after our careful evaluation of the data and information shared by the firm, we have no further

the FDA on the origins of the stem cells used, as well as the safety procedures undertaken in the early stages of the process, and the testing operations.[33]

At the stage of harvest of the cells, oversight tasks transition to USDA-FSIS, which is competent for inspecting processing, packaging and labelling, ultimately being responsible for ensuring the safety of the final cultured meat products and labelling standards. The Agreement also commits both agencies to share information, enforce regulations in their respective domains, and collaborate to address safety concerns.[34] The agencies have established specific working groups with the purpose of enforcing the agreed supervision framework. These include an FDA-led premarket assessment working group, a labelling group under the responsibility of the USDA and a group on the transfer of jurisdiction, which is led by the FDA and USDA together.[35]

As a result, while the regulation of the development of cultured meat in the early stages of the process falls within the competence of the FDA, the USDA-FSIS is responsible for the processing stages and the transformation of cells into edible meat products. Conversely, when cell-based cultured seafood is involved, the FDA is the competent authority for both the development process and the subsequent processing phases leading to the final product.[36]

The agreement has also raised criticism in scholarly debate, especially in light of the risk of influence of private interests over the process and the threat of the USDA being "captured" by the traditional meat industry,[37] although other authors have hailed it as a practical solution to provide a clear regulatory framework.[38]

Based on the 2019 agreement, in November 2022, the first premarket consultation was completed by the FDA for Upside Foods' cultivated chicken. The FDA's assessment diffusely built on the information provided by Upside Foods, which was checked to verify the adequacy of the controlled environment and bio-reactors used in the first phases. The procedure was completed with a "no further questions" statement issued by the FDA.[39] A similar process was finalized in March 2023 for Good

---

questions at this time about the firm's safety conclusion". FDA (2022), p. 1. See Formici (2024), p. 186.

[33] See DaVault and Sinha (2025), p. 31. Pursuant to Section 4(A) of the Agreement between the FDA and the USDA cited at footnote 27, "HHS-FDA will: 1. Conduct premarket consultation processes to evaluate production materials/processes and manufacturing controls, to include oversight of tissue collection, cell lines and banks, and all components and inputs".

[34] CRS (2023), pp. 8–11.

[35] Government Accountability Office (GAO) (2020), p. 18.

[36] CRS (2023), p. 9.

[37] Bernstein (2021), p. 26 argues that the arrangement "is far from ideal" and that "the USDA is subject to agency "capture" by the meat industry, meaning that the meat industry is able to leverage its influence to induce the USDA to consistently act in accordance with the industry's goals rather than acting to protect the public interest". See Etheridge (2022), p. 1736.

[38] DaVault and Sinha (2025), p. 33. According to the authors "The joint approach between FDA and USDA appears to be a well-reasoned and innovative approach to regulating an otherwise difficult-to-regulate novel product".

[39] FDA (2022). For more, see Formici (2024), p. 185.

Meat's cultivated chicken.[40] This second premarket clearance also led to a "no further questions" statement by the FDA. Both companies then submitted additional facility information to USDA-FSIS, which carried out inspections and approved the sites, thus enabling commercial sales in San Francisco and Washington D.C. in 2023.[41]

Yet regulatory challenges in the US extend beyond authorisation. As discussed in Sect. 1, definitions and labelling for cell-cultivated meat products remain critical issues.[42] While these products fall under the definition of "food" according to Section 201 of the Federal Food, Drug and Cosmetic Act (FFDCA), their classification within the definitions of "meat" and "meat by-products",[43] that describe food grown or produced from muscle of animals and products derived or prepared in whole or in substantial part from portions of animals, respectively, remains a matter of debate.[44] Additionally, the Poultry Products Inspection Act (PPIA) provides definitions of "poultry"[45] and "poultry product"[46] that are relevant to Upside Foods and Good Meat cultured chicken. Unlike other jurisdictions analysed in this chapter, the PPIA does not strictly require that poultry products be derived solely from slaughtered animals, and thus cultured chicken may arguably be considered to fall within the definition of poultry product.[47] However, scholars such as DaVault and Sinha have advocated introducing a new dedicated classification for these products in order to clarify existing doubts, arguing however for the possibility of including, *de iure condito*, the cultured meat products within the definition of meat.[48]

This allows us to tackle another key regulatory issue, i.e. cultured meat labelling. Even in this case, as no specific laws have been introduced so far by the Congress to regulate cultured meat products,[49] they are therefore subject to general food

[40] FDA (2023) and CRS (2023), p. 10.

[41] CRS (2023), p. 11.

[42] See *The Regulatory Landscape of Cultured Meat and Fish: Terminology and Consumer Information Provisions* by Sirsi and Rolandi in this volume.

[43] Code of Federal Regulations (CFR) para 301.2, available at https://www.govinfo.gov/content/pkg/CFR-2024-title9-vol2/pdf/CFR-2024-title9-vol2-sec301-2.pdf. Accessed 24 February 2025.

[44] DaVault and Sinha (2025), p. 35.

[45] "Poultry" is defined as "any domesticated bird, whether live or dead". TITLE 21 United State Code (USC), Chapter 10, para 453(e).

[46] "Poultry product" is defined as "any poultry carcass, or part thereof; or any product which is made wholly or in part from any poultry carcass or part thereof". TITLE 21 USC, Chapter 10, para 453(f).

[47] This seems to be the conclusion reached by the United States District Court in the Northern District of Florida, Tallahassee Division (11 October 2024) case no. 4:24-cv-00316-MW-MAF *Upside Foods, Inc. v. Simpson et al.* available at https://ij.org/wp-content/uploads/2024/10/Upside-PI-decision.pdf. Accessed on 31 March 2025.

[48] DaVault and Sinha (2025), p. 36: "While this paper does argue, contrary to some prior literature, that the final product produced by the cultivation process is meat rather than a meat alternative, a new category of food classification must still be created".

[49] See USDA-FSIS (2023).

labelling rules that require a clear and non-deceptive description of the food items.[50] The USDA-FSIS is competent to oversee the process of labelling with a pre-approval procedure deemed to verify whether cell-cultured meat is labelled in compliance with general labelling principles as well as truthfully, for consumers.[51] The alternative protein industry initially advocated using the term "clean meat", meeting criticism from the livestock industry.[52] Ultimately, in June 2023, Good Meat reported that USDA-FSIS approved "cell-cultivated chicken" as the official label for products to be sold in the US by the company.[53]

Recent developments also seem to confirm the controversial nature of cultured meat regulation in the US. First, several states have enacted laws restricting the use of terms such as "meat," "hamburger" or "sausage" for cultured products. As of 2018, states such as Arkansas,[54] Kansas,[55] South Carolina[56] and Wyoming[57] have passed legislation limiting or prohibiting the use of the term "meat" for non-traditionally sourced products, potentially in conflict with federal labelling regulations.[58]

A related trend is the introduction of bans on cultured meat.[59] In 2024, Florida and Alabama enacted laws criminalizing the production, sale and distribution of cultivated meat, including penalties for violations.[60] In 2025, Mississippi became the third state to follow the trend.[61] These pieces of legislation, which have inspired other similar initiatives currently under discussion in other States such as Arizona,[62] have also raised constitutional concerns from early commentators regarding the Commerce Clause.[63] Notably, in 2024 Upside Foods filed a lawsuit challenging Florida's ban SB 1084 as unconstitutional before the United States District Court, Northern District of Florida, Tallahassee Division.[64] The Court has so far denied the

[50] Such as the Fair Packaging and Labelling Act (FPLA), TITLE 15 USC, Chapter 39, para 1451–1461.

[51] Notably, in September 2021, the USDA-FSIS has issued an Advance Notice of Proposed Rulemaking (ANPR) in order to gather public commentaries and input on labelling regulation for cultured meat and poultry products. USDA (2021).

[52] See Hall (2023).

[53] Good Meat (2023); CRS (2023), p. 12.

[54] Arkansas, H.B. 1407, passed on 18 March 2019 (especially Section 2-1-302).

[55] Kansas, S.B. 261, passed on 5 May 2022 (especially Section 2, 65-665).

[56] South Carolina H.B. 4245, passed on 9 May 2019 (especially Article 5, Section 47-17-510).

[57] Wyoming S.B. 68, signed into law on 26 February 2019 (especially Section 1, 35-7-111).

[58] CRS (2023), p. 13.

[59] National Agricultural Law Center (2024).

[60] Florida SB 1084 passed on 2 May 2024 and Alabama SB 23, enacted on 8 May 2024.

[61] Mississippi HB 1006, signed into law on 21 March 2025.

[62] See Arizona HB21121 (2024).

[63] Mousa (2025), p. 157; Fortin (2024).

[64] See Institute for Justice (2024), available at https://ij.org/press-release/federal-judge-denies-preliminary-injunction-in-upside-foods-lawsuit-against-floridas-ban-on-cultivated-meat/. Accessed 31 March 2025.

request for a preliminary injunction to suspend the effects of the ban.[65] Significantly, the Court underscored that even though cultured meat may fall within the definition of poultry products under the PPIA,[66] Upside Foods failed to demonstrate that Florida's ban on cultured meat is inconsistent with federal law.[67]

In addition, at federal level these tensions have been echoed in recent legislative proposals, such as Bill H.R. 8757 (2024), which seeks to prohibit the use of federal funds to sustain the research and development of cultured meat products.[68] This trend aligns, however, with trends common in other jurisdictions, such as the 2023 Italian ban[69] on the production and commercialization of "synthetic food", including cultured meat,[70] and France's attempts to limit use of terms such as "meat", as well as advanced proposals of statutes including bans on cultured meat products.[71]

4 Singapore

In 2020, Singapore became the first country to authorize the commercialization and sale of cultured chicken, produced by the California-based start-up Eat Just Inc.[72] This was the result of an expedited food safety assessment and authorization process by the Singapore Food Agency (SFA), an authority established by the SFA Act of 2019 (SFAA) under the former Singapore Ministry of the Environment and Water Resources (MEWR), now renamed Ministry of Sustainability and the Environment

---

[65] United States District Court Northern District of Florida, Tallahassee Division (11 October 2024) case no. 4:24-cv-00316-MW-MAF *Upside Foods, Inc. v. Simpson et al.*

[66] See CRS (2023), pp. 14–15.

[67] *Verbatim* from page 17 of the order of the Court "To be clear, Plaintiff has identified no federal authority that requires the sale or use of cultivated meat in any derivative poultry product. Nor has Plaintiff pointed to anything other than the statutory definition of "poultry product" to suggest that cultivated chicken must be permitted as an ingredient in any derivative poultry product. Moreover, although Plaintiff's cultivated chicken arguably falls within the plain meaning of the PPIA's "poultry product" definition, federal regulations identifying the standards for kinds and classes, and for cuts of raw poultry available in the market do not appear to include cultivated chicken as a "kind" or "class" of poultry".

[68] Formici (2024), p. 187.

[69] Law No. 172 of 1 December 2023, *Disposizioni in materia di divieto di produzione e di immissione sul mercato di alimenti e mangimi costituiti, isolati o prodotti a partire da colture cellulari o di tessuti derivanti da animali vertebrati nonche' di divieto della denominazione di carne per prodotti trasformati contenenti proteine vegetali.* (23G00188) (Official Gazette General Series No. 281 of 1 December 2023), https://www.gazzettaufficiale.it/eli/id/2023/12/01/23G00188/sg. Accessed 31 March 2025. See *Cell-Based Meat in the European Union: A Regulatory Crossroads* by Corazza and Formici in this volume.

[70] As noted by Formici (2024), p. 167. See also Lanzoni et al. (2024), p. 3.

[71] *Proposition de Loi* No. 1965, 5 December 2023 includes a ban on production, process and commercialization of cultured meat in the interests of human health. See Paganizza (2025), p. 63.

[72] Hall (2023), p. 5 who notes that the Singapore restaurant "1880" became the first establishment in the world to serve cultured meat to the public.

(MSE).[73] Pursuant to the SFAA, the Agency is entrusted *inter alia* with: implementation of a comprehensive food safety framework in Singapore, Section 12(b); regulation of the safety and wholesomeness of the food supply, Section 11(a); and provision of strategic recommendations to the government concerning the agri-food system, Section 11(f).

The regulatory environment that enabled the authorization of Eat Just Inc. cultured chicken relied on a Novel Food normative framework introduced in 2019,[74] which was later reformed in January 2025. In this Section I briefly discuss the legal framework that enabled the authorisation of cultured meat, insofar as granted, and then focus on the recently introduced 2025 discipline.

At the time of the authorisation granted to Eat Just Inc., the above framework aligned with broader normative strategies such as the Singapore Green Plan 2030[75] of 2021, a government policy aimed at fostering sustainable development, and the Singapore Food Security Plan "30 by 30", which included actions to enhance food security and promote the use of alternative protein sources, with the ultimate goal of reducing the country's high dependence on imported food.[76]

The 2019 Guidelines on Novel Food applied to foods that lacked a history of safe consumption,[77] and adopted a chronological criterion. Specifically, based on these provisions, a food was to be considered novel if it had not been consumed as part of the diet for at least 20 years.[78] In the absence of such a history, cultured meat products were considered Novel Foods and thus subject to these provisions.[79] This approach, albeit taking into the history of use in a manner comparable to the European Novel Food framework,[80] introduced a "mobile chronological reference",

---

[73] SFAA 2019. Similar functions were previously conducted by the former Agri-Food and Veterinary Authority of Singapore regulated by the Agri-food and Veterinary Authority Act, https://sso.agc.gov.sg/Act-Rev/AFVAA2000/Published/20121231?DocDate=20121231&ViewType=Pdf&_=20250308235322. Accessed 22 April 2025.

[74] SFA (2023b).

[75] Singapore Government Press Centre (2021).

[76] Singapore Ministry of Sustainability and the Environment (2024) and Teng and Montesclaros (2019).

[77] SFA (2023c); Formici (2024), p. 189.

[78] SFA (2023b).

[79] Miyake et al. (2023), pp. 4–6.

[80] According to Regulation (EU) 2015/2283, the current discipline in force after reform of Regulation (EC) no 258/97, the regulatory definition of novel food has two components. Article 3 of the Regulation envisages a chronological criterion similar to the SFA requirement, according to which novel foods are any food that was not used in the EU to a significant degree before 15 May 1997. Nevertheless, according to Article 3 (2) of EU Regulation 2015/2283, the scope of the definition extends only to novel foods which are included in one of the expressly mentioned categories. These include *inter alia* "food with a new or intentionally modified molecular structure, where that structure was not used as, or in, a food within the Union before 15 May 1997" and "food consisting of, isolated from or produced from material of mineral origin". See Volpato (2022), p. 18; Scaffardi (2020), pp. 44–65; see *Cell-Based Meat in the European Union: A Regulatory Crossroads* by Corazza and Formici in this volume.

thereby differing substantially from the European model, which fixes a cut-off date (15 May 1997).

Companies interested in commercializing, importing, distributing or selling Novel Foods and ingredients such as cultured meat in Singapore had to submit a comprehensive dossier demonstrating the safety of the products, including aspects such as toxicity, allergenicity, as well as compliance with additional requirements set by relevant legislation.[81] A safety risk assessment was then conducted by the SFA in light of the scientific data and information provided. The safety assessment was assisted by the SFA-established Novel Food Safety Expert Working Group, which comprises 11 experts (mainly academics) specialized in food sciences, and provides advice on scientific matters.[82] The evaluation process, conducted on a case-by-case basis, included a joint dialogue with the companies involved, supported by the presence of permanent platforms for stakeholder and government discussions, such as the Novel Food Virtual Clinics.[83] The SFA had also introduced self-assessment checklists specifically designed for cultured meat and seafood products. These guidelines were intended to help applicants address all necessary safety considerations before submitting their applications.[84]

As far as the labelling of cultured and cell-based meat products is concerned, Singaporean authorities require labelling that reflects the true nature of the products. Labels must include descriptors such as "cultured" or "cell-based" to ensure transparency for consumers and to prevent misleading claims or confusion with conventional meat.[85]

In 2024, a proposal for a comprehensive reform of food safety and security legislation was advanced.[86] In January 2025, as previously mentioned, the Food Safety and Security Act 2025 (FSSA) was passed by Singapore Parliament.[87] The Act repealed the previous Sale of Food Act 1973 and the Wholesome Meat and Fish Act, which had also served as the legal basis for the pre-market assessment of cultured chicken by Eat Just Inc. This new comprehensive regulatory framework is intended to strengthen Singapore's food security by maintaining essential food reserves, promoting sustainable domestic production and mitigating disruptions in the food supply. It also aims to ensure food safety, to increase local production of primary food

[81] Particularly the Sale of Food Act of 1 May 1973, https://sso.agc.gov.sg/Act/SFA1973, and the Wholesome Meat and Fish Act of 10 December 1999, https://sso.agc.gov.sg/Act/WMFA1999, as well as implementing acts. All accessed 22 April 2025.

[82] SFA (2025). See also Vlčko et al. (2023), p. 931.

[83] Formici (2024), p. 189.

[84] SFA (2023a).

[85] SFA (2023c); Vlčko et al. (2023), p. 931.

[86] The proposal was accompanied by a public consultation on the draft from which the MSE and SFA were seeking feedback, especially on the category of "defined food" and the pre-market approval process. See MSE and SFA (2024); Paganizza (2025), p. 65.

[87] Republic of Singapore, Government Gazette, Acts supplement No. 1, 7 February 2025, FSSA 2025, passed by Parliament, 8 January 2025, https://sso.agc.gov.sg/Act/FSSA2025?ViewType=Advance&Phrase=Novel+food&WiAl=1. Accessed 24 April 2025.

products in a sustainable manner, and to uphold standards for both imported and exported food. Additionally, it seeks to promote public health through precise food information, regulation of agri-food inputs and promotion of national dietary guidelines.[88]

Section 3 of the Act includes an extensive list of normative definitions, including the newly introduced category of "defined food",[89] specifically envisaged by Section 13(1), along with other definitions relevant to the present chapter such as "Novel Food", "meat",[90] "meat product",[91] "meat processing",[92] "fish"[93] and "fish products".[94]

Under the FSSA, cultured meat falls within the definition of novel foods, a definition primarily based on a chronological criterion. This includes categories of food that have not been significantly consumed inside or outside Singapore, as well as other criteria related to the product's characteristics and origin, or the processes used in its production. Novel Foods include: (1) foods lacking significant use for a period of at least 20 years inside or outside Singapore, expressly mentioning food isolated or produced by "*cell culture or tissue culture derived from animals, plants, bacteria or yeast, fungi, algae or other micro-organism*" (Section 13(a)); (2) foods produced with innovative processes not applied for at least 20 years (Section 13(b)); (3) food derived from materials of mineral origin (Section 13(c)); and (4) foods that contain or consist of engineered nanomaterials (Section 13(d)).

A novelty introduced by the FSSA, aimed at establishing a clear pre-market assessment framework for food businesses, is the category of "defined food". A food product is referred to by Section 13(1) as "defined food" when it "(a) is, consists of or has as an ingredient a novel food in respect of which no pre-market approval is granted; (b) is, consists of or has as an ingredient a genetically modified

[88] FSSA, Part 1, Section 2.

[89] See Paganizza (2025), p. 66.

[90] Meat is defined as "raw food that is the whole or any part of the carcass of a slaughtered animal (including edible offal) after it is dressed, whether fresh or after freezing or chilling, but does not include any hide, tallow, meat product, fish or fish product". See FSSA Section 3(1), p. 34.

[91] This includes a broader category of food intended for human consumption including: "(a) any product wholly or partially prepared or made from meat; (b) any product in paragraph (a) which is partially processed, prepared for further processing or fully processed, by any method mentioned in paragraph (a) of the definition of "meat processing"". See FSSA Section 3(1), p. 35.

[92] "Meat processing" means "(a) the production of meat products from slaughtered animals (i) by mixing meat with another meat or any other substance; or (ii) by cutting, mincing, curing, boning, drying, smoking, preserving, salting, fermenting or otherwise treating, meat after slaughter; (b) the storage or packing of any meat or meat product; or (c) any other activity that is prescribed by the Minister, by order in the Gazette, for the purpose of this definition where the activity involves meat". See FSSA Section 3(1), p. 34.

[93] "Fish" means "any species of fish (whether marine or freshwater), and includes (a) crustaceans, shellfish, echinoderms, mollusks or any other form of aquatic life; and (b) the roe or young of any fish". See FSSA Section 3(1), p. 27.

[94] These are defined as "(a) any part of any fish; (b) any product derived from processing or preserving fish; (c) any product containing fish". See FSSA Section 3(1), p. 28.

food in respect of which no pre-market approval is granted; or (c) is, consists of or has as an ingredient, in any form (whether whole or in parts and whether fresh, chilled, frozen, dried, smoked, salted or in brine, or as flour) an edible insect-like species which is not a catalogued insect-like species". With this classification, the FSSA restructured the approval process for novel food, insect-based products and genetically-modified ingredients, which is now governed by Part 5 of the Act.[95] Pre-market approval is required before such products can be made available to consumers, based on health and safety assessments conducted by the Singapore Food Agency (SFA).[96] The Agency may grant or refuse approval, impose conditions and even revoke granted approvals.[97]

As far as cultured meat is concerned, it is worth noting that the definition of "meat" under Section 3 of the FSSA appears to limit its scope to parts of slaughtered animals, whereas the production process typical of cultured meat involves extracting stem cells from living animals. The reference to slaughter is also present in the definitions of "meat products" and "meat processing". Cultured meat therefore does not seem to fall under the FSSA's definition of "meat"[98]. Nevertheless, it is worth adding that the Act seems to allow inclusion of other products within the scope of "meat processing" for other activities prescribed by the Minister.[99]

Furthermore, labelling requirements are governed by Section 309 FSSA, which authorizes the SFA, with Ministerial approval, to establish regulations on information, safety standards for agri-food production inputs and labelling contents.[100]

Additional references to cultured and cell-based foods are also found in the First Schedule of the Act.[101] Section 6(1) of the Act defines licensable food businesses as those operating in Singapore under retail and non-retail food categories specified in the First Schedule. In this regard, the competent Minister is empowered to amend the classes listed in the Schedule by adding or removing business types and updating their descriptors, based on criteria such as risks for public health and type of food and consumers.[102] The current version of the First Schedule distinguishes retail (Part 1) and non-retail (Part 2) licensable food businesses, and provides definitions

[95] Pre-market approval, as defined in FSSA Section 105, is the authorisation intended for novel or genetically modified food to be consumed by the public, after assessment by the Agency of health and safety risks. It includes potential adverse effects, composition, preparation process, source, consumption patterns, and other relevant factors.

[96] FSSA Section 105 specifies that the assessment is carried out having regard to adverse effects on humans, the composition of the novel food, and the processes and sources used to prepare the novel food. Supplying defined foods without approval is an offence and offenders are punished with fines or imprisonment, according to Sections 106–107. See Mahira et al. (2025).

[97] See FSSA Sections 110–113.

[98] Vlčko et al. (2023), p. 931.

[99] "Minister" refers to the minister responsible for food safety, supply resilience and sustainability or for the purposes of Part 9, the Minister responsible for the promotion of health pursuant to Section 3.

[100] Section 183 also sets out the offences and penalties for mislabelling food.

[101] FSSA, First Schedule, Licensable Food Business.

[102] See FSSA Section 6(3).

and examples of the businesses considered licensable. According to Section 4(g) of Part 2 of the Schedule, a food business that, for the purpose of sale or export, involves "cell-based food production, using cell or tissue culture of animals, plants, micro-organisms, fungi or algae" is considered a non-retail licensable food business. Food business licences are granted to applicants by the SFA, taking into account factors set out in Section 92(2) of the Act,[103] and remain in force for a period specified in the license, although not exceeding 5 years.[104]

5 Israel

Like Singapore, Israel also takes a generally favourable approach to cultured and cell-based food products, part of a broader trend aimed at encouraging research and innovation in alternative proteins. In 2024, Israel became the first country to authorise the commercialization of cultured beef by the company Aleph Farms.[105] Additionally, while the company SuperMeat has opened a test restaurant in Tel Aviv serving cultured chicken, the food business Future Meat recently established an industrial-scale production facility in Israel.[106]

Like the EU, with which Israel shares commercial interests in food production and thus partly aligns with its food regulatory standards,[107] the latter has established a legal framework on Novel Foods that encompasses cultured meat products. In this regard, the authorization granted to Aleph Food was regulated by the Novel Food Directive 2006 (004-08)[108] and the Public Health Protection (Food) Law 2015-5576 (PHPFL).[109]

Novel Foods are defined by Section 2 PHPFL as any food or food component (excluding supplements, additives, flavourings or processing aids) that meets at least one of the following criteria: it has a modified primary molecular structure with no established safety record in Israel before February 2006; it contains genetically-modified organisms; it includes or is derived from plants, animals, microorganisms, fungi or algae on which there is insufficient information in Israel with regard to their safety, or is produced using an unconventional manufacturing

[103] See FSSA, Section 92(2), letters (a)–(j). These include the applicant's qualifications, past disqualifications or offences, ongoing or prior regulatory actions, the condition and legal status of the premises, the existence of business plans, compliance history, and the overall public interest of Singapore.

[104] This is specified by FSSA Section 94(2): "Every food business licence granted under this Part is in force for the period (not exceeding 5 years) specified in the licence — (a) except when it is wholly suspended under section 96(2); or (b) unless it is earlier revoked under section 96".

[105] Formici (2024), p. 188.

[106] Vlčko et al. (2023), p. 934.

[107] Letterhead Food Research (2024).

[108] Miyake et al. (2023), p. 5.

[109] Artom (2025), p. 55; Vlčko et al. (2023), p. 934.

process that alters the food's composition, structure or nutritional value and is not commonly consumed in Israel.[110] Foods that meet the definition of novel foods, including cultured meat and alternative proteins,[111] are subject to a premarket Novel Food Registration Procedure.[112] The process is overseen by the National Food Service (NFS), under the Ministry of Health, after a proposal by a dedicated body, the Novel Food Committee (NFC).[113] The process, which applies both to imported goods and novel foods developed in Israel, aims to assess the safety and nutritional value of such products prior to their commercialization on the domestic market.[114]

Interested food companies may submit an online application in Hebrew or English to register new foods, jointly with a safety dossier.[115] The documents provided are reviewed during the safety assessment conducted by the Food Risk Management Department (FRMD) of the NFS, which refers to the Novel Food Committee. The latter subsequently provides a final recommendation for registration to the Director of the NFS.[116] Approved petitions are then notified to the applicant, including any conditions of approval, and are published on the NFS website. As far as the safety assessment is concerned, this involves evaluating risks such as undesirable changes resulting from genetic modifications, the toxicogenic or allergenic potential of the cell lines used in the production, and potential long-term effects on consumers from frequent consumption of the food products under review.[117]

Pursuant to the PHPFL of 2015, the Ministry of Health also holds competence regarding the labelling of Novel Foods.[118] The proposed labelling must be included in the applicant's dossier, and should detail the product and ingredient names,

[110] Public Health Protection (Food) Law 2015-5576 (PHPFL), Section 2 (definitions), available in English at https://www.fao.org/fileadmin/user_upload/gmfp/docs/Israeli%20public%20health%20protection%202015.pdf. Accessed 14 August 2025. For more, see Food Control Service (2024), p. 2; FAO (2022), p. 7.

[111] According to the Guidance Document on Submission of Applications for Approval of Novel Proteins as Novel Food, foods that include "novel proteins" (alternative proteins) produced by precision fermentation, and "cell-based foods" such as "cultured meat" produced by cell culture (henceforth "novel proteins", the subject of this guide), meet the definition of novel food in Israel pursuant to Section 2 of the PHPFL. See Food Control Service (2024), pp. 1–2.

[112] Novel Food Registration Procedure (Procedure No. 004-08). The Procedure, in effect since 2006, has received the status of secondary legislation pursuant to Article 321(d) of the PHPFL. See Food Control Service (2024), p. 2.

[113] Vlčko et al. (2023), p. 934; Formici (2024), p. 190.

[114] Food Control Service (2024), p. 1.

[115] Regarding alternative proteins including cultured meat, the dossier should include information on the identity, origin and characteristics of the product. Full description of the production process is also required, including the cell extraction process, main biological processes, recovery of raw materials, physical and chemical processes and storage conditions as for cultured meat. See Food Control Service (2024), p. 17.

[116] Food Control Service (2024), p. 1.

[117] *Ibid*, pp. 25–26.

[118] Gross (2021).

instructions for use, allergens, and a description of alleged environmental benefits.[119] Labelling of food products in Israel is also governed by the Protection of Public Health (Food) (Nutritional Labelling) Regulations 5778-2017 (PPFNL),[120] which establish general requirements for the accessibility of information to consumers as well as the nutritional labelling of pre-packed food.[121]

A closely related issue in the Israeli context concerns the eligibility of cultured meat for *kosher* certification, which would allow its consumption by observant Jews.[122] In this regard, some authorities maintain that the *kosher* certification can only be granted to meat derived from animals slaughtered according to religious rules.[123] Conversely, in 2023, poultry meat cultured by Super Meat was deemed *kosher* by the US-based Orthodox Union Kosher, an influential *kosher* certification body.[124] Moreover, a position recently expounded by the Chief Rabbi David Lau supports the classification of beef cultured by Aleph Farm as *kosher* and *pareve*, i.e. a food which is neither meat nor dairy.[125] However, other rabbinic opinions tend to advocate stricter criteria for *kosher* certification, emphasizing the need for a slaughtered animal source.[126]

In general terms, various factors underscore the strong interest of Israeli authorities in the cultured meat market, which is viewed as a strategic field to advance the country's leadership in the area of novel foods and alternative proteins.[127] This is exemplified by the establishment of a task force dedicated to Novel Food in the risk management Unit of the National Food Service, "in order to preserve Israel's position as a leader in this area".[128] Israel's support to the industry, driven by the country's dependence on imported food, is also demonstrated by significant government-led investments aimed at fostering research and development of alternative protein sources.

A central role, in this regard, is played by the Israeli Innovation Authority (IIA), a public body responsible for formulating and implementing national innovation policy as well as strengthening the Israeli innovation environment. Notably, the IIA has developed funding platforms to support innovative protein industry operators and has recently promoted the creation of a dedicated consortium, which involves

---

[119] Food Control Service (2024), p. 22; Vlčko et al. (2023), p. 934.

[120] Available in English at https://faolex.fao.org/docs/pdf/isr217950.pdf. Accessed 31 March 2025.

[121] Gross (2021).

[122] Vlčko et al. (2023), p. 934.

[123] Soleymani et al. (2024), p. 16. Cf., *amplius*, *Food Justice in Religious Laws: Hypotheses on Cultivated Meat* by Anello in this volume.

[124] Yun et al. (2024), p. 331.

[125] Opinion of the Chief Rabbi of Israel David Lau of January 2023, at https://a7.org/?file=20230118144302.pdf. Accessed 14 August 2025. For more, see Wrobel (2023), Artom (2025), p. 55; Wee (2024).

[126] Wrobel (2023); Artom (2025), pp. 55–56.

[127] This is also evident from the endorsement of the Netanyahu administration noted by Formici (2024), p. 190.

[128] Ministry of Health of Israel (2024); Artom (2025), p. 56.

leading world business operators, academic laboratories and even *kosher* food manufacturers.[129] The Consortium aims to develop innovative, industrial-scale methods for producing cultured meat. The total estimated investment amounts to approximately 600 million US dollars, helping to place Israel as the second-largest global investor (after the US) in the alternative protein sector.[130]

6 Concluding Remarks

The present comparative analysis shows shared elements and regulatory divergence. To properly address them, it seems useful to rely on the methodology described in Sect. 2.

Considering the "who-question" regarding cultured meat, all the jurisdictions considered involve premarket assessments which are entrusted to agencies or governmental bodies, sometimes supported by experts, and consist mainly of scientific risk assessments, as they are confined to the evaluation of food safety aspects.[131] In the US, following the 2019 inter-agency agreement, this function, which consists of a voluntary premarket consultation is divided between the FDA, which oversees the initial phases, and FSIS-USDA, which supervises post-harvest activities. Notably, both agencies operate within the executive branch of the federal government and are subject to political appointment and oversight.[132] In Singapore, the safety risk assessment is conducted by the SFA, a statutory board under the Ministry of Sustainability and the Environment (MSE) of Singapore, with scientific assistance from the Novel Food Safety Expert Working Group. The 2025 FSSA, in continuity with the previous regulatory framework, confirmed the role of the SFA in market approval and the relevant safety assessment of defined food, as well as the registration of food business licences to firms intending to sell or export cell-based food products. Similarly, the Israeli National Food Service—again an internal division of the Ministry of Health of Israel—alongside its Novel Food Committee, oversees the premarket registration procedure in Israel. While all three jurisdictions show a significant presence of start-ups and private investment, Israel and Singapore have added dedicated public support. In particular, the dialogue between the SFA and interested stakeholders is facilitated by platforms such as the Novel Food Virtual Clinics and the introduction of self-assessment checklists, aimed at attracting and encouraging investment in the country's cultured food industry. The Israeli Innovation Authority, on the other hand, acts as a catalyst for advancing the cultured

[129] Artom (2025), p. 57.

[130] GFI Israel (2023), p. 8.

[131] See *amplius* on this aspect Formici (2024), p. 194.

[132] In this regard, see scholarly arguments regarding risk of "capture" of these agencies by Bernstein (2021), p. 26.

meat industry and supports the creation of synergies between business operators, academic institutions and policy-makers.

Considering the "what-question", i.e. the identification of objectives, concepts and principles underlying food laws, in Singapore and Israel the supportive regulatory environment of cultured meat is coupled with policy trends aligning investments in alternative proteins with broader national priorities related to sustainability (especially in the case of Singapore, with the Green Plan 2030) and food security, as well as the purpose of reducing reliance on food imports, a common problem of these countries. Moreover, these two countries have not so far experienced the resistance and opposition described in the section on the United States and common to some European countries such as Italy, Hungary and France. Criticism of cultured meat in the US has in fact reached the regulatory environment at state and federal level, with state laws limiting use of the term "meat" for cultured products, and even prohibiting their commercialization, alongside legislative proposals aimed at excluding the use of federal funds to support the industry.

It is also worth underlining that unlike Israel, Singapore and the EU, the US legal framework does not anchor the authorization of cultured meat products to a novel food regulatory regime.[133] Conversely, Israel and Singapore have developed similar criteria to identify novel foods, generally referring to a lack of history of use (20 years in Singapore, and lack of records of safe use before February 2006 in Israel), ingredient characteristics, origin and production processes.[134] A main difference between the two jurisdictions, however, is that the Israeli framework emphasizes the availability of safety information on the products, while Section 13 of Singapore's FSSA explicitly refers to food isolated from or produced by cell or tissue cultures.

This issue relates to the possibility of considering cultured meat within the regulatory classifications of meat. Section 3 of Singapore's 2025 FSSA, which defines meat as food derived from slaughtered animals and classifies the products of cell agriculture as novel foods, excludes this possibility. Conversely, in the US, the possibility of considering cultured products to be meat under the FFDCA and the PPIA, although widely debated, remains open, despite the clarifications provided by the 2019 Agreement.[135] A similar situation to Singapore may be appreciated under the Israeli legal framework, where foods containing novel proteins produced by cell culture are considered to be novel food and not meat.[136] The Israeli case is also noteworthy in this context due to ongoing debate about the possibility of considering cultured meat to be *kosher*. Indeed, recent openings by *kosher* certification authorities may actually lead to the result of declaring *kosher* cell-based products.[137]

[133] Formici (2024), pp. 193–195.

[134] See Sects. 4 and 5 above.

[135] DaVault and Sinha (2025), p. 36.

[136] See Food Control Service (2024), p. 1.

[137] See Sect. 5 above. For broader perspectives on the connections between cultured food and religion, see *Food Justice in Religious Laws: Hypotheses on Cultivated Meat* by Anello in this volume.

Parallel to this is the Islamic debate on whether cultured meat can be deemed *halal*, which could significantly expand the market potential of cell-based meat, since the forecasted global market value of *halal* food is estimated at around 2043.2 billion US dollars by 2027.[138]

This brings us to the "how-question". An interesting specific dimension of cultured meat regulation concerns labelling. In the US, labelling falls under the competence of the USDA-FSIS, although the agency shares food safety responsibilities with the FDA under the 2019 agreement. Following the 2023 USDA-FSIS decision, the official label approved must clearly state "cell-cultivated chicken".[139] In Singapore, labelling is the responsibility of the SFA, which requires the inclusion of descriptors such as "cultured" or "cell-based" to avoid confusion with conventional meat.[140] In addition, under Section 309 of the 2025 FSSA, the SFA is authorized (with ministerial approval) to issue additional regulations concerning labelling and food information. The Ministry of Health is also involved in the labelling process in Israel, where food businesses are required to detail the proposed labels and, if prepacked food are concerned to comply with the general requirements of the 2017 PPFNL.

Regarding the specific procedures in each jurisdiction, it is worth underlining that in all three case studies, cultured meat is subject to a premarket review or consultation primarily focused on a food safety assessment[141] that takes into account potential threats to consumers health.[142] These assessments are conducted on a case-by-case basis, although the composition of the evaluation committees varies in the jurisdictions considered, and in some cases, such as in the United States and Singapore, specific working groups may be established to oversee single segments of the assessment or provide scientific advice.[143] In all three case studies, the assessment primarily relies on technical dossiers submitted by the applicants, although it may also involve inspections by public authorities, as in the US, where site inspections are conducted by the USDA-FSIS, which also grants approval for production sites.

In conclusion, the recent interest in cultured meat among scholars, stakeholders and policymakers in the three jurisdictions appears to be connected to its potential advantages in addressing major challenges of the Anthropocene, such as global food security and sustainability, while also reducing the suffering caused by intensive animal farming. On the other hand, this emerging frontier of alternative proteins is also becoming a central area of competition among global leaders in food innovation. This broader economic dimension appears crucial for understanding the

---

[138] Shahbandeh (2024), p. 29.

[139] See Sect. 3 above.

[140] SFA (2023c).

[141] See Formici (2024), p. 193 who discusses differences with respect to the EU, where factors other than food safety may be considered.

[142] See, for instance, Section 105 of the Singapore 2025 FSSA.

[143] See Sects. 3 and 4.

regulatory challenges that cultured meat entails. Besides being the only countries where cultured meat has so far been authorized for commercialization, the US, Israel and Singapore are major global investors in the sector,[144] and view alternative protein production as a field of strategic technological and economic leadership. At the same time, these jurisdictions face significant regulatory challenges. The outlook for US regulation remains uncertain due to a fragmented normative framework and growing political opposition, including legal disputes already brought before courts. The recent inauguration of the second Trump administration may be an additional factor of uncertainty. Arguably, the deregulatory stance announced by the Trump Administration may actually facilitate approval processes for the alternative protein industry, reducing compliance burdens and encouraging private investment in biotechnology. On the other hand, it is worth recalling the recent appointment of Robert F. Kennedy Jr., an open critic of cultured meat, as Secretary for Health and Human Services.[145] Cultural opposition to cell-based meat is also widespread among Republicans, as testified by the legislative bans enacted in Republican-led states such as Florida, Alabama and Mississippi,[146] and the impact on the industry of the application of US global tariffs and the resulting "commercial war" seems difficult to predict.

While Israel and Singapore have intentionally designed supportive regulatory environments to attract foreign investment, they are now confronted with the results achieved and broader market dynamics, such as energy costs and the complex regulatory nature of a technology that intersects multiple legal domains. In this context, Israel has opted for a more incremental approach, making limited adjustments to the existing framework to accommodate novel foods, without pursuing systemic reform. By contrast, Singapore's 2025 Food Safety and Security Act establishes a broad and ambitious regulatory framework, explicitly aimed at increasing accessibility for foreign operators and combining the objectives of food innovation, sustainability, food safety and food security in a unified legislative strategy. Whether such an ambitious and integrated regulatory model will prove effective in addressing global competition in the cultured meat sector remains to be seen.

References

Artom A (2025) Oltre l'Unione europea: Brevi note sulla produzione di carne coltivata in Israele. Quaderni della Rivista di Diritto Alimentare 1:54–61

Bernstein ZA (2021) The fight over Frankenmeat: the FDA as the proper agency to regulate cell-based "clean meat". Brooklyn Law Rev 86:593–626

---

[144] GFI Israel (2023), p. 8.

[145] The FDA, which is responsible for overseeing the pre-market assessment of cultured meat together with USDA-FSIS under the inter-agency agreement as outlined in Sect. 3, is an agency of the Department of Health and Human Services (HHS).

[146] Fortin (2024).

Bruno F (2017) Il diritto alimentare nel contesto globale. USA e UE a confronto: food safety, food health e food security. CEDAM, Padova
Cheli F, Tedeschi G, Lanzoni D, Selvi S, Baldi A (2024) Carne a base cellulare: Una nuova frontiera della ricerca. BioLaw J 4:151–164
Choudhury D, Tseng TW, Swartz E (2020) The business of cultured meat. Trends Biotechnol 38(6):573–577
CRS Report (2023) Cell-cultivated meat: an overview. https://www.congress.gov/crs-product/R47697. Accessed 31 Mar 2025
DaVault J, Sinha MS (2025) Meat, the future: the role of regulators in the lab-grown revolution. Am J Law Med 51:69–99. https://ssrn.com/abstract=4782080. Accessed 31 Mar 2025
Etheridge TS (2022) What's the beef? The FDA, USDA, and cell-cultured meat. Wash Lee Law Rev 78:1729–1741
FAO (2018) Sustainable food systems: concept and framework. https://openknowledge.fao.org/server/api/core/bitstreams/b620989c-407b-4caf-a152-f790f55fec71/content. Accessed 31 Mar 2025
FAO (2022) Food safety aspects of cell-based food. Background document three – regulatory frameworks. Rome. https://doi.org/10.4060/cc2353en. Accessed 22 Apr 2025
FDA (2022) FDA completes first pre-market consultation for human food made using animal cell culture technology. https://www.fda.gov/food/hfp-constituent-updates/fda-completes-first-pre-market-consultation-human-food-made-using-animal-cell-culture-technology. Accessed 25 Apr 2025
FDA (2023) Cell Culture Consultation Notification File: CCC 000001 (Good Meat Inc), Letter from the Office of Food Additive Safety Center for Food Safety and Applied Nutrition. https://www.fda.gov/media/166347/download. Accessed 14 Aug 2025
Ferri M, Schiavo N, Cofelice MG, Lombardo F (2024) La carne coltivata: Stato dell'arte e criticità. https://www.veterinariapreventiva.it/wp-content/uploads/2024/03/Maurizio-Ferri.-La-carne-coltivata.-8-Marzo-2024_.pdf. Accessed 31 Mar 2025
Food Control Service of Israel (2024) Guidance document on the submission of an application for the approval as "novel food" of "novel proteins". https://www.gov.il/BlobFolder/news/guide-for-submitting-applications/he/units_fcs_Guide%20for%20Submitting%20Applications%20for%20the%20Approval%20of%20Novel%20Proteins-%20English.pdf. Accessed 31 Mar 2025
Formici G (2024) Un assaggio di futuro? La carne a base cellulare e il vivace dibattito regolatorio dentro e fuori i confini dell'Unione europea. BioLaw J 4:165–197
Fortin N (2024) Florida bans cultivated meat. Michigan State University Institute for Food Law and Regulation, 8 May 2024. https://www.canr.msu.edu/news/cultivated-meat-laws-andregulations. Accessed 31 Mar 2025
FSANZ (2023) Vow FSANZ Dossier, executive summary. https://www.foodstandards.gov.au/sites/default/files/food-standards-code/applications/Documents/Vow%20FSANZ%20dossier%20-%20Executive%20summary_Redacted.pdf. Accessed 14 Aug 2025
FSANZ (2025a) Approval report – application A1269, cultured quail as a novel food. https://www.foodstandards.gov.au/sites/default/files/2025-04/A1269%20Approval%20report.pdf. Accessed 14 Aug 2025
FSANZ (2025b) A1269 - cultured quail as a novel food. https://www.foodstandards.gov.au/food-standards-code/applications/A1269-Cultured-Quail-as-a-Novel-Food. Accessed 14 Aug 2025
FSIS-USDA (2023) Human food made with cultured animal cells. https://www.fsis.usda.gov/inspection/compliance-guidance/labelling/labelling-policies/human-food-made-cultured-animal-cells#:~:text=for%20products%20to%20receive%20a,must%20be%20preapproved%20by%20FSIS. Accessed 31 Mar 2025T

GAO (2020) Report to the chairwoman, subcommittee on labor, health and human services, education, and related agencies, committee on appropriations, House of Representatives. https://www.gao.gov/assets/gao-20-325.pdf. Accessed 31 Mar 2025

GFI Israel (2023) State of alternative proteins. https://gfi.org.il/wp-content/uploads/2023/02/SOIR-2023.pdf. Accessed 31 Mar 2025

Good Food Institute (2023) State of the industry report: cultivated meat and seafood. gfi.org/resource/cultivated-meat-eggs-and-dairy-state-of-the-industry-report/. Accessed 31 Mar 2025

Good Meat (2023) Press release "GOOD Meat Receives First-Ever USDA Label Approval for Cultivated Meat." https://www.goodmeat.co/all-news/good-meat-receives-first-ever-usda-label-approval-for-cultivated-meat. Accessed 14 Aug 2025

Gross T (2021) Novel food regulation in Israel – from directive to regulation. GSAP Blog. https://www.gsap.co.il/novel-food-regulation-in-israel-from-directive-to-regulation/. Accessed 31 Mar 2025

Hall K (2023) High steaks: how USDA involvement will spoil lab-grown meat for Americans. Alabama Law Rev 75:275–297

Lanzoni D, Rebucci R, Formici G, Cheli F, Ragone G, Baldi A, Violini L, Sundaram TS, Giromini C (2024) Cultured meat in the European Union: legislative context and food safety issues. Curr Res Food Sci 8:100722

Letterhead Food Research Group (2024) Israel set to adopt European food regulations. https://www.leatherheadfood.com/news-detail/blog-israel-adopts-european-food-regulations/. Accessed 31 Mar 2025

Mahira A, Nadya C, Vergano PR (2025) Singapore's new food safety and security act: implications for Singapore's food industry and for traders. Lexology. https://www.lexology.com/library/detail.aspx?g=e3d1e8df-db9a-4350-8000-e3f51c80f9d7. Accessed 31 Mar 2025

Michaels R (2019) The functional method of comparative law. In: Reimann M, Zimmermann R (eds) The Oxford handbook of comparative law, 2nd edn. Oxford University Press, Oxford

Ministry of Health of Israel (2024) Novel food in Israel. https://www.gov.il/en/pages/novel-food. Accessed 31 Mar 2025

Miyake Y, Tachikawa M, Kohsaka R (2023) Policy frameworks and regulations for the research and development of cell-based meats: systematic literature review. Food Res Int 167:112599

Monateri PG (2022) The structures of comparative law: metaphors and methods. ASCL Blog, 15 November 2022. https://ascl.org/the-structures-of-comparative-law-metaphors-and-methods/?utm_source=chatgpt.com. Accessed 3 Feb 2025

Mousa N (2025) Regulating cultured meat labels in the United States. Colorado Environ Law J 36(1):137–166

MSE and SFA (2024) Consultation on draft provisions in the food safety and security bill – defined food and pre-market approval. https://www.reach.gov.sg/latest-happenings/public-consultation-pages/2024/consultation-on-draft-provisions-in-the-food-safety-and-security-bill%2D%2D-defined-food-and-pre-market-approval. Accessed 25 Mar 2025

National Agricultural Law Center (2024) Cell cultured meat updates: state bans, labelling requirements, and regulatory clarifications. https://nationalaglawcenter.org/cell-cultured-meat-updates-state-bans-labeling-requirements-and-regulatory-clarifications/. Accessed 25 Feb 2025

Olenic M, Thorrez L (2023) Cultured meat production: what we know, what we don't know and what we should know. Italian J Anim Sci 22(1):749–751

Paganizza V (2025) Carne coltivata: Una comparazione tra Australia e Unione europea, passando da Singapore. Quaderni della Rivista di Diritto Alimentare 1:62–75

Saija R (2025) Carni e pesci in vitro. Quali regole in vista della sostenibilità? Quaderni della Rivista di Diritto Alimentare 1:33–44

Sanchez M (2018) Food law and regulation for non-lawyers: a US perspective. Springer, Cham

Scaffardi L (2020) I novel food, un futuro ancora da definire. BioLaw J 2:43–66
SFA (2023a) Requirements for the safety assessment of novel foods and novel food ingredients. https://www.sfa.gov.sg/docs/default-source/regulatory-standards-frameworks-and-guidelines/requirements-for-the-safety-assessment-of-novel-foods-and-novel-food-ingredients.pdf. Accessed 31 Mar 2025
SFA (2023b) Guidelines on novel food. https://www.sfa.gov.sg/regulatory-standards-frameworks-guidelines/novel-food-framework/guidelines-on-novel-food. Accessed 31 Mar 2025
SFA (2023c) Safety of alternative proteins. https://www.sfa.gov.sg/food-safety-tips/food-risk-concerns/risk-at-a-glance/safety-of-alternative-protein. Accessed 31 Mar 2025
SFA (2025) Overview of novel food regulatory framework. https://www.sfa.gov.sg/regulatory-standards-frameworks-guidelines/novel-food-framework/overview-of-novel-food-regulatory-framework. Accessed 24 Apr 2025
Shahbandeh M (2024) Halal food market revenue worldwide 2018–2027. Statista (Campus Licence, University of Genoa). https://www.statista.com/statistics/785499/halal-food-market-revenue-global/. Accessed 31 Mar 2025
Singapore Government Press Centre (2021) Singapore green plan 2030. https://www.greenplan.gov.sg/. Accessed 25 Mar 2025
Singapore Ministry of Sustainability and the Environment (2024). https://www.mse.gov.sg/policies/food. Accessed 31 Mar 2025
Soleymani S, Naghib SM, Mozafari MR (2024) An overview of cultured meat and stem cell bioprinting: how to make it, challenges and prospects, environmental effects, society's culture and the influence of religions. J Agric Food Res 18(2):101307
Teng P, Montesclaros JMLP (2019) Singapore's '30 by 30' strategy: can food self production be achieved? RSIS Commentaries 054. Nanyang Technological University, Singapore
Treich N (2021) Cultured meat: promises and challenges. Environ Resour Econ 79:33–61
USDA (2021) USDA seeks comments on labelling meat and poultry products derived from animal cells. https://www.usda.gov/about-usda/news/press-releases/2021/09/02/usda-seeks-comments-labelling-meat-and-poultry-products-derived-animal-cells. Accessed 25 Jan 2025
Van der Meulen B, Wernaart B (2023) Comparative food law. In: Roberts MT (ed) Research handbook on international food law. Edward Elgar, Cheltenham
Vlčko T, Bokwa K, Jarosz I, Szymkowiak A, Golian J, Antoniak M, Kulawik P (2023) Cell-based meat labelling. Current worldwide legislation status. A review. Ann Anim Sci 23(4):927–938
Volpato A (2022) Novel foods in the EU integrated administrative space: an institutional perspective. In: Formici G, Scaffardi L (eds) Novel foods and edible insects in the European Union. An interdisciplinary analysis. Springer, Cham, pp 15–36
Wee SL (2024) The country where you can buy meat grown in a lab. The New York Times, 24 July 2024. https://www.nytimes.com/2024/07/24/world/asia/singapore-cultivated-lab-meat.html. Accessed 25 Apr 2025
Wrobel S (2023) Kosher, now maybe halal: Aleph farms gears up for rollout of cultivated steak. https://www.timesofisrael.com/kosher-now-maybe-halal-aleph-farms-gears-up-for-rollout-of-cultivated-steak/. Accessed 25 Apr 2025
Yun SH, Lee DY, Lee J, Mariano E Jr, Choi Y, Park J, Han D, Kim JS, Hur SJ (2024) Current research, industrialization status, and future perspective of cultured meat. Food Sci Anim Resour 44(2):326–355
Zweigert K, Kötz H (1998) An introduction to comparative law. Oxford University Press, Oxford

The Regulatory Landscape of Cultured Meat and Fish: Terminology and Consumer Information Provisions

Eleonora Sirsi and Silvia Rolandi

Abstract This chapter delves into cultured meat and fish dynamics and evolving field. From a literature review, it explores the terminology used and the information provisions, specifically those for consumers and businesses (Business to Consumer, B2C and Business to Business, B2B). A comparative analysis of the relevant legislation, particularly concerning information and labelling in the European Union, the United States of America and Singapore, focuses on the regulatory frameworks necessary to trade and market cultured meat and fish. It offers points for reflection that contributes to the broader discourse on the future of food, regulation, consumer choices and transparency in trade, providing insights for researchers, policymakers and industry stakeholders.

Keywords Neo-meat · Terminology · Legal name · Information

1 Introduction

From the perspective of alternatives to meat obtained from slaughtered or captured animals, legislators aim to address key issues related to safety, fair trade, and providing accurate information to ensure consumers can make informed choices. In a context of abundant information, this requires effective marketing communication,

E. Sirsi
University of Pisa, Department of Law, Pisa, Italy
e-mail: eleonora.sirsi@unipi.it

S. Rolandi (✉)
University of Pisa, Department of Agriculture, Food and Environment, Pisa, Italy

Wageningen University and Research, Law Group, Wageningen, The Netherlands
e-mail: silvia.rolandi@agr.unipi.it

G. Formici et al. (eds.), *Cell-Based Meat in the European Union and Beyond*,
https://doi.org/10.1007/978-3-032-14624-3_6

properly defined product names in accordance with regulations, and correctness in presentation.

The introduction of products that represents alternatives to the meat obtained from the slaughter of raised or captured animals, created many issues, in the agri-food domain, about safety, fair trade, and correct information. Particularly when there is an abundance yet not clear presence of information, consumers rights about making informed choices depend on appropriate marketing communication terminology and product names defined according to the regulatory context.

While rules exist for meat obtained from plant ingredients ("non-meat"), in part due to the intervention of the European Court of Justice and National Courts, "neo-meat"[1] is still awaiting a definition. In the case of meat obtained by laboratory cultivation of cells from animals, resulting in a product essentially corresponding to that obtained from the slaughter of the animals, it is necessary to adopt the relevant framework. In the European Union, there is the Novel Food Regulation, accompanying horizontal and vertical regulations, and eventually national provisions from member states. Several European member states[2] have manifested their willingness to address the matter by introducing specific national regulations.[3] In different countries, such as the United States of America (USA), the players are realising the importance of an agreement within the Codex Alimentarius, in order to establish a reference point and guidance for terminology, nomenclature, and information.

2 The Terminology

The noun "terminology" refers to "the special words and phrases that are used in a particular subject"[4] and its role is of unique relevance when dealing with a topic that is still quite new in the international research agenda and requires further investigation. Nevertheless, given the pervasiveness of global trade, shared terminology is needed for clear understanding and communication with and for consumers and to anticipate new regulations. The importance of communication and thus information was recently remarked by the European Court of Auditors[5] in a report on the role of

[1] Sirsi (2020), p. 670.

[2] Including France, Austria, Hungary and Italy. In particular, it is worth mentioning that the Hungarian Government, independently from the Commission Opinion 2777/2024, about the notification of a draft text, presented a law (T/11151) about the ban and commercialization of cultured meat.

[3] See *Cell-Based Meat in the European Union: A Regulatory Crossroads* by Corazza and Formici in this volume.

[4] Cambridge Dictionary. https://dictionary.cambridge.org/dictionary/learner-english/terminology#google_vignette. Accessed 12 February 2025.

[5] European Court of Auditors (2024). https://www.eca.europa.eu/en/publications?ref=SR-2024-23. Accessed 12 February 2025.

information to consumers for avoiding confusion and misleading practices.[6] The Court not only underlined the lack of clarity of communication but also the lack of effectiveness of controls, sanctions and reporting. All these aspects emerge when there is a lack of correspondence between terminology and definitions. On an international market scale, common shared terminology is important to achieve collaboration between different agencies for controls on matters such as safety,[7] but also to build debate and discussion on matters such as environmental impacts, animal welfare, ethical question and related issues.

Research on cultured meat began more than 25 years ago; the first prototype was presented in 2013 at a press conference in London as a "lab-grown beef burger".[8] On the international scene, consumers had to wait until 2020 to purchase these lab-grown products in grocery stores. This became possible when the Singapore Food Agency granted authorization to market "meat developed from animal cell culture". European Union consumers have not yet had this possibility.

In the meantime, debate on the terms to use has not been a stylistic exercise but concerns commercial opportunities and consumer protection. The grey literature indicates that various terms are used by authorities, the food industry, developers, academia and the media to indicate meat and fish produced without slaughtering.[9] Only a few of these terms, namely cell-based, cell-cultured, cultivated and cultured, are shared by all sectors, while the media have shown the greatest fantasy (with terms such as Frankenmeat).[10] Recently, interest has been shown in words that suggest less impact on the environment, such as "no cow" or "clean cow".[11] The debate is usually about meat,[12] but the products in question also concern fish,[13] which has the same legislative grounds as meat.[14]

What matters most is how public authorities and legislation choose to use and define terms such as cell-based, and whether and to what extent they can be used to describe meat or fish. This is addressed later in the present chapter. So far the different terms can affect people's perceptions of characteristics that these products do not have, confusing consumers and inhibiting market competitiveness.

A point to clarify concerns the use of the terms "cells" and "cellular agriculture." Regarding "cells," since most food products consist of cells, the terms "cell-based" or "cell-cultured" lack scientific clarity and could be misunderstood by consumers; not all products indicated as cell-cultured contain cells. A method is used to make

[6] For recent contributions on these aspects, see Albisinni (2025) and Di Lauro (2025).

[7] See Broucke et al. (2023).

[8] FAO and WHO (2023a), p. 4.

[9] See Mayhall (2019), Lähteenmäki-Uutela et al. (2021) and Lanzoni et al. (2024).

[10] FAO and WHO (2023a), p. 5.

[11] Carolan (2025), p. 17.

[12] See, among others, Post et al. (2020), Miyake et al. (2023), Planchenstainer (2024), Formici (2024), Monaco (2025) and Paganizza (2025).

[13] See, among others, Rubio et al. (2019), De Amstalden (2022) and Xu et al. (2023).

[14] For further information on fish-related terminology, see Malerich and Bryant (2022), p. 56.

acellular and cellular products "where acellular products are made of organic molecules like proteins and fats and contain no cellular or living material" in their final form, whereas "cellular products are made of living or once living cells." This illustrates a risk of unclear information to consumers, risking to cause confusion on their overall understanding of the products. "Cellular agriculture" on the other hand, suggests a direct link with agriculture, a word derived from the Latin *ager* (field) and *colo* (cultivate), meaning tillage of the land.[15] Agriculture has acquired different meanings in the course of time, but it could be controversial to accept complete absence of a field-related activity, replaced by an *in vitro* one with the use of bioreactors.

The different shades of meaning of words in different languages add a further level of complexity.[16] An interesting example from the grey literature is found in a document published before Israel authorized a cultured product as a Novel Food. The need for appropriate terminology emerged to meet needs such as product information in Hebrew, Arabic and English.[17] On 17 January 2024, the Israel Ministry of Health announced approval of "cattle-based cultivated meat".[18]

In 2021, the Codex Alimentarius Commission sought guidance on a list of topics,[19] including "cell-culture-based food products (meat, fish, dairy)", which was followed by discussion[20] and several recommendations.[21] The latest document is on "food safety aspects of cell-based food and precision-fermentation-derived food products" which is also quite useful for terminology considerations as it proposes agreed use of nouns, expressions and definitions.[22] While setting the background, the commission indicated that "cell-based food production involves using cells iso-

[15] See Harris and Fuller (2014), p. 64.

[16] See *ibid*, p. 10. The need to share common terminology also emerges from private initiatives such as the Asia-Pacific Society for Cellular Agriculture that signed a memorandum of understanding where it was agreed to use "cultivated" as the preferred English-language term for cell-cultivated products, 36 companies agreed. See https://www.foodanddrinkbusiness.com.au/news/apac-cellular-agriculture-announces-mou. Accessed 12 February 2025.

[17] FAO and WHO (2023a), p. 34. Interestingly, the report states: "Several common terms are used in the media in Hebrew and are to some extent known and are familiar to the public. These terms include: "[modifier] Meturbat – מתורבת []" Hebrew for cultivated, cultured, refined or domesticated [meat, poultry, etc.]; "[modifier] Maabada – מעבדה [ ]" Hebrew for lab (grown) [meat, poultry, etc.]. Less common terms include "[modifier] Synteti – סינטטי []" Hebrew for synthetic [meat, poultry, etc.]"; "[modifier] Naki – נקי [ ]" Hebrew for clean [meat, poultry, etc.]. In addition to the Hebrew-speaking population, about 20% of the Israeli public speak Arabic and are informed by Arabic-speaking media. Common terms known by the Arabic-speaking population include: "[modifier] Masna مصنّع [ ]" Arabic for processed [meat, poultry, etc.]; "[modifier] Mazru – مزروع [ ]" Arabic for cultivated [meat, poultry, etc.]; and "[modifier] Fi'l Muhtabar – في المختبر [ ]" Arabic for lab (grown) [meat, poultry, etc.]".

[18] For the announcement, see https://www.gov.il/en/pages/17012024-02. Accessed 13 February 2025.

[19] Codex Alimentarius Commission (2021).

[20] The discussion was held during the 46th session of the Codex Alimentarius Commission (2023).

[21] Codex Alimentarius Commission (2023).

[22] Codex Alimentarius Commission (2024).

lated from animals, plants or microorganisms to produce food products, ingredients or additives. Often the products can be seen as comparable to existing animal-source products, such as meat, poultry, aquatic products, dairy products and eggs". It seems important to underline that in "cell-based food", the cell is "utilized" and not "cultured". Also, the claim that the products in question are comparable to other existing ones has stimulated debate that is covered in the next section, and likewise the combination of words such as "meat" or "fish" with "cell-based".

At the time of writing, the Food and Agriculture Organization (FAO) was editing a new report[23] that provides an overview of the regulatory frameworks for cell-based food production as an aid for navigating the current regulations.

3 The Information Regulatory Framework in the European Union

In the writing and amending of national, European and international rules and contracts, nomenclature and information are crucial to establish a common shared meaning in a specific language. In the European Union (EU), especially in the agri-food sector, clarity is sought for consumer protection, safety and right to be informed. Regulation (EC) no 178/2002[24] (General Food Law, GFL) at "Whereas 22" indicates that food safety must be ensured to "increment consumer confidence and the confidence of trading partners through the open and transparent development of food law" and the aim expressed in Article 1[25] is to provide "a high level of protection of human health and consumers' interest in relation to food, taking into account in particular the diversity in the supply of food including traditional products, whilst ensuring the effective functioning of the internal market".

Information and therefore product names are important for achieving two fundamental aspects of the GFL Regulation: the "protection of human health" and "consumers' interests concerning food". The former concerns food safety including product traceability in relation to hygiene thus protecting human health. Regarding the latter, consumers need to understand what they are purchasing. Information is therefore essential and is subject to specific rules. Article 8 of (EC) no 178/2002 concerns the protection of consumers' interests: "Food law shall aim at the protection of the interests of consumers and shall provide a basis for consumers to make informed choices in relation to the foods they consume. It shall aim at the prevention of: (a) fraudulent or deceptive practices; (b) the adulteration of food; and (c) any other practices which may mislead the consumer." The regulatory framework

[23] Previous reports of interest are FAO and WHO (2023a, b) and FAO (2023).

[24] Regulation (EC) no 178/2002 of the European Parliament and of the Council of 28 January 2002 laying down the general principles and requirements of food law, establishing the European Food Safety Authority and laying down procedures in matters of food safety.

[25] Article 1 of Regulation (EC) no 178/2002 also indicates other relevant aspects, such as the stages of production to which it applies.

aims to ensure that consumers receive clear, accurate and non-misleading information about food products to facilitate informed choices. Building on letter (c), Regulation (EU) no 1169/2011[26] concerns the provision of food information to consumers. It sets detailed labelling requirements, including lists of ingredients, nutritional values and allergen declarations. It also seeks to prevent deceptive practices by regulating product names and descriptions. Directive 2005/29/EC[27] on unfair commercial practices provides further safeguards against misleading advertising and deceptive food-labelling. Focusing on the European Union food information regulation, Article 9 indicates what information is mandatory. Article 1(a) concerns "the name of the food", which is then specified in Article 17: "1. The name of the food shall be its legal name. In the absence of such a name, the name of the food shall be its customary name, or if there is no customary name or the customary name is not used, a descriptive name of the food shall be provided".[28] No legal name exists in the EU for "cell-culture-based food products (meat, fish, dairy)", and given the characteristic of the products itself, that is framed within the application of the EU Novel Food Regulation,[29] we can affirm that there is neither a customary name, so a descriptive name, at some point, shall be provided. Before moving to a punctual indication concerning the Novel Food Regulation and the aspect relates to information, it is worth noting that these aspects belong to the horizontal rules and that we

[26] Regulation (EU) no 1169/2011of the European Parliament and of the Council of 25 October 2011 on the provision of food information to consumers, amending Regulations (EC) no 1924/2006 and (EC) no 1925/2006 of the European Parliament and of the Council and repealing Commission Directive 87/250/EEC, Council Directive 90/496/EEC, Commission Directive 1999/10/EC, Directive 2000/13/EC of the European Parliament and of the Council, Commission Directives 2002/67/EC and 2008/5/EC and Regulation (EC) no 608/2004.

[27] Directive 2005/29/EC of the European Parliament and of the Council of 11 May 2005 concerning unfair business-to-consumer commercial practices in the internal market and amending Council Directive 84/450/EEC, Directives 97/7/EC, 98/27/EC and 2002/65/EC of the European Parliament and of the Council and Regulation (EC) no 2006/2004 of the European Parliament and of the Council ('Unfair Commercial Practices Directive').

[28] The Article also states: "2. The use in the Member State of marketing of the name of the food under which the product is legally manufactured and marketed in the Member State of production shall be allowed. However, where the application of the other provisions of this Regulation, in particular those set out in Article 9, would not enable consumers in the Member State of marketing to know the true nature of the food and to distinguish it from foods with which they could confuse it, the name of the food shall be accompanied by other descriptive information which shall appear in proximity to the name of the food. 3. In exceptional cases, the name of the food in the Member State of production shall not be used in the Member State of marketing when the food which it designates in the Member State of production is so different, as regards its composition or manufacture, from the food known under that name in the Member State of marketing that paragraph 2 is not sufficient to ensure, in the Member State of marketing, correct information for consumers. 4. The name of the food shall not be replaced with a name protected as intellectual property, brand name or fancy name. 5. Specific provisions on the name of the food and particulars that shall accompany it are laid down in Annex VI".

[29] About this specific aspect and the possible controversy regarding application of the GMO Regulation, see *Cell-Based Meat in the European Union: A Regulatory Crossroads* by Corazza and Formici in this volume.

also have to consider the vertical rules that apply to our product or ingredients. Regulation (EU) no 1379/2013 on the Common Organisation of the Markets in Fishery and Aquaculture Products[30] sets marketing standards and labelling rules for seafood. If cultivated fish products enter the market, they may require distinct labelling like those imposed on other Novel Foods. The same applies to meat products as indicated in Regulation (EU) no 1308/2013 on the common organization of markets for agricultural products.[31] The vertical regulation and specific definitions also indicated in Regulation (EC) no 853/2024[32] on hygiene, especially Annex I, provide definitions of "meat", "mechanically separated meat", "meat preparations", "fishery products" and "meat products".

The regulation of Novel Foods in the EU falls under Regulation (EU) 2015/2283,[33] which defines Novel Foods as those that are not significantly consumed in the EU before 15 May 1997 and fall under one of the listed categories.[34] As emerges from "Whereas 8" that acknowledges the scientific and technological developments in the sector, the categories of food that constitute Novel Food are updated. The Novel Food Regulation now includes food products, as indicated in Article 3 (VI) of the Regulation, "consisting of, isolated from or produced from cell culture or tissue culture derived from animals, plants".[35]

The European Food Safety Authority (EFSA) plays a key role in assessing the safety of Novel Foods before they can be authorized for the EU market. Consumer information requirements for Novel Foods emphasize transparency and safety, ensuring that such products undergo rigorous scientific assessment before market approval. The labelling of Novel Foods must not be misleading and must provide necessary information on the nature and composition of the product.

The Novel Food Regulation concerns possible specific labelling information, as indicated in Article 9.3 (b). When the Union list is updated, the indication of Novel Foods in the list has to include "additional specific labelling requirements to inform the final consumer of any specific characteristic or food property, such as the

---

[30] Regulation (EU) no 1379/2013 of the European Parliament and of the Council of 11 December 2013 on the common organisation of the markets in fishery and aquaculture products, amending Council Regulations (EC) no 1184/2006 and (EC) no 1224/2009 and repealing Council Regulation (EC) no 104/2000.

[31] Regulation (EU) no 1308/2013 of the European Parliament and of the Council of 17 December 2013 establishing common organisation of the markets in agricultural products and repealing Council Regulations (EEC) no 922/72, (EEC) no 234/79, (EC) no 1037/2001 and (EC) no 1234/2007.

[32] Regulation (EC) no 853/2004 of the European Parliament and of the Council of 29 April 2004 laying down specific hygiene rules for food of animal origin.

[33] Regulation (EU) 2015/2283 of the European Parliament and of the Council of 25 November 2015 on Novel Foods, amending Regulation (EU) no 1169/2011 of the European Parliament and of the Council and repealing Regulation (EC) no 258/97 of the European Parliament and of the Council and Commission Regulation (EC) no 1852/2001.

[34] See Rubino and Rossi Dal Pozzo (2024).

[35] Note that "Whereas 8" of the Novel Food Regulation also refers to "certain micelles or liposomes", to which the regulation applies.

composition, nutritional value or nutritional effects and intended use of the food, which renders a Novel Food no longer equivalent to an existing food or of implications for the health of specific groups of the population". It is still unknown whether and when requests for authorization will be accepted and what the information provisions will be. Currently, two requests have been filed, the first in 2024 by Gourmey CEO, a French start-up, that has also applied to the United States Food and Drug Administration (FDA), the Singapore Food Agency (SFA), the Food Standards Agency (FSA) of the United Kingdom and the Swiss Federal Food Safety and Veterinary Office (FSVO). The outcomes will therefore be of particular interest.

The second request was filed by Mosa Meat, which announced on 22 January 2025 that the request concerned the authorization of "cultivated meat fat" and that the future product would "combine cultivated beef fat with plant-based ingredients". The reason for this choice was that the EU Novel Food Regulation foresees authorization of ingredients.[36] This is interesting if observed in conjunction with the proceedings of EFSA Scientific Colloquium 27—Cell Culture-derived Foods and Food Ingredients,[37] which indicate that critical concerns about the need for clear labelling were expressed in a survey of 138 actors. A specific concern was raised by "industry associations in the fields of livestock, dairy production, food and beverage and meat processing"[38] on how to ensure transparency and increase consumer choice. Concern emerged about future utilization of specific names and terms, when and if EU approval is granted.

The recent EU jurisprudence is of even greater doctrinal interest in light of these items from the EFSA Colloquium and the recent declarations regarding lab-grown meat fat. A recent ruling by the Court of Justice of the European Union (CJEU) discussed the principles governing food labelling, particularly in the case of meat- and plant-based alternatives. The decision follows the precedent set by the *Tofu Town* case,[39] where the Court ruled that designations traditionally associated with animal products (e.g. "milk", "butter", "cheese") cannot be used for plant-based alternatives unless explicitly envisaged by EU law. This ruling was based on Regulation (EU) no 1308/2013, which protects the use of specific denominations for animal-derived products.

The latest ruling in *Protéines France et al.*[40] responded to a request for a preliminary ruling concerning the interpretation of Regulation (EU) no 1169/2011 and in particular, Articles 7, 17 and 38 and point 4 of Part A Annex VI. On 18 July 2022,

---

[36] See the company website "Submitting Our First EU Market Authorisation Request" https://mosameat.com/blog/submitting-our-first-eu-market-authorisation-request. Accessed 10 March 2025.

[37] EFSA (2024) https://efsa.onlinelibrary.wiley.com/doi/abs/10.2903/sp.efsa.2024.EN-8664. Accessed 14 February 2025.

[38] *Ibid*, p. 32.

[39] CJ Judgement (14 June 2017) Case C-422/16 *Verband Sozialer Wettbewerb eV* v. *TofuTown.com GmbH*.

[40] CJ Judgement (4 October 2024) Case C-438/23 *Protéines France et al.* v. *Ministre de l'Économie, des Finances et de la Souveraineté industrielle et numérique*.

Protéines France, which represents the interests of companies in France's vegetable protein market, brought an action before the referring court, *Conseil d'État* (Council of State, France), seeking annulment of Decree No. 2022-947, later supported by other applications, all reunited by the CJEU. It addressed the legality of a French regulation, Decree No. 2022-947 of 29 June 2022, that sought to prohibit plant-based products from using meat-related terminology. To draw some reflections, we follow the line of reasoning of the ruling. The Court first indicated the legal framework, referring to the legislation mentioned above, in particular Regulation (EC) no 178/2002, Regulation (EC) no 853/2004 and Regulation (EU) no 1169/2011. We previously mentioned the importance of the general food law and the law on the regulation of food information. The court did not address Regulation (EC) no 853/2004, just generically mentioned the need to use the name of the product. Since the product in question was meat, Regulation (EC) no 853/2004 was included in the legal framework because it defines "meat", "mechanically separated meat", "meat preparations", "fishery products" and "meat products" at points 1.1, 1.14, 1.15, 3.1 and 7.1 of Annex I.

The Court's reasoning opens the door to innovative food practices while interpreting the current legislation, which attempts to maintain harmonization between different member states and different sensibilities in relation to specific matters, particularly innovation. The CJEU ruled that the restrictions introduced with the French Decree must align with EU law and avoid creating unnecessary barriers to trade within the European market. In any case, the judgment affirms that it "does not preclude a Member State from laying down administrative penalties for failure to comply with the requirements and prohibitions resulting from the provisions of that regulation and from national measures which are in conformity with it. However, that specific harmonisation precludes a Member State from adopting a national measure setting the proportions of vegetable proteins below which the use of names, other than legal names, consisting of terms from the butchery and charcuterie sectors to describe, market or promote foods containing vegetable proteins is still authorised".[41] The Court emphasized the importance of preventing consumer confusion while allowing fair competition in the market, and stated that while the *Tofu Town* case was about a product with a legal name, there is no legal name for the processed meat products indicated in the new French legislation and therefore those names can be utilized to "market or promote foods containing vegetable proteins".[42] At the same time, exceptions were indicated in the *Tofu Town* case, referring to the provisions of Annex VII, Part III, point 5, second subparagraph, to Regulation (EU) no 1308/2013.[43]

Considering the future implications of Novel Food authorisation, if it is approved, it is essential to recognize that products composed solely of cell-based meat or fish might

---

[41] CJ Judgement (4 October 2024) Case C-438/23 *Protéines France et al.* v. *Ministre de l'Économie, des Finances et de la Souveraineté industrielle et numérique*, para 106.

[42] CJ Judgement (4 October 2024) Case C-438/23 *Protéines France et al.* v. *Ministre de l'Économie, des Finances et de la Souveraineté industrielle et numérique*, Ruling 2.

[43] As highlighted in Sirsi (2020), p. 671.

not be classified as such under existing legal definitions. Nevertheless, given that lab-grown cells would be utilized as ingredients, the final product could potentially be designated as sausage. This scenario hinges on the interpretation of the term "edible parts of the animal" as outlined in the legal nomenclature, which may require a distinction between what is deemed edible and the anatomical origin of the substance. Provided that the authorized product is indeed consumable, the connection to the animal source may be acknowledged even if it involves laboratory-mediated replication of cells. Further elucidating this point, in instances referring to the Moose Meat case, where the lab-grown substrate is combined with plant-based proteins, meat-related terminology is more likely to be accepted. Conversely, in the case of food products like *foie gras* (under consideration by the European Food Safety Authority) that could be protected by a quality scheme, even in the absence of a legal definition, how should we proceed?

In the meantime, a vibrant debate has arisen in several member states, including France and Italy. In particular, Italy approved a law (Legge 1 dicembre 2023, n. 172) banning the production and marketing of food and feed consisting of, isolated from, or produced from cell cultures or tissues derived from vertebrate animals, as well as the designation of processed products containing vegetable proteins as "meat". This is of particular interest, as it refers to use of a specific term "meat" and consumer information.[44] It also demonstrates the confusion surrounding the question in the EU.[45]

4 The Information Regulatory Framework in Third Countries

Only focusing on the terminology and information dimensions for a chronological overview of the situation in third countries, where authorizations have been issued for "cell-culture-based food products (meat, fish, dairy)". We consider only Singapore and the United States where current debate is lively and legislative measures have been taken.[46]

4.1 Singapore

In 2019, Singapore introduced the Novel Food regulatory framework that indicates what products, including cell-culture-based foods, should be called.[47] Companies are required to label such products with terms like "mock", "cultured" or "plant-

[44] See *The Legal Debate on Cultured Meat Around the World: Comparative Insights from the United States, Singapore and Israel* by Pitto in this volume for further considerations.

[45] See Rubino (2024) and Spoto (2024).

[46] To clarify this aspect, Israel has approved regulatory measures for the production of beef without using animals, but there is no specific provision concerning food information-related aspects and nomenclature as yet; comprehensive labelling guidelines are still being formulated.

[47] See Kit Lim (2023).

based" to differentiate them from the products consumers are familiar with. According to the regulation, establishments selling alternative protein products, food traders and the like that misrepresent the nature of the food sold may be convicted as indicated in the Singapore Sale of Food Act. In January 2025, the Food Regulation (Amendment)[48] 2025 was published and will come into force on 30 January 2026. It mostly deals with the list of ingredients but also contains definitions of specific products that imply need for a further interpretation of how the regulation will apply to Novel Foods. Another legislative reform, still in progress, will affect the communication of information on food products and will apply to cell-culture-based food: the Food Safety and Security Act 2025 will repeal the Sale of Food Act 1973 and the Wholesome Meat and Fish Act 1999. The new act introduces the concept of "defined foods" that include Novel Foods and therefore also culture-based food. It will be interesting to follow how and whether the new regulation is interpreted in light of the existing regulation on information requirements for Novel Foods where information standards shall meet detailed information about ingredients, nutritional content, instructions for use and storage recommendations.[49]

4.2 *United States of America*

In 2022, the US Food and Drug Administration FDA[50] issued two companies with a premarket review and approval to produce and market cell-cultivated chicken in the US.[51] In 2023, the US Department of Agriculture (USDA) FSIS issued inspection grants to the two companies and decided that their products should be labelled "cell-cultivated chicken".[52] What appeared to be a settled question is currently embroiled in an interesting phase of debates and confusion, where some States have introduced

[48] It amends the Sale of Food Act 1973, see https://sso.agc.gov.sg/SL-Supp/S92-2025/Published/20250131?DocDate=20250131. Accessed 14 February 2025.

[49] According to the Novel Food Regulatory Framework of Singapore.

[50] Congressional Research Service (2023).

[51] The companies are Good Meat and Upside Foods. In July 2023, Upside Foods sold cell-cultured chicken at restaurants in San Francisco, and Good Meat in Washington, DC.

[52] The reason for the way this particular matter is addressed in the United States, with joint activity of the FDA and USDA-FSIS, is based in a formal agreement signed on 7 March 2019 (https://www.usda.gov/about-usda/news/press-releases/2018/11/16/statement-usda-secretary-perdue-and-fda-commissioner-gottlieb-regulation-cell-cultured-food-products). Under this agreement, the FDA's remit includes oversight of cell collection, cell banks, cell growth and differentiation. The transition of oversight to the United States Department of Agriculture—Food Safety and Inspection Service (USDA-FSIS) occurs during the harvesting stage of the cell culture process. The USDA-FSIS assumes responsibility for the subsequent production and labelling of these products. In accordance with the agreement, the FDA and USDA-FSIS are collaborating to formulate comprehensive procedures aimed at streamlining the coordination of their respective regulatory oversight functions pertaining to the harvesting of cultured livestock and poultry cells (Commissioner 2020).

laws to restrict the use of specific terms such as "meat", "hamburger" and "sausage" except for products of animals that were raised and slaughtered. This implies that the terms cannot be used for cell-cultivated products. Since 2018, certain state governments have enacted laws to prohibit use of the term "meat" for cell-based products or to consider use of the term *meat* for such products misbranding. In particular, as indicated in the Congressional Research Service Report, Arkansas "prohibits the labeling of cell-cultivated food as meat";[53] Kansas requires that "alterative meat products sold in its state using meat terms be considered misbranded if those products do not include a disclaimer that they do not contain meat";[54] Kentucky[55] "states that food made of cultivated animal tissue produced from in vitro cultivated animal cells cannot be labeled as meat or a meat product"; Mississippi[56] "prohibits the labeling of cell-cultivated meat using meat terms"; Missouri[57] "prohibits the making of false claims of meat if the meat was not derived from the harvested production of livestock or poultry"; North Dakota[58] "states that cell-cultivated meat products cannot be labeled as meat"; Oklahoma[59] "defines meat as "any edible portion of livestock or part thereof" and declares it to be misrepresentation to label a product as meat that is not derived from harvested production of livestock"; South Carolina[60] "makes it unlawful to advertise, sell, label, or misrepresent as "meat" or "clean meat" cell-cultivated meat that is not derived from harvested production livestock, poultry, fish, or crustaceans"; and Wyoming[61] "prohibits the use of meat terms for the labeling, advertising or selling of cell-cultivated meat products and requires that cell-cultivated products be labeled as "containing cell-cultured products"". Nine out of 50 states introduced such bans with laws dated before 2023, setting the stage for the debate mentioned above. At the time of writing, Indiana is enacting a bill[62] entitled "lab-grown meat labeling" that "prohibits a person from intentionally misbranding a cultivated meat product as a meat product." It provides that the cultivated meat product is misbranded as a meat product if the cultivated meat product is: (1) advertised; (2) labelled; or (3) offered for sale or sold in a manner that does not clearly indicate that it is a cultivated meat product or not in accordance with rules adopted by the Board of Animal Health (BOAH). It requires BOAH to administer

[53] H.B. 1407, enacted 18 March 2019. Later on a lawsuit was filed against Arkansas. The case *Turtle Island Foods SCP v. Soman*—USDC E.D. Arkansas, No. 4:19-cv-514—was closed with a judgment in favor of plaintiff.

[54] S.B. 261, enacted 5 May 2022.

[55] H.B. 311, enacted 21 March 2019.

[56] S.B. 2922, enacted 12 March 2019.

[57] S.B. 627, enacted 1 June 2018.

[58] H.B. 140, enacted 12 March 2019.

[59] H.B. 306, enacted 19 May 2020.

[60] H.B. 4245, enacted 16 May 2019.

[61] S.B. 68, enacted 10 January 2018.

[62] H.B. 1425, enacted 1 July 2025.

the chapter, while Florida,[63] Alabama[64] and Mississippi[65] introduced a ban on the manufacture, sale and distribution of cultured meat in the state. Nebraska and Dakota have also introduced proposals to ban lab-grown meat. In the meantime (March 2025), the FDA completed pre-market consultation for a human food made with cultured pork fat cells.[66]

In conclusion, while a growing number of countries have authorized or are working towards approving cultivated meat for commercial consumption,[67] the regulatory landscape continues to develop. Each country has its own distinct regulatory approach, many focusing on food safety assessments, transparency and developing clear guidelines for labelling and nomenclature. As technology advances, it is anticipated that more countries will introduce regulations and authorizations for the sale of cultivated meat and fish, reflecting the growing interest in this emerging sector.

References

Albisinni F (2025) Quale carne? Rivista di Diritto Alimentare 1:104–112

Broucke K, Van Pamel E, Van Coillie E, Herman L, Van Royen G (2023) Cultured meat and challenges ahead: a review on nutritional, technofunctional and sensorial properties, safety and legislation. Meat Sci 195:109006

Carolan M (2025) Sustainable protein transitions or transformations: contested agrifood frames across "no cow" and "clean cow" futures. Sustainability 17:2637

Codex Alimentarius Commission (2021) Joint FAO and WHO Food Standards Programme Codex Alimentarius Commission, Forty-Fourth Session, 8–17 November 2021. New food sources and production systems: need for codex attention and guidance? https://www.fao.org/fao-who-codexalimentarius/sh-proxy/jp/?lnk=1&url=https%253A%252F%252Fworkspace.fao.org%252Fsites%252Fcodex%252FMeetings%252FCX-701-44%252FWorking%2BDocuments%252Fcac44_15.Add.1e.pdf. Accessed 15 Feb 2025

Codex Alimentarius Commission (2023) Joint FAO and WHO Food Standards Programme Codex Alimentarius Commission, Forty-Sixth Session, FAO Headquarters, Rome, Italy 27 November – 2 December 2023. https://www.fao.org/fao-who-codexalimentarius/sh-proxy/en/?lnk=1&url=https%253A%252F%252Fworkspace.fao.org%252Fsites%252Fcodex%252FMeetings%252FCX-701-46%252F%25E2%2598%2585Final%252520Report%252FREP23_CACe.pdf. Accessed 15 Feb 2025

Codex Alimentarius Commission (2024) Joint FAO and WHO Food Standards Programme Codex Alimentarius Commission, Forty-Seventh Session, Geneva, Switzerland, CICG, 25–30

[63] S.B. 1084, enacted 1 May 2024.

[64] S.B. 32, enacted 7 May 2024.

[65] H.B. 1006, enacted 1 July 2025.

[66] See the information released about the consultation at https://www.fda.gov/food/hfp-constituent-updates/fda-completes-pre-market-consultation-human-food-made-cultured-pork-fat-cells. Accessed 13 February 2025.

[67] As of March 2025, the sale of cultivated meat has been approved in Singapore, Israel, the US, Hong Kong and the United Arab Emirates (UAE). Conversely, countries such as South Korea, Japan, Canada and the EU are in the process of establishing regulatory frameworks for cultivated meat or evaluating proposals. The United Kingdom has approved the use of lab-grown meat in pet food, while Brazil and New Zealand are exploring the possibility of implementing regulations.

November 2024. Food safety aspects of cell-based food and precision fermentation derived food products. https://www.fao.org/fao-who-codexalimentarius/sh-proxy/fr/?lnk=1&url=https%253A%252F%252Fworkspace.fao.org%252Fsites%252Fcodex%252FMeetings%252FCX-701-47%252FCRDs%252Fcac47_crd04x.pdf. Accessed 15 Feb 2025
Commissioner (2020) Statement from USDA Secretary Perdue and FDA Commissioner Gottlieb on the regulation of cell-cultured food products from cell lines of livestock and poultry. https://www.fda.gov/news-events/press-announcements/statement-usda-secretary-perdue-and-fda-commissioner-gottlieb-regulation-cell-cultured-food-products. Accessed 15 Feb 2025
Congressional Research Service (2023, September 19) Cell-cultivated meat: an overview (R47697). https://crsreports.congress.gov. Accessed 15 Feb 2025
De Amstalden M (2022) Seafood without the sea: Article 20 of the agreement on trade-related aspects of intellectual property rights, the 'justifiability test' and innovative technologies in a sustainable blue economy. J World Invest Trade 23(1):68–94
Di Lauro A (2025) Carne "artificiale" e meatsounding: un orizzonte cannibale possibile e la retorica del "senza". Rivista di Diritto Alimentare 1:4–16
EFSA, Afonso AL, Gelbmann W, Germini A, Fernández EN, Parrino L, Precup G, Ververis E (2024) EFSA Scientific Colloquium 27: cell culture-derived foods and food ingredients. EFSA Supporting Publications 21(3):8664E
European Court of Auditors (2024) Special Report, food labelling in the EU. Consumers can get lost in the maze of labels. https://www.eca.europa.eu/en/publications?ref=SR-2024-23. Accessed 16 Feb 2025
FAO (2023) Cell-based food: its safety and its future role – stakeholder roundtable meeting report. Tel Aviv, 7 September 2022
FAO and WHO (2023a) Food safety aspects of cell-based food. https://doi.org/10.4060/cc4855en. Accessed 16 Feb 2025
FAO and WHO (2023b) Nine things to know about food safety aspects of cell-based food. https://www.fao.org/documents/card/en/c/cc6419en. Accessed 16 Feb 2025
Formici G (2024) Un assaggio di futuro? La carne a base cellulare e il vivace dibattito regolatorio dentro e fuori i confini dell'Unione europea. BioLaw J 4:165–197
Harris DR, Fuller DQ (2014) Agriculture: definition and overview. In: Smith C (ed) Encyclopedia of global archaeology. Springer, New York
Kit Lim LW (2023) Cultivated meat in Singapore: the road to commercialization. Int J Zool Anim Biol 6(4):1–5
Lähteenmäki-Uutela A, Rahikainen M, Lonkila A, Yang B (2021) Alternative proteins and EU food law. Food Control 13:108336
Lanzoni D, Rebucci R, Formici G, Cheli F, Ragone G, Baldi A, Violini L, Sundaram TS, Giromini C (2024) Cultured meat in the European Union: legislative context and food safety issues. Curr Res Food Sci 8:100722
Malerich M, Bryant C (2022) Nomenclature of cell-cultivated meat and seafood products. NPJ Sci of Food 6(1):56
Mayhall TA (2019) The meat of the matter: regulating a laboratory-grown alternative. Food Drug Law J 74:151–169
Miyake Y, Tachikawa M, Kohsaka R (2023) Policy frameworks and regulations for the research and development of cell-based meats: systematic literature review. Food Res Int 167:112599
Monaco A (2025) A perspective on the regulation of cultivated meat in the European Union. NPJ Sci Food 9(1):21
Paganizza V (2025) Carne coltivata: una comparazione tra Australia e Unione europea, passando da Singapore. Rivista di Diritto Alimentare 1:62–67
Planchenstainer F (2024) Meat me in Italy: the Italian ban on sounding names and cell-cultured meat. Eur Food Feed Law Rev 19(2):66–71
Post MJ, Levenberg S, Kaplan DL, Genovese N, Fu J, Bryant CJ, Negowetti N, Verzijden K, Moutsatsou P (2020) Scientific, sustainability and regulatory challenges of cultured meat. Nat Food 1(7):403–415

Rubino V (2024) La battaglia della carne coltivata dalle aule parlamentari a quelle di giustizia? Considerazioni a margine della legge 172/2023 fra armonizzazione, leale cooperazione e margini di autonomia degli Stati membri. Rivista di Diritto Alimentare 1:34–55

Rubino V, Rossi Dal Pozzo F (2024) The regulatory framework for the authorisation to produce and market cultured meat in the EU. Eur Food Feed Law Rev 19:199–203

Rubio N, Datar I, Stachura D, Kaplan D, Krueger K (2019) Cell-based fish: a novel approach to seafood production and an opportunity for cellular agriculture. Front Sustain Food Syst 3:43

Sirsi E (2020) Il nome delle cose e i paradigmi dell'innovazione nel mercato agro-alimentare. Il caso delle carni alternative. In: Carmignani S, Lucifero N (eds) Le regole del mercato agroalimentare tra sicurezza e concorrenza. Editoriale Scientifica, Napoli, pp 665–708

Spoto G (2024) Libertà di ricerca e paternalismo giuridico in Italia: La carne coltivata in laboratorio. Rivista di Diritto Alimentare 1:56–61

Xu E, Niu R, Lao J, Zhang S, Li J, Zhu Y, Shi H, Zhu Q, Chen Y, Jiang Y, Wang W, Yin J, Chen Q, Huang X, Chen J, Liu D (2023) Tissue-like cultured fish fillets through a synthetic food pipeline. NPJ Sci Food 7:17

Cell-Based Meat and Food Policy

Maria Cecilia Mancini

Abstract The global food system faces significant challenges, including food security, environmental sustainability and fair economic opportunities. Tackling these issues demands innovation, particularly in protein production. Among the emerging alternatives, cell-based meat is a disruptive technology that may lessen climate impact, reduce land and water usage and improve food security. However, widespread adoption relies on overcoming technical, economic and regulatory obstacles. This study examines the role of public policy in steering cell-based meat innovation, highlighting the need for a coherent regulatory framework. It looks at global policy approaches and the economic instruments propelling alternative protein development. The shift to alternative protein sources raises concern about market concentration and social inequalities that call for inclusive governance. A unified, long-term strategy that aligns regulatory policies, public investments and sustainability goals is essential for integrating cell-based meat into global food systems. This study underscores the importance of adaptive policymaking to promote innovation while ensuring ethical and equitable food production.

Keywords Cell-based meat · Governance · Public policy · Innovation · Sustainability

---

M. C. Mancini (✉)
Department of Economics and Management, University of Parma, Parma, Italy
e-mail: mariacecilia.mancini@unipr.it

G. Formici et al. (eds.), *Cell-Based Meat in the European Union and Beyond*,
https://doi.org/10.1007/978-3-032-14624-3_7

1 Cell-Based Meat and Innovation

According to the Organisation for Economic Co-operation and Development (OECD), the global food system faces a triple challenge that demands immediate intervention.[1] Firstly, it must ensure food security and provide adequate nutrition to meet the needs of a growing world population. Secondly, it needs to secure livelihoods and equitable incomes for those involved in agricultural production and individuals working in the broader agri-food system. Thirdly, these goals must be achieved while promoting environmental sustainability, which entails minimising the ecological footprint of food production and mitigating the adverse effects of climate change.

Addressing these interrelated challenges calls for a comprehensive and integrated approach that combines scientific innovation with strong political commitment. This transformation is not merely incremental: it entails potentially disruptive changes that require a reconfiguration of existing paradigms of food production, distribution and consumption.

A pressing concern is rising world demand for protein sources, driven by population growth in Africa and Asia, increasing incomes in the Middle and Far East, and rapid urbanisation.[2] A sustainable response to this demand entails balancing social imperatives—such as ensuring equitable access to food—with the urgent need to tackle environmental issues like land degradation, water scarcity and GreenHouse Gas (GHG) emissions.

Alternative protein sources are a promising avenue for addressing these challenges. They encompass microorganisms (such as fungi and bacteria), algae, microalgae, insects[3] and lab-grown proteins, including cell-based meat.[4] Members of the scientific community and policymakers are increasingly emphasising these novel protein sources as potential solutions to environmental impacts and global food security problems. In this emerging landscape, meat analogues—products designed to replicate the sensory characteristics of conventional animal-derived meat—have attracted substantial attention.[5] Indeed, the fact that cell-based meat provides real animal protein places this product in a distinct position. Further scientific research and structural changes are nevertheless required before cultivated meat can alter current production and market patterns.[6]

Cultivated meat is a radical innovation, a transformative technological breakthrough rooted in entirely new scientific and engineering principles.[7] This was

[1] OECD (2021).

[2] FAO (2020).

[3] Pojić et al. (2018).

[4] The terms cell-based meat, cultured meat, and cultivated meat will be used interchangeably with the same meaning.

[5] Lusk et al. (2022).

[6] Dean et al. (2022).

[7] Dess and Beard (1984).

clearly articulated at the World Economic Forum,[8] where it was asserted that transitioning to safe, affordable, sustainable protein cannot be achieved solely through incremental improvements in the sustainability of conventional livestock farming and new protein sources, such as cell-based meat, are needed. While such innovations create new market opportunities and reshape commercial landscapes, they also challenge established firms entrenched in traditional business models.[9] Indeed, radical innovations enable the entry of new market participants and may result in a redefinition of production sectors. Research highlights the fact that industries embedded in dominant socio-technical systems often resist significant technological shifts, primarily due to entrenched economic, institutional and social structures.[10] However, when external pressures—such as policy shifts, evolving social norms and macroeconomic changes—reach a critical threshold, established industries relinquish outdated practices.[11] This process of destabilisation is crucial for the widespread adoption of innovations, and facilitates the introduction of more sustainable technologies.

Public policy plays a crucial role in guiding innovation towards broader societal objectives. This influence is exerted in increasingly interconnected ways. Firstly, governments can support research and development through public investments. Such support is vital in addressing market failures, particularly in sectors characterised by high uncertainty or where long-term return horizons deter private investment. By providing essential funding and resources, governments catalyse the creation of new knowledge and technologies that might otherwise remain untapped.[12]

Secondly, effective public policies strengthen national and regional innovation ecosystems by fostering collaborative networks and forging connections between different stakeholders, including research institutions, industry players and civil society organisations. By facilitating exchange of knowledge and promoting the spread of innovations, governments can cultivate an environment conducive to ground-breaking solutions. Traditional innovation policies have largely emphasised enhancing competitiveness and stimulating economic growth.[13] However, these policies are often insufficient to address the complex social and environmental challenges inherent in the global food system.[14] There is consequently growing consensus that innovation should be directed towards broader societal issues and aligned with international frameworks such as the United Nations Sustainable Development Goals (SDGs).[15]

---

[8] World Economic Forum (2019).

[9] Tushman and Nelson (1990).

[10] Galli et al. (2018).

[11] Turnheim and Geels (2012).

[12] Marques et al. (2024) and Haddad et al. (2022).

[13] Haddad et al. (2022).

[14] Diercks et al. (2019).

[15] Haddad et al. (2022) and Schot and Steinmueller (2018).

This chapter contends that policymaking and regulatory frameworks need revision to effectively address disruptive innovations and facilitate the transition towards sustainability in agri-food systems. These frameworks ought to support stakeholders and protect consumer interests throughout the agri-food system. This means coordinated integration of scientific research, technological innovation and policy reform aimed at reshaping global food production processes in response to the complex challenges of the twenty-first century.

The following section provides a brief overview of the leading innovative approaches proposed to meet the growing demand for alternative protein sources, with a focus on cell-based meat. Section 3 gives an update on current public policies, regulatory frameworks and economic instruments concerning alternative protein sources worldwide. Section 4 identifies key issues in existing public food policies, particularly at European level, and offers recommendations for a long-term framework to promote sustainability. Section 5 draws some conclusions.

2 Scenarios for Sustainability: An Overview

There is growing global consensus on the urgent need to transform agriculture and food systems to achieve a more sustainable agri-food system.[16] However, debate continues on the approaches to prioritise in order to achieve this goal. Proposed scenarios range from holistic frameworks combining ecological and social principles—such as agroecology—to technological innovations, including alternative protein sources and cellular agriculture.

While all these approaches share the objective of promoting sustainability, they diverge significantly in their principles, research background and implementation. Agroecology relies primarily on ecological and social innovation, emphasising local knowledge, biodiversity and sustainable resource management.[17] By contrast, cellular agriculture and alternative protein technologies[18] focus on cutting-edge scientific advancements to achieve sustainability goals. The different pathways therefore generate distinct impacts and side effects on socioeconomic structures, ecosystems and consumption patterns.

Specific approaches, particularly agroecology, are already widely implemented and their impacts are increasingly documented. Studies indicate that agroecological

---

[16] Iqbal et al. (2025) and Haddad et al. (2022).

[17] Agroecology seeks to optimize the interactions between plants, animals, humans and the environment while also addressing the need for socially equitable food systems where people can exercise choice over what they eat and how and where it is produced. See Altieri (2018).

[18] Alternative sources of proteins are obtained from novel sources, including microorganisms (fungi and bacteria), algae, microalgae and insects—see Pojić et al. (2018)—or animal cells grown in laboratories. Cellular agriculture refers to the production of agricultural goods through animal or plant cell cultures, or microbes, rather than employing animals or crop plants, see Nyyssölä et al. (2022).

practices significantly contribute to enhancing the three pillars of sustainability in agricultural and food systems when compared to conventional industrial farming systems.[19] Agroecology improves biodiversity and soil health and strengthens rural economies by supporting smallholder farmers and promoting equitable resource distribution.[20]

Similarly, plant-based meat alternatives have emerged as promising solutions for building more sustainable food systems. Over the past few decades, technological improvements have led to substantial enhancements in the sensory qualities of plant-based meat products, facilitating their widespread availability in supermarkets and food service establishments. Major global fast-food chains, such as McDonald's, Burger King and KFC, now offer plant-based alternatives to traditional meat products, reflecting the growing consumer demand for sustainable dietary options. Furthermore, substantial investments in the plant-based meat sector, particularly in countries such as the Netherlands, the United Kingdom (UK) and Germany, show rapid market growth and suggest strong potential for a shift toward plant-based diets.[21]

In addition to plant-based alternatives, edible insects have been recognised as a highly sustainable protein source. They provide significant nutritional value, including essential amino acids, vitamins and minerals,[22] while demonstrating considerably lower environmental impacts compared to traditional livestock farming.[23] Although Western consumers often resist consuming insects,[24] the expanding market for farmed insects indicates growing potential to address future food requirements.

On the other hand, the environmental, social and economic implications of large-scale cell-based meat production remain largely underexplored due to the nascent stage of commercial development and very limited global availability.[25] Cultivated meat for human consumption is currently approved for sale in three countries (Singapore, United States of America—USA—and Israel),[26] limiting opportunities for comprehensive, data-driven analysis of its broader impacts. Early-stage studies and pilot projects suggest that cell-based meat could significantly reduce GHG emissions, land use and water consumption with respect to conventional animal agriculture.[27] However, these findings largely relied on hypothetical models and small-scale experimental data. The recent literature has re-evaluated the environ-

---

[19] Mouratiadou et al. (2024).

[20] Guareschi et al. (2023), Van der Ploeg et al. (2019) and Boone et al. (2019).

[21] Andreani et al. (2023) and Petersen et al. (2021).

[22] Ordoñez-Araque and Egas-Montenegro (2021).

[23] FAO (2013).

[24] Sogari et al. (2023) and Ardoin and Prinyawiwatkul (2021).

[25] Wood et al. (2023).

[26] See also *Cell-Based Meat in the European Union: A Regulatory Crossroads* by Corazza and Formici, as well as *The Legal Debate on Cultured Meat Around the World: Comparative Insights from the United States, Singapore and Israel*, by Pitto in this volume.

[27] Tuomisto and Teixeira de Mattos (2011); see also *What Is Cell-Based Meat? Food Safety and Sustainability Aspects* by Cheli et al. in this volume.

mental benefits of cultivated meat, suggesting they may be less substantial and with more nuanced prospects than early Life Cycle Assessments (LCA) suggested.[28] Moreover, large-scale production introduces variables such as energy requirements for bioreactors, sourcing of growth media and waste management, which could offset environmental benefits.[29]

The literature also suggests that cultured and plant-based meat products could confer public and individual health benefits by reducing antibiotic use and lessening the likelihood of foodborne illness.[30] These alternatives also address ethical concerns related to animal welfare by removing the need for conventional animal slaughter.[31]

The socio-economic impacts of scaling up cell-based meat production are particularly challenging to predict. Currently, production costs for cultivated meat remain significantly higher than those for conventional meat, primarily due to the expense of growth medium and the complex bioreactor infrastructure required for production. Estimates suggest that producing one kilogram of cultured meat in a large-scale facility could cost approximately $63/kg.[32] However, technological advances, economies of scale and public-private partnerships could reduce costs over time, enhancing the market competitiveness of cultivated meat.

The global landscape for cell-based meat production is evolving rapidly. Worldwide, the number of publicly announced cultivated meat companies, which focus primarily on cultivated meat inputs or end products, increased to 174 in 2023, mostly in the United States (n. 45), Israel (n. 19) the United Kingdom (n. 16) and Singapore (n. 11).[33] An increasing number of companies are concentrating on the cultivated meat value chain, beyond merely manufacturing end products. Some scholars[34] envision opportunities for small enterprises to operate like microbreweries in the cultured-meat landscape[35] and argue that the growth of the alternative protein sector could create new market opportunities across various segments of the supply chain, including suppliers of cell culture medium, growth factors and other essential inputs such as cell nutrition and genetics.[36] However, most funding comes from venture capital and large meat processing companies,[37] and large-scale adoption of cell-based meat could disrupt traditional livestock farming communities by

[28] Tuomisto et al. (2022), Heller and Keoleian (2018) and Tuomisto et al. (2015).

[29] Tuomisto and Ryynänen (2024).

[30] Newton and Blaustein-Rejto (2021), Santo et al. (2020) and Mayhall (2019).

[31] See *Animal Welfare as a Constitutional Value: The Debate in Comparative Law and the Emerging Role of Cultivated Meat*, by Piciocchi in this volume.

[32] See Humbird (2021). The three major costs of production are the cell-culture medium, bioreactors and labor. These costs make up over 80% of the overall cost of production.

[33] Good Food Institute (GFI) (2024).

[34] Van der Weele and Tramper (2014).

[35] Stephens et al. (2018).

[36] Newton and Blaustein-Rejto (2021), Choudhury et al. (2020) and Reis et al. (2020a).

[37] Santo et al. (2020).

consolidating market power in the hands of a few dominant firms.[38] As already observed, cultured meat is a disruptive innovation with a vertically integrated value chain that could centralise economic benefits among a handful of multinational corporations.[39] Many of the largest meat and consumer packaged goods companies across the globe—including JBS, Tyson, Cargill, Nestlé and Danone—are engaged in the cultivated meat industry through investments, acquisitions, partnerships and/or research, development and manufacturing.[40] The global protein transition—shifting from traditional animal-sourced proteins to alternatives like plant-based and cell-based meats—could therefore significantly affect rural economies heavily reliant on livestock production. In regions, such as certain USA States, where intensive animal farming is widely practised, declining demand for livestock products could lead to job losses, rural depopulation and economic disintegration.[41] The rise of cell-based meat technologies could also require reconfiguring workforce skills and responsibilities in the agri-food sector.[42]

However, rising global meat demand in developing countries may allow conventional meat industries to coexist with emerging alternatives.[43] Premium meat products, such as those bearing Geographical Indications, could also benefit from being premium price alternatives to mass-produced cultured meats.[44]

The other side of the issue is consumer acceptance of cell-based meat, which is influenced by a complex interplay of factors, including perceptions of safety, taste and the ethical dimensions of lab-grown food.[45] Research findings show general uncertainty among consumers, particularly in Western societies, driven by concerns about the unnaturalness of the technology and its long-term health implications.[46] Moreover, since most consumer surveys have been conducted using hypothetical products without tasting, the role of sensory attributes—such as flavour and texture—remains largely underexplored.

From a broader perspective, cellular agriculture has the potential to enhance global food security by providing a stable and scalable source of protein.[47] However, the authors indicate that the success of this potential relies heavily on the production

---

38 Mylan et al. (2023), Treich (2021) and Bryant (2020).

39 Reis et al. (2020b).

40 GFI (2024).

41 Chriki and Hocquette (2020).

42 Goodman (2023).

43 Stephens et al. (2018).

44 Mancini and Antonioli (2022).

45 See *Socio-Demographic and Behavioural Determinants of Consumer Attitudes to Cell-Based Meat and Insect-Flour-Based Products: An Empirical Study of Italians' Propensity to Accept New Foods for Human Nutrition and Pet Food*, by Degli Antoni and Faillo; *Navigating Consumer Acceptance of Cultured Meat: Insights from Social Science Research* by Lin Hi and Burdorf, and *Who Hates Cultured Meat? Examining Public Opposition to Cellular Agriculture* by Broad and Glenna, all in this volume.

46 Pakseresht et al. (2022) and Bryant and Barnett (2020).

47 Soice and Johnston (2021).

model adopted by companies. Many cell agriculture firms are pursuing a centralised supply system, where production is concentrated in a limited number of large facilities. While this model offers economies of scale, it also introduces vulnerabilities, particularly regarding supply chain disruptions—an issue highlighted by the impact of the COVID-19 pandemic on centralised meat processing systems. Alternative supply models that favour decentralised production could help mitigate these vulnerabilities by distributing production across a broader geographic area, thereby enhancing resilience and reducing dependence on large-scale industrial infrastructure.[48]

Significantly, the alternative protein sector experienced a decline in private funding in 2023. While alternative protein firms secured £15.7 billion from 2014 to 2023, over half of this sum was amassed in 2020 and 2021, and investments dropped from £2.9 billion in 2022 to £1.6 billion in 2023.[49] Given the tighter private funding landscape expected to extend into the coming years, the alternative protein companies best positioned to attract equity investment will be those demonstrating pathways to revenue and profitability. Simultaneously, long-term debt, grants and governmental incentives are essential to enable companies to reduce their production costs and achieve price parity as they scale up their operations. To assist alternative protein companies in accessing such funding, innovative solutions and leveraging of governmental and philanthropic funding will be required.[50]

3 Policies and Regulatory Frameworks for Cell-Based Meat and Alternative Protein Sources

Alternative protein foods have become a significant focus for key international organisations and stakeholders in the global food governance framework. Institutions such as the Food and Agriculture Organization (FAO), the World Health Organization (WHO) and the United Nations Environment Programme (UNEP) have increasingly emphasised the relevance of alternative proteins in advancing sustainable food systems. In April 2023, the FAO and the WHO published a milestone report entitled *Food Safety Aspects of Cell-Based Food*,[51] which established the first global framework for identifying food safety hazards related to cell-based products. Similarly, UNEP released a landmark report evaluating the environmental, social and food security benefits of alternative proteins, and offering guidance on how governments could foster their development through appropriate regulatory frameworks and policy interventions.

[48] Soice and Johnston (2021).

[49] GFI (2024).

[50] *Ibid.*

[51] FAO and WHO (2023).

Some federal government policies have incorporated alternative protein sources into coherent regulatory frameworks. A significant milestone in regulatory cooperation was achieved in 2019 when the United States Department of Agriculture (USDA)'s Food Safety Inspection Service (FSIS) and the USA Food and Drug Administration (FDA) formalised a joint agreement on regulating cell-based meats. Under this arrangement, the FDA oversees the production phase up to the point of processing, while the FSIS governs the subsequent stages, including processing, packaging and labelling.[52]

The U.S. Defense Advanced Research Projects Agency (DARPA) launched the Cornucopia program in February 2023, an initiative focused on developing novel microbial-based foods using minimal inputs—air, water and electricity—without additional supplementation.

In tandem, alternative protein research has been explicitly incorporated into the mandates of the USDA and the Department of Energy (DOE). These agencies have prioritised research on alternative proteins, formed public-private partnerships, and created agriculture-focused biomanufacturing ecosystems.

On the European Union (EU) front, the European Commission launched the European Green Deal as a comprehensive strategy to achieve carbon neutrality and conserve biodiversity.[53] Two pivotal strategies emerged from this framework: the Biodiversity Strategy and the Farm-to-Fork Strategy,[54] which aim to promote food-system sustainability, mitigate climate change and provide equitable access to food. In this context, the Farm-to-Fork Strategy explicitly encourages research into alternative proteins, including insect-derived and cell-based meat alternatives.

In line with these objectives, the EU Horizon 2020 Programme awarded funding to the *Meat4All* project,[55] marking the European Union's first significant investment in cell-based meat research. This initiative supports the development of regulatory, technological and consumer acceptance frameworks.

Under Novel Food Regulation, Regulation (EU) 2015/2283,[56] cell-based meat requires premarket approval. This regulation also mandates precise labelling requirements to safeguard consumer health and provide accurate information about food products. Cell-based products utilizing Genetically Modified Organisms

[52] See *The Legal Debate on Cultured Meat Around the World: Comparative Insights from the United States, Singapore and Israel* by Pitto in this volume.

[53] European Commission (2023a).

[54] European Commission (2023b).

[55] European Commission (2020).

[56] Regulation (EU) 2015/2283 of the European Parliament and of the Council of 25 November 2015 on Novel Foods, amending Regulation (EU) no 1169/2011 of the European Parliament and of the Council and repealing Regulation (EC) no 258/97 of the European Parliament and of the Council and Commission Regulation (EC) no 1852/2001.

(GMOs) are regulated under the European GMO Regulation,[57] requiring traceability and labelling measures throughout the distribution process.[58]

At the national level, countries such as Singapore and Israel have emerged as pioneers in establishing regulatory frameworks for cell-based foods. The Singapore Food Agency (SFA) authorised the marketing of cell-based chicken in 2020, categorising it as a Novel Food—defined as any food product without a history of prior consumption. The Singaporean government has also made substantial investments in research and development through initiatives such as the Singapore Food Story 2.0,[59] a $117 million program to advance research in future foods and food safety. In 2023, Enterprise Singapore, a government-backed agency, introduced a Food Technology Program to assist local start-ups specialising in plant-based proteins in accessing the Chinese market.[60] Singapore has also introduced preferential visa rules to attract foreign experts in alternative protein research.

In Israel, governmental support for alternative proteins has been robust, with considerable investment in both research and commercialisation. Israel has also launched initiatives to cultivate human capital by fostering expertise in alternative proteins and food technology. In 2024, the Israel Ministry of Health authorised the marketing of cell-based beef under the country's Novel Foods regulatory framework.[61]

In 2024, the UK authorised the marketing of cell-based meat as pet food,[62] and the same year the French company Gourmey filed for the authorisation of cell-based meat ("foie gras") with the European Commission.[63] In the Netherlands, cultivated meat is advocated at research and institutional level, and the tasting of the meat grown is authorised by the government under controlled conditions defined by a code of conduct.[64]

---

[57] Regulation (EC) no 1830/2003 of the European Parliament and of the Council of 22 September 2003 concerning the traceability and labelling of genetically modified organisms and the traceability of food and feed products produced from genetically modified organisms and amending Directive 2001/18/EC.

[58] See *Cell-Based Meat in the European Union: A Regulatory Crossroads* by Corazza and Formici in this volume.

[59] Singapore Food Story 2.0 is financed through allocations designated in the government's annual Supply Act, which authorizes the government to use resources of the Consolidated Fund and the Development Fund.

[60] https://www.enterprisesg.gov.sg/grow-your-business/innovate-with-us/market-access-and-networks/global-innovation-alliance/cip/foodtech-project-call-on-alternative-proteins. Accessed 1 March 2025.

[61] https://www.gov.il/en/pages/17012024-02. Accessed 1 March 2025. See also *Cell-Based Meat in the European Union: A Regulatory Crossroads* by Corazza and Formici in this volume.

[62] Meatly received regulatory approval from the Animal & Plant Health Agency (APHA)—in collaboration with the Food Standards Agency (FSA) and DEFRA—on 2 July 2024.

[63] https://www.eitfood.eu/news/gourmey-files-the-first-eu-novel-food-submission-for-cultured-meat?utm_source=chatgpt.com. Accessed 1 March 2025.

[64] The Dutch government formally issued the Code of Practice for Safely Conducting Tastings of Cultivated Foods Prior to EU Approval on 5 July 2023. See also *Cell-Based Meat in the European Union: A Regulatory Crossroads* by Corazza and Formici in this volume.

At the time of writing, five cultivated meat products have been approved in five countries. Good Meat received approval for a cell-based chicken hybrid in Singapore in December 2020[65] and for cultivated chicken in the USA in December 2023.[66] Upside Foods gained approval for a cultivated chicken fillet in the USA in June 2023.[67] Aleph Farms secured approval for a thin-cut beef steak in Israel in January 2024[68] and Vow received approval for cultivated Japanese quail in Singapore in April 2024 and in Hong Kong in November 2024.[69] Further approvals are expected in Australia, New Zealand, China, South Korea and Canada in the coming years.

Despite the growing global momentum, several jurisdictions have resisted the commercialization of cell-based meat. In certain States of the U.S., including Florida[70] and Alabama,[71] legislation prohibits the production, sale and distribution of cultivated meat due to concerns regarding economic disruption and food safety. Other U.S. States[72] have approved or are in the process of adopting legislation aimed at preventing the use of the term "meat" for cell products. Label censorship laws either ban terms such as "burger" or "sausage" on products that are not made from an animal carcass, or compel language on cultivated meat labels to include terms such as "lab-grown" in a specific font size.[73] A similar decision has emerged internationally, with countries like Uruguay[74] and some EU Member States discussing or enacting legislation to restrict or outright ban the sale of cell-based meat. Notably, Italy approved a national law[75] in December 2023 prohibiting cell-based foods. The resistance of some European countries has led to a formal "Note from the Austrian, French and Italian delegations—Common Agricultural Policy (CAP)'s role in safeguarding high-quality and primary farm-based food production", in which

[65] https://www.sfa.gov.sg/food-safety-tips/food-risk-concerns/risk-at-a-glance/safety-of-alternative-protein. Accessed 1 March 2025.

[66] https://www.goodmeat.co/all-news/good-meat-receives-first-ever-usda-label-approval-for-cultivated-meat?utm_source=chatgpt.com. Accessed 1 March 2025.

[67] https://upsidefoods.com/blog/the-upshot-how-do-the-fda-and-usda-regulate-cultivated-meat. Accessed 1 March 2025.

[68] https://www.gov.il/en/pages/17012024-02. Accessed 1 March 2025.

[69] GFI (2024).

[70] Senate Bill 1084 (SB 1084), signed into law by Governor Ron De Santis on 1 May 2024.

[71] Gov. Kay Ivey signed the Alabama Bill, which prohibits "the manufacture, sale, or distribution of food products made from cultured animal cells," into law on 7 May 2024.

[72] Arkansas, Georgia, Kansas, Kentucky, Louisiana, Maine, Mississippi, Missouri, Montana, North Dakota, Oklahoma, South Carolina, South Dakota, Texas, Wyoming. Formici (2024).

[73] See *The Legal Debate on Cultured Meat Around the World: Comparative Insights from the United States, Singapore and Israel* by Pitto, in this volume.

[74] Uruguay took the first measures against cell-based meat in 2020, with the Budget Law for 2020–2024. Article 292 of the Regulation states that "food containing animal culture cells artificially produced in a laboratory" cannot be identified as meat.

[75] Italian Law No. 172 of 1 December 2023 (Official Gazette No. 281, 1 December 2023).

lab-grown artificial cell-based foods are claimed to be a threat for the primary farm-based approach and genuine production methods.[76]

Far from being comprehensive, Table 1 is meant to show how active and sometimes contradictory the global regulatory and public funding framework is.

3.1 Economic Instruments for Alternative Protein Sources

Governments are currently employing a variety of economic instruments—primarily financial mechanisms—to address the development and adoption of alternative protein sources. Economic instruments are tools[77] in the economic policy framework designed to influence economic behaviour by incentivising or discouraging specific activities, such as consuming particular goods or utilising certain resources. These instruments are effective mechanisms for achieving targeted policy objectives, particularly in promoting sustainable food systems. However, economic instruments extend beyond mere monetary interventions to encompass a range of elements that influence the effectiveness of these policies. Indeed, the implementation and efficacy of economic intervention depend on several factors, including complementary measures aimed at promoting action and facilitating verification processes.[78]

These interventions can be categorised into two domains: supply-side and demand-side measures. Most interventions focus on supporting the supply side, aiming to facilitate alternative protein-source production, research and scalability. This includes funding research through grants, investing in start-ups, offering tax breaks, and supporting education and training programs to build a skilled workforce for this sector. Other policies are consumer-focused, aimed at shaping consumer perceptions, market acceptance and affordability. This is the case of labelling policies, which clarify how products can be marketed, and tax incentives, which make

---

[76] The note was supported by ten States: Czech Republic, Cyprus, Greece, Hungary, Luxembourg, Malta, Poland, Romania, Slovakia and Spain. See CAP's role in safeguarding high-quality and primary farm-based food production—Note from the Austrian, French and Italian Delegations, 5469/2024-REV1, https://data.consilium.europa.eu/doc/document/ST-5469-2024-REV-1/en/pdf. Accessed 1 March 2025. See also *Cell-Based Meat in the European Union: A Regulatory Crossroads* by Corazza and Formici, in this volume.

[77] There are seven major categories of economic instrument: (1) property rights, market creation, fiscal instruments; (2) taxes and subsidies, charge systems; (3) fines and fees, financial instruments; (4) incentives, liability instruments; (5) socially responsible behaviour; (6) performance bonds and (7) deposit refund systems.

[78] This comprehensive ecosystem includes components such as collection of data, monitoring, reporting and verification frameworks; integration of environmental legislation and compliance mechanisms; government-led policy initiatives that align with sustainability goals; deployment of technological solutions to enhance production and efficiency; active stakeholder participation and engagement to ensure industry collaboration; public awareness and educational campaigns to shift societal perceptions and foster acceptance of alternative protein sources. See Srinivas (2023).

Table 1 Main examples of regulations and public funding on alternative protein sources

Country	Regulations	Public funding
United States	FDA pre-market consultation and USDA grant of inspection. USDA approval of labels case-by-case.	Federal and state-level grants; USDA-backed loans; Department of Energy and Department of Defense public funding opportunities.[a]
Singapore	Regulatory approvals for serum-free media and multiple forms of cultivated chicken; Food Safety and Security Bill.	Singapore Food Story 2.0 Research Programme ($165 million); business grants for start-ups.[b]
Israel	The National Food Service in Ministry of Health has regulatory authority over cultivated meat produced and marketed in Israel.	In 2022, the Israel Innovation Authority granted $18 million to establish a consortium of 14 companies and 10 academic laboratories (MAGNET Consortium Program).[c]
Netherlands	First EU country to enable pre-market tastings for cultivated meat and seafood. [d]	€1 million awarded for research on scalability and cost-effectiveness of precision fermentation.[e]
Switzerland	Separate Novel Food approval process.	Swiss National Science Foundation funding for alternative protein research.[f]
United Kingdom	Retained EU Novel Food Regulation; first applications for cultivated meat submitted in 2023.	£12 million for Cellular Agriculture Research Hub. £3.4 million for seven research projects on cultivated meat. Additional support for domestic cultivated meat industry through business grants (e.g. a £2.5 million grant to Accelerator program growth medium start-up).[g]
Germany	Federal funding for 2024.	€38 million in federal funding towards sustainable protein transition, including funding for innovation in alternative protein production and aid to farmers and companies transitioning to plant-based, cultivated or fermented protein production.[h]
France	Regulatory pushback on plant-based labelling; no explicit cultivated meat policy.	€7.4 million awarded for alternative protein processing and research.
Spain	Some regulatory support but not explicitly defined for cultivated meat.	€335,000 grant for mycelium-based meat;[i] €7 million for Center for Innovation in Alternative Proteins.[j]
South Korea	Regulation-free zone for cultivated meat R&D; approval process introduced in 2023[k].	National plan for plant-based foods; $6.7 million Cellular Agriculture Support Center.

(continued)

Table 1 (continued)

Country	Regulations	Public funding
Australia and New Zealand	Joint Novel Foods Standard; first cultivated meat regulatory application by Vow Foods in 2023.	NZD 1 million for cultivated meat technology research.
China	No formal regulatory framework for cultivated meat.	Government-funded cultivated meat research as part of the 5-year plan.[l]
Japan	Prime Minister endorsed cultivated meat; work underway on regulatory framework.	Grant to cultivated meat company ($13.1 million).[m]
Brazil	Developing regulatory framework for cultivated meat under National Health Surveillance Agency and Ministry of Agriculture.	Research funding for hybrid and plant-based as well as cultivated sausage and fat at Brazilian state research agency.[n]
Italy	Banned cultivated meat in 2023.	Projects funded under the National Recovery and Resilience Fund.[o]

Source: Author's elaboration based on various sources (see footnotes)

[a]DOE announcement (January 2025) regarding calls for proposals/funding in the Department of Energy's website (https://www.energy.gov/eere) under the Office of Energy Efficiency & Renewable Energy. Accessed 1 March 2025

[b]https://www.sfa.gov.sg/docs/default-source/recognition-programmes-and-grants/sfs/sfs2-0_seeds-grant-call_info-sheet_tracked_180434.pdf. Accessed 4 November 2024

[c]https://stip.oecd.org/stip/interactive-dashboards/policy-initiatives/2023%2Fdata%2FpolicyInitiatives%2F99993435.4. Accessed 3 November 2024

[d]The approval originated from a motion passed by the Dutch House of Representatives (Tweede Kamer) on 8 March 2022—Parliamentary document 27 428, No. 383

[e]Dutch Research Council (NWO https://www.nwo.nl/en/researchprogrammes/perspectief/perspectief-grants/fermi-fermentations-optimized-by-rational-models-to-improve-plant-based-substrates). Accessed 3 November 2024

[f]https://data.snf.ch/grants/. Accessed 3 November 2024

[g]Grant EPSRC Ref: EP/X038114/1

[h]Allocation in the 2024 Federal Budget—decision of the Bundestag/Budget Committee (2024 federal budget documents)

[i]Libre Foods, a Barcelona-based biotech company, was awarded a €335,000 public R&D grant by Neotec, a program supporting technology-based companies

[j]In 2023, the regional government of Catalonia invested €7 million in the Center for Innovation in Alternative Proteins (CiPA). This centre aims to promote research and technology related to co-products, by-products, waste, and surplus from the agri-food sector, focusing on sustainable food and feed production

[k]https://worldbiomarketinsights.com/south-korea-establishes-special-regulatory-free-zone-for-cultivated-meat-industry/?utm_source=chatgpt.com. Accessed 3 November 2024

[l]Chinese Ministry of Agriculture and Rural Affairs—14th Five-Year Agricultural Plan (2021–2025)

[m]Small and Medium-sized Enterprises Innovation Promotion Fund Project, managed by the Ministry of Agriculture, Forestry and Fisheries

[n]Novo Arranjo de Pesquisa e Inovação em Proteínas Alternativas, NAPI-PA

[o]See, for instance, CELLtoFOOD project, funded under Italy's National Recovery and Resilience Plan (PNRR) by the Ministry of University and Research (MUR). The project focuses on developing innovative technologies for sustainable food production, with a particular emphasis on cellular agriculture

alternative proteins more affordable for consumers (e.g. reduced Valued Added Tax (VAT) on plant-based milks in Germany and Czechia). Demand-side measures are fewer but crucial for consumer acceptance and market penetration, primarily through clear labelling and financial incentives (Table 2).

As useful as these instruments may be in supporting the development of alternative protein sources, it is also true that they are limited unless used as part of a coherent and comprehensive strategy. The weaknesses of such fragmentation are the subject of the next section.

4 The Role of Public Policy in Governing the Alternative Protein Value Chain

Analysis of the public policies adopted on alternative protein sources clearly shows expectations ranging from environmental benefits to food security, thus making value chain governance a multifaceted challenge.

Several factors contribute to the complexity of this issue, highlighting the need for comprehensive and forward-thinking public policies. Below are the main aspects proposed.

Although still in its early stages of development, cellular agriculture, including cell-based meat, could significantly contribute to reshaping food production and tackling global sustainability challenges. Governments that choose to ban cellular agriculture or refrain from intervening in its regulation and political direction are taking a risky stance. Inaction results in a regulatory moratorium that may cause significant issues. Without clear guidelines, there is a higher chance of unregulated practices, poor safety controls and monopolization by a few powerful corporations. The absence of government oversight may also hinder innovation by failing to provide the necessary support for research and fair competition. Given the potential

Table 2 Main policy types and examples

Policy type	Description
Research grants	Public funding for R&D in alternative proteins, including cultivated meat and precision fermentation.
Regulatory frameworks	Government-defined processes for approval, labelling and safety of cultured meat products.
Public funding and investment	Government investments, subsidies and loans for alternative protein start-ups and scale-ups.
Labelling policies	Rules on how cultivated and plant-based meat can be marketed and labelled.
Trade and export support	Policies supporting alternative proteins in national trade strategies.
Tax incentives	Lower tax rates or subsidies for alternative protein products.
Workforce and education initiatives	Programs to train workers and develop expertise in the alternative protein sector.

Source: Author's elaboration

of cellular agriculture in addressing food security and sustainability challenges, a lack of political direction and regulation could result in missed opportunities, ethical concerns and unforeseen consequences. Proactive public governance is therefore essential to ensure that cell agriculture develops in a way that is safe, equitable and beneficial for society as a whole. Historical precedents, such as the European Union's delay on GMO regulations and patents, show that legislative moratoria and avoiding technological support can result in commercial and legal disputes, particularly in the framework of international trade agreements under the World Trade Organization (WTO), as well as economic disadvantage.[79] The GMO experience also highlights that avoiding regulation does not necessarily prevent the entry of new technologies into domestic markets, as seen in European Union, where various GM agricultural commodities are marketed despite regulatory resistance.

The need for intervention is particularly pressing as the development of cultured meat demands substantial investments due to its high entry costs and the need for constant or increasing returns to scale.[80] In this context, smaller agribusinesses may struggle to secure government subsidies or access the highly technical expertise necessary for competing with well-capitalized multinational corporations, thus further entrenching the dominance of existing agribusiness giants. Additionally, the industry requires a highly skilled workforce, which tends to be consolidated in multinational corporations. Empirical evidence supports these concerns, as most start-ups operating in the cultured meat sector are in industrialized nations, often backed by established corporations such as Tyson Foods and Cargill.[81] This trend raises fears of increasing disparities between the global North and South, exacerbating existing inequalities.

Thus there is a risk of market concentration, where a few large firms gain disproportionate power to influence suppliers, consumers and industrial practices. This risk is exacerbated by the expected consolidation of companies, as some exhaust their funds or are acquired.[82] To avoid market failures, policymakers have to find a balance between different trade-offs, such as legal protection of inventions versus the need for access to innovation, or technological progress versus consumer health protection. The former trade-off is sustained by economic evidence that through patents, exclusive rights boost firms to invest in research and development, which in turn leads to greater market power, reducing consumer surplus and distorting investment incentives.[83] The latter concerns the dyscrasia arising from the gap between the time required for research to assess the food safety of new technologies and the time taken for their adoption by industries. Indeed, while research and regulatory approval processes are essential to ensure consumer safety, they can be

[79] See WTO—Dispute DS291/DS292/DS293. This case concerns the *de facto* moratorium imposed by the EU on GMO approvals between 1999 and 2003. The WTO ruled that the moratorium caused unjustified delays in the approval process, violating international trade rules.

[80] Rubio et al. (2020) and Treich (2021).

[81] Reis et al. (2020a), Santo et al. (2020), Mouat and Prince (2018) and Hocquette (2016).

[82] Tavan et al. (2025).

[83] Arrow (1962) and Bloom et al. (2019).

time-consuming. On the other hand, industries often push for rapid adoption to capitalize on innovation and market opportunities. Balancing these timelines is crucial for safety as well as for efficient technological progress in the food sector.

Again, recent history shows us that when governments fail to provide robust regulatory frameworks, the private sector often takes a dominant role in governance. This is exemplified by the rise of private standards such as Global Good Agricultural Practice (GAP) and Global Food Safety Initiative (GFSI), which have reshaped global food systems by influencing social, political and economic relationships on a transnational scale.[84] The growing influence of private entities in areas traditionally governed by the public sector—such as food safety, environmental stewardship, and ethical standards—raises critical concerns about transparency and ethical governance. For cellular agriculture, the shift from public to private governance could undermine the capacity of regulatory institutions to ensure an equitable distribution of benefits and foster inclusive participation in decision-making processes. Zimmermann[85] states that improving the quality of research on the environmental impact of cultured meat requires research that is at least partly publicly funded. Public policies must therefore be designed to pre-emptively address these gaps in governance by establishing precise accountability mechanisms and ensuring that ethical considerations are embedded in regulatory frameworks.

Nor can sustainability in the agri-food system be achieved without addressing the conflicting principles and practices spanning different technological, economic and ecological paradigms. The EU has acknowledged the need for an integrated governance model for food systems. In 2016, the European Economic and Social Committee (EESC) adopted an opinion titled "More sustainable food systems" advocating a comprehensive food policy to ensure better coherence and integration of food-related policy objectives.[86] This opinion emphasized the need for a unified approach to food policy, linking agricultural objectives to nutrition and the creation of supply chains that safeguard public health. Current policy responses have so far been inadequate in addressing the complex and interconnected challenges faced by European food systems. This inadequacy stems from fragmented governance structures and the proliferation of contradictory objectives across policy domains, such as agriculture, trade, environment, food safety, education and competition. A notable example is EU Common Agricultural Policy, which has well-documented limitations concerning sustainability and equity.[87] Gaps and inconsistencies between policies are widespread rather than isolated occurrences, often arising from conflicting objectives and priorities across various sectors. These contradictions create significant challenges in policy-making, particularly without effective mechanisms to reconcile competing goals. Addressing such complexities requires a comprehensive

---

[84] Busch and Bain (2004).

[85] Zimmermann et al. (2022).

[86] https://www.eesc.europa.eu/en/news-media/news/eesc-calls-comprehensive-food-policy-eu. Accessed 2 November 2024.

[87] IPES (2019).

policy framework capable of coordinating diverse actions over extended periods, free from the limitations imposed by short-term political cycles, such as those dictated by electoral terms.[88]

In this context, a well-designed food policy framework would clearly distinguish between short-term and long-term objectives, ensuring that immediate actions align with broader, long-term sustainability goals. It would also establish a structured approach to evaluating trade-offs, weighing the benefits and costs of different policy options over time.[89] This would involve assessing externalities[90] such as environmental degradation, public health impacts and social inequalities.

Without such a comprehensive and forward-looking approach, policy-making risks becoming reactive and fragmented, addressing only immediate concerns while leaving deeper systemic problems unresolved. As a result, societies remain locked in cycles of short-term fixes that fail to produce meaningful, long-lasting solutions.

The EU, by adopting a systems-based approach, should also address the heterogeneity of European food landscapes—from industrial agriculture to smallholder systems with rich cultural traditions—through policies that are context-specific and capable of balancing food security, environmental sustainability and economic viability.[91]

Indeed, a successful transition to sustainable food systems requires acknowledgement of local socio-economic and cultural specificities.[92] Rather than adopting a single, one-size-fits-all approach, it is more plausible to develop a pluralistic strategy that combines multiple approaches—including agroecology, plant-based proteins and emerging technologies, such as cell-based meat—based on context-specific factors on the supply and demand sides.

It is also important to highlight that while some scholars argue that alternative protein products inherently compete with one another,[93] the author believes that these innovations may play complementary roles, depending on local contexts and production capacities. Embracing only one approach is unrealistic; cell agriculture may rather be a component of a composite framework, provided that food policy can address it coherently.

A holistic food policy framework must accommodate this diversity while ensuring transparency and consumer trust through clear regulations and labelling practices. A harmonised and transparent EU-wide labelling framework would help resolve these discrepancies and ensure consumer protection and market regulation consistency. Currently, labelling is just one component of the wider debate regarding the adequacy of Regulation (EU) 2015/2283 on Novel Foods. The efficacy of this regulation has been questioned by several Member States in the Council of the

[88] De Schutter et al. (2020).

[89] Béné et al. (2019).

[90] Costs or benefits that affect third parties and are not reflected in market prices.

[91] Kanter et al. (2018) and Horlings and Marsden (2011).

[92] Béné et al. (2019).

[93] Sexton et al. (2019).

EU as well as by the EU Parliament,[94] thus rendering the development of a regulatory framework that effectively balances innovation and inclusivity a significant challenge to accomplish.

At a time when the alternative protein sector suffers a decline in private funding and most cultivated meat companies are in pre-revenue stage, still not having transitioned from pilot to commercial production, it is increasingly important for governments to step up and address disruptive innovations as tools for tackling societal needs, namely the sustainability challenge.

5 Conclusions

The transition towards alternative protein sources presents both significant opportunities and multifaceted challenges for reshaping global food systems. The rising demand for sustainable protein solutions underscores the urgency of developing innovative alternatives. Among these, cell-based meat emerges as a potentially transformative solution, capable of addressing critical issues such as greenhouse gas emissions, land use and animal welfare. However, the widespread adoption of this disruptive technology hinges on several factors, including technological advancements, regulatory frameworks, market acceptance and public policies.

Policymaking plays a pivotal role in steering the development of alternative protein sources towards broader societal goals. The integration of public funding, regulatory oversight and collaborative networks is crucial to fostering innovation while ensuring safety, inclusivity and equitable market access. However, the fragmented and often inconsistent regulatory landscape across countries presents a barrier to the harmonised advancement of these technologies. Addressing this requires the establishment of comprehensive policy frameworks that not only promote scientific research and economic incentives but also safeguard consumer interests and prevent market concentration.

The socioeconomic implications of scaling up cell-based meat production demand careful consideration. While the technology holds promise for enhancing food security and reducing environmental footprints, its potential to disrupt traditional livestock farming systems and rural economies cannot be overlooked. Policies must therefore balance technological progress with the protection of vulnerable communities, fostering a pluralistic approach that accommodates diverse food production models, including agroecology, plant-based proteins and cellular agriculture.

Ultimately, the future of alternative proteins depends on a coordinated, long-term governance strategy that aligns technological innovation with sustainability objectives and social inclusivity. By adopting integrated, forward-thinking policies, governments can harness the transformative potential of alternative protein sources to build more resilient, equitable and environmentally sustainable food systems in the decades to come.

---

[94] General Secretariat of the Council of the European Union (2024), p. 2. See also *Who Hates Cultured Meat? Examining Public Opposition to Cellular Agriculture* by Broad and Glenna in this volume; Formici (2024).

References

Altieri MA (2018) Agroecology: the science of sustainable agriculture, 2nd edn. CRC Press, Boca Raton
Andreani G, Sogari G, Marti A, Froldi F, Dagevos H, Martini D (2023) Plant-based meat alternatives: technological, nutritional, environmental, market, and social challenges and opportunities. Nutrients 15(2):452
Ardoin R, Prinyawiwatkul W (2021) Consumer perceptions of insect consumption: a review of Western research since 2015. Int J Food Sci Technol 56(10):4942–4958
Arrow K (1962) Economic welfare and the allocation of resources for invention. In: Universities-National Bureau Committee for Economic Research, Committee on Economic Growth of the Social Science Research Council (ed) The rate and direction of inventive activity: economic and social factors. Princeton University Press, Princeton, pp 609–626
Béné C, Oosterveer P, Lamotte L, Brouwer ID, de Haan S, Prager SD, Khoury CK (2019) When food systems meet sustainability – current narratives and implications for actions. World Dev 113:116–130
Bloom N, Van Reenen J, Williams H (2019) A toolkit of policies to promote innovation. J Econ Perspect 33(3):163–184
Boone L, Roldán-Ruiz I, Muylle H, Dewulf J (2019) Environmental sustainability of conventional and organic farming: accounting for ecosystem services in life cycle assessment. Sci Total Environ 695:133841
Bryant CJ (2020) Culture, meat, and cultured meat. J Anim Sci 98(8):1–7
Bryant CJ, Barnett J (2020) Consumer acceptance of cultured meat: an updated review (2018–2020). Appl Sci 10(15):5201
Busch L, Bain C (2004) New! improved? The transformation of the global agrifood system. Rural Sociol 69(3):321–346
Choudhury D, Tseng TW, Swartz E (2020) The business of cultured meat. Trends Biotechnol 38(6):573–577
Chriki S, Hocquette JF (2020) The myth of cultured meat: a review. Front Nutr 7:7
De Schutter O, Jacobs N, Clément C (2020) A 'Common Food Policy' for Europe: how governance reforms can spark a shift to healthy diets and sustainable food systems. Food Policy 96:101859
Dean T, Zhang H, Xiao Y (2022) The role of complexity in the Valley of Death and radical innovation performance. Technovation 109:102160
Dess GG, Beard DW (1984) Dimensions of organizational task environments. Acad Manag Rev 29(1):52–73
Diercks G, Larsen H, Steward F (2019) Transformative innovation policy: addressing variety in an emerging policy paradigm. Res Policy 48(4):880–894
European Commission (2020) Industrialization and commercialization of a competitive, sustainable and consumer oriented alternative animal protein source. https://cordis.europa.eu/project/id/958660. Accessed 09 Nov 2024
European Commission (2023a) A European green deal striving to be the first climate-neutral continent. https://commission.europa.eu/strategy-andpolicy/priorities2019-2024/europeangreen-deal_en. Accessed 29 Nov 2024
European Commission (2023b) Farm to fork strategy for a fair, healthy and environmentally-friendly food system. https://eur-lex.europa.eu/legal-content/EN/TXT/?uri=CELEX:52020DC0381. Accessed 29 Nov 2024
FAO (2013) Edible insects: future prospects for food and feed security. https://www.fao.org/4/i3253e/i3253e.pdf. Accessed 30 Dec 2024
FAO (2020) The state of agricultural commodity markets. Agricultural markets and sustainable development: global value chains, smallholder farmers and digital innovations. FAO, Rome
FAO and WHO (2023) Food safety aspects of cell-based food. Rome. https://doi.org/10.4060/cc4855en. Accessed 30 Dec 2024

Formici G (2024) Un assaggio di futuro? La carne a base cellulare e il vivace dibattito regolatorio dentro e fuori i confini dell'Unione europea. BioLaw J 4(24):165–197
Galli F, Favilli E, D'Amico S, Brunori G (2018) A transition towards sustainable food systems in Europe: food policy blueprints copying study. Laboratorio di Studi Rurali Sismondi, Pisa
General Secretariat of the Council of the European Union (2024) The CAP's role on safeguarding high-quality and primary farm-based food production. https://data.consilium.europa.eu/doc/document/ST-5469-2024-REV-1/en/pdf. Accessed 20 Mar 2025
Good Food Institute (2024) State of the industry report: cultivated meat and seafood. https://gfi.org/resource/cultivated-meat-and-seafood-state-of-the-industry-report/. Accessed 4 Nov 2024
Goodman D (2023) Transforming agriculture and foodways: the digital-molecular convergence. Bristol University Press, Bristol
Guareschi M, Mancini MC, Lottici C, Arfini F (2023) Strategies for the valorization of sustainable productions through an organic district model. Agroecol Sustain Food Syst 47(1):100–125
Haddad CR, Nakić V, Bergek A, Hellsmark H (2022) Transformative innovation policy: a systematic review. Environ Innov Soc Trans 43:14–40
Heller MC, Keoleian GA (2018) Beyond meat's beyond burger life cycle assessment: a detailed comparison. Center for Sustainable Systems, University of Michigan
Hocquette JF (2016) Is in vitro meat the solution for the future? Meat Sci 120:167–176
Horlings LG, Marsden TK (2011) Towards the real green revolution? Exploring the conceptual dimensions of a new ecological modernisation of agriculture that could 'feed the world'. Glob Environ Chang 21(2):441–452
Humbird D (2021) Scale-up economics for cultured meat: techno-economic analysis of future factory farming. Biotechnol Bioeng 118(8):3239–3250
IPES-Food (2019) Towards a Common Food Policy for the European Union: the policy reform and realignment required to build sustainable food systems in Europe. International Panel of Experts on Sustainable Food Systems, Brussels
Iqbal B, Alabbosh KF, Jalal A, Suboktagin S, Elboughdiri N (2025) Sustainable food systems transformation in the face of climate change: strategies, challenges, and policy implications. Food Sci Biotechnol 34:871–883
Kanter DR, Musumba M, Wood SL, Palm C, Antle J, Balvanera P, Andelman S (2018) Evaluating agricultural trade-offs in the age of sustainable development. Agr Syst 163:73–88
Lusk JL, Blaustein-Rejto D, Shah S, Tonsor GT (2022) Impact of plant-based meat alternatives on cattle inventories and greenhouse gas emissions. Environ Res Lett 17(2):024035
Mancini MC, Antonioli F (2022) The future of cultured meat between sustainability expectations and socio-economic challenges. In: Bhat R (ed) Future foods. Cambridge Academic Press, Cambridge, pp 331–350
Marques M, Morais-da-Silva RL, Biscarra-Bellio JC, Ueta MH, Maiolino Molento CF (2024) Global and regional policies for cultivated meat. In: Soccol CR, Maiolino Molento CF, Glufke Reis G, Karp SG (eds) Cultivated meat. Springer Nature, Cham, pp 359–384
Mayhall TA (2019) The meat of the matter: regulating a laboratory-grown alternative. Food Drug Law J 74:151
Mouat MJ, Prince R (2018) Cultured meat and cowless milk: on making markets for animal-free food. J Cult Econ 11(4):315–329
Mouratiadou I, Wezel A, Kamilia K, Marchetti A, Paracchini ML, Bàrberi P (2024) The socio-economic performance of agroecology. A review. Agron Sustain Dev 44(2):19
Mylan J, Andrews J, Maye D (2023) The big business of sustainable food production and consumption: exploring the transition to alternative proteins. Proc Natl Acad Sci 120(47):e2207782120
Newton P, Blaustein-Rejto D (2021) Social and economic opportunities and challenges of plant-based and cultured meat for rural producers in the US. Front Sustain Food Syst 5:624270
Nyyssölä A, Suhonen A, Ritala A, Oksman-Caldentey KM (2022) The role of single cell protein in cellular agriculture. Curr Opin Biotechnol 75:102686
OECD (2021) Agricultural policy monitoring and evaluation 2021: addressing the challenges facing food systems. OECD Publishing, Paris

Ordoñez-Araque R, Egas-Montenegro E (2021) Edible insects: a food alternative for the sustainable development of the planet. Int J Gastron Food Sci 23:100304

Pakseresht A, Kaliji SA, Canavari M (2022) Review of factors affecting consumer acceptance of cultured meat. Appetite 170:105829

Petersen T, Hartmann M, Hirsch S (2021) Which meat (substitute) to buy? Is front of package information reliable to identify the healthier and more natural choice? Food Qual Prefer 94:104298

Pojić M, Mišan A, Tiwari B (2018) Eco-innovative technologies for extraction of proteins for human consumption from renewable protein sources of plant origin. Trends Food Sci Technol 75:93–104

Reis GG, Heidemann MS, de Matos KHO, Maiolino Molento CF (2020a) Cell-based meat and firms' environmental strategies: new rationales as per available literature. Sustainability 12(22):9418

Reis GG, Heidemann MS, Borini FM, Maiolino Molento CF (2020b) Livestock value chain in transition: cultivated (cell-based) meat and the need for breakthrough capabilities. Technol Soc 62:101286

Rubio NR, Xiang N, Kaplan DL (2020) Plant-based and cell-based approaches to meat production. Nat Commun 11(1):6276

Santo RE, Kim BF, Goldman SE, Dutkiewicz J, Biehl EMB, Bloem MW, Nachman KE (2020) Considering plant-based meat substitutes and cell-based meats: a public health and food systems perspective. Front Sustain Food Syst 4:134

Schot J, Steinmueller WE (2018) Three frames for innovation policy: R&D, systems of innovation and transformative change. Res Policy 47(9):1554–1567

Sexton AE, Garnett T, Lorimer J (2019) Framing the future of food: the contested promises of alternative proteins. Environ Plann E Nat Space 2(1):47–72

Sogari G, Riccioli F, Moruzzo R, Menozzi D, Sosa DAT, Li J, Mancini S (2023) Engaging in entomophagy: the role of food neophobia and disgust between insect and non-insect eaters. Food Qual Prefer 104:104764

Soice E, Johnston J (2021) How cellular agriculture systems can promote food security. Front Sustain Food Syst 5:753996

Srinivas H (2023) Economic instruments for environmental policy: solving global environmental problems with local economic action. GDRC Research Output - Policy Analysis Series E-190. Global Development Research Center, Kobe. https://www.gdrc.org/sustbiz/econ-instruments/index.html. Accessed 27 Nov 2024

Stephens N, Di Silvio L, Dunsford I, Ellis M, Glencross A, Sexton A (2018) Bringing cultured meat to market: technical, socio-political, and regulatory challenges in cellular agriculture. Trends Food Sci Technol 78:155–166

Tavan M, Smith NW, McNabb WC, Wood P (2025) Reassessing the sustainability promise of cultured meat: a critical review with new data perspectives. Crit Rev Food Sci Nutr:1–9

Treich N (2021) Cultured meat: promises and challenges. Environ Resour Econ 79:33–61

Tuomisto H, Ryynänen T (2024) Environmental impacts of cultivated meat. In: Soccol CR, Maiolino Molento CF, Glufke Reis G, Karp SG (eds) Cultivated meat. Springer Nature, Cham, pp 277–297

Tuomisto HL, Teixeira de Mattos MJ (2011) Environmental impacts of cultured meat production. Environ Sci Technol 45(14):6117–6123

Tuomisto HL, Ellis MJ, Haastrup P (2015) Environmental impacts of cultured meat: alternative production scenarios. In: Schenck R, Huizenga D (eds) Conference proceedings. Proceedings of the 9th international conference on life cycle assessment in the agri-food sector. ACLCA, Vashon, pp 1360–1366

Tuomisto HL, Allan SJ, Ellis MJ (2022) Prospective life cycle assessment of a bioprocess design for cultured meat production in hollow fiber bioreactors. Sci Total Environ 851:158051

Turnheim B, Geels FW (2012) Regime destabilisation as the flipside of energy transitions: lessons from the history of the British coal industry (1913–1997). Energy Policy 50:35–49

Tushman ML, Nelson RR (1990) Introduction: technology, organizations, and innovation. Adm Sci Q 35(1):1–8

Van der Ploeg JD, Barjolle D, Bruil J, Brunori G, Madureira LMC, Dessein J, Wezel A (2019) The economic potential of agroecology: empirical evidence from Europe. J Rural Stud 71:46–61

Van der Weele C, Tramper J (2014) Cultured meat: every village its own factory? Trends Biotechnol 32(6):294–296

Wood P, Thorrez L, Hocquette JF, Troy D, Gagaoua M (2023) Cellular agriculture: Current gaps between facts and claims regarding "cell-based meat". Anim Front 13(2):68–74

World Economic Forum (2019) Meat: the future – a roadmap for delivering 21st-century protein. White paper. https://www3.weforum.org/docs/WEF_White_Paper_Roadmap_Protein.pdf. Accessed 29 Nov 2024

Zimmermann AW, Langhorst T, Moni S, Schaidle JA, Bensebaa F, Bardow A (2022) Life-cycle and techno-economic assessment of early-stage carbon capture and utilization technologies: a discussion of current challenges and best practices. Front Clim 4:841907

Animal Welfare as a Constitutional Value: The Debate in Comparative Law and the Emerging Role of Cultivated Meat

Cinzia Piciocchi

Abstract The chapter examines the process of legal recognition in the protection of animals, with a particular focus on constitutional law. After a brief analysis of comparative constitutional jurisprudence, especially regarding the potential to grant rights to animals, the various approaches to animal rights are explored. Specifically, the welfarist and dignitarian approaches are discussed, with a specific focus on their relevance to the current debate over cultivated meat and their implications for animal rights. The potential impact of cultivated meat in challenging traditional food systems and prompting a re-evaluation of the ethics surrounding animal exploitation and animal welfare is considered.

Keywords Constitutional law · Animals' Habeas Corpus · Constitutional jurisprudence · Welfarism/dignitarism · Cultivated meat

1 Introduction

The law does not regulate all behaviours and potential human interactions. In our daily lives, for example, we follow various rules: we greet one another, dress and speak in ways appropriate to the context, behaviours (for the most part) not governed by legal systems but by social norms. The perception of whether a behaviour is public or private evolves over time, and as such, what is regulated by law can also change. Consider, for instance, the abolition of so-called "honour crimes," alongside the increasing influence of scientific advances in genetics, reproductive health and social sciences, which have reshaped our understanding of human rights and individual autonomy. Certain matters are now regulated, though they were not in the

C. Piciocchi (✉)
Trento University, Faculty of Law, Trento, Italy
e-mail: cinzia.piciocchi@unitn.it

G. Formici et al. (eds.), *Cell-Based Meat in the European Union and Beyond*,
https://doi.org/10.1007/978-3-032-14624-3_8

past, simply because they did not exist before: it's not hard to find examples when considering scientific and technological advances, which create new realities and new choices for individuals that require legal frameworks. There are many concrete examples: from options in scientific research to the possibilities enabled by advances in artificial intelligence and progress in medicine. These advances have become embedded in daily life, from the most routine aspects (such as the opportunities created by apps) to the most existential (such as reproductive choices or end-of-life decisions), presenting a clear need for regulation. The legal literature has documented all these changes: by reviewing recent scholarly works, one can observe an increasingly interdisciplinary approach to analysing many topics, with a growing awareness of the need for a comprehensive perspective to fully grasp the various complexities involved.

Sometimes, the "new" areas where the law intervenes do not emerge from events that previously did not exist, but rather from existing phenomena that begin to be "seen" in light of changing social conditions. Examples can be drawn from various fields: regulation of the legal consequences of choices in affective matters (such as marriage, cohabitation, parenthood, etc.) are products of their time, largely influenced by social acceptance of relationships, choices and statuses that, if not recognized, could face ostracism and marginalisation. Similarly, many anti-discrimination laws protecting groups, minorities or regulations addressing individual vulnerabilities do not arise from the creation of categories that did not exist before, but rather from acknowledgment of their existence, and crucially, their needs.

The legal regulation of animals is not a recent development: animals have been an integral part of human history. They have stood alongside humans since the earliest cave paintings, which are evidence of their presence.[1] The need for regulatory norms is therefore intrinsic, animals being an integral part of human life. What has evolved over time is the recognition of animal welfare as a value to be protected for its own sake. This has led to the development of laws that prioritize the protection of animals, irrespective of their use or utility to humans. This cultural shift initially took root in philosophy and was later progressively adopted in legal frameworks.[2] Initially, the legal recognition of enhanced animal protection appeared in specific sectors (such as farming, scientific use and transport) before gradually pervading international and supranational law from a broader perspective. In this context, international declarations, particularly the Treaty on the Functioning of the European Union (TFEU) as amended by the Lisbon Treaty, are noteworthy.[3] The TFEU

---

[1] Regarding the relationship between animals and the law in antiquity, see for example Oylan and Rosenblum (2021).

[2] Cao and White (2016), in particular p. 111.

[3] See Article 13 of the Treaty on the Functioning of the European Union (TFEU, consolidated version, in OJ C 202 7.6.2016, p. 47): "In formulating and implementing the Union's agriculture, fisheries, transport, internal market, research and technological development and space policies, the Union and the Member States shall, since animals are sentient beings, pay full regard to the welfare requirements of animals, while respecting the legislative or administrative provisions and customs of the Member States relating in particular to religious rites, cultural traditions and

notably enshrined the recognition of animals as sentient beings, a pivotal definition that underpins their legal status.

At state level, normative sources addressing specific issues typically preceded constitutional provisions. The inclusion of animals in constitutional texts is in fact a relatively recent development, representing a significant milestone that elevates the protection of animal welfare to its highest form at the top of legal sources.

Here I examine the constitutional significance of animal welfare from a comparative law perspective, exploring potential links to the factual reality of cultivated meat, made possible by scientific advancement. This topic could have profound implications for the legal protection of animals. In Sect. 2, I explore the constitutional recognition of animals, analysing the values protected (such as dignity and welfare) and how this is done. In Sect. 3, I assess how the issue of cultivated meat aligns with these values (or otherwise), offering a critical review of the relevant doctrinal perspectives.

2 Animals in Constitutions

As previously mentioned, the inclusion of animals in constitutional texts is a relatively recent development. This does not imply that animals were not subject to legal regulation in the past, but rather that such regulation typically occurred through primary and secondary sources, often focused on specific areas. One of the earliest examples of constitutional intervention on this issue is Switzerland, where a reference to animals was added to the Constitution in 1973 following a referendum.[4] Since then, several other countries have followed suit, including Egypt (2014), Italy (2022) and Germany (2002).[5] It is also worth noting that both newdrafts of the Chilean Constitution (2022 and 2023; which ultimately did not pass) included reference to animals.[6]

Constitutional provisions on animal protection vary across countries, reflecting different structures and approaches. Some associate animal protection with the preservation of ecosystems, while others include the rights of future generations,

regional heritage". Regarding the protection of animals at international level, see Cao and White (2016), Kelch (2017) and Chapaux et al. (2023).

[4] See the *Arrêté fédéral concernant un article sur la protection des animaux qui remplace l'article 25bis actuel de la constitution fédérale* (27 June 1973), FF 1973 I 1633 on which Bolliger (2016). https://www.fedlex.admin.ch/eli/fga/1973/1_1685_1633_1397/fr. Accessed 4 April 2025.

[5] Examples of constitutional provisions regarding animals can be found at https://animal.law.harvard.edu/. Accessed 4 April 2025.

[6] Regarding the protection of animals in the 2022 and 2023 draft Constitutions of Chile, see Cerda-Guzman and Insunza (2024). From a more general perspective, with regard also to the constitutional amendments concerning animal protection, see Le Bot (2024).

sometimes combining all three elements in a single provision.[7] The scope of protection also differs in the constitutional articles of various countries: some refer to general animal protection (e.g. Austria and Brazil), while others emphasize promoting and safeguarding animal welfare, such as Article 41 of the Luxembourg Constitution, which focuses on the protection of their welfare.

The Slovenian Constitution (1991) Article 72(4) specifically protects animals from torture, stating, "The protection of animals against torture shall be regulated by law." The Egyptian Constitution calls for "kind treatment" of animals, which is similar in phrasing to the "compassion for living creatures" in Article 51A of the Indian Constitution. Likewise, the Russian Constitution (2020, Article 114(f5)) mandates that the federal government "shall undertake measures aimed at forming a responsible attitude in society towards animals".[8]

It is also important to note that some of these constitutional provisions have been amended, sometimes affecting the values being protected. For example, the Swiss Constitution was amended in 1992 after a referendum to include reference to dignity of living beings, a concept that has since been interpreted in various ways in its application.[9] The same principle (dignity) was later reaffirmed in Swiss law with the adoption of the Animal Welfare Act in 2005, or even earlier in the Swiss Civil Code, which explicitly stated in 2003 that animals are not things. However, legal scholars

---

[7] This being the case, for example, of the Italian constitution, whose Article 9, para 3 was amended by Constitutional Law of 11 February 2022, No. 1, (in Official Journal no. 44 of 22 February 2022): "The Republic [...] shall safeguard the environment, biodiversity and ecosystems, also in the interest of future generations. State law shall regulate the methods and means of safeguarding animals" (English translation available at https://www.cortecostituzionale.it/. Accessed 4 April 2025). See also Article 20a of the German Constitution: "Protection of the natural foundations of life and animals. Mindful also of its responsibility towards future generations, the state shall protect the natural foundations of life and animals by legislation and, in accordance with law and justice, by executive and judicial action, all within the framework of the constitutional order" (English translation available at https://www.gesetze-im-internet.de/. Accessed 4 April 2025). See also Van Hoorick and Veriners (2024), p. 227.

[8] Eisen (2017) and Le Bot (2023). English translations taken from https://animal.law.harvard.edu/what-we-do/projects/constitutional-law/. Accessed 4 April 2025.

[9] Art. 120 of the Swiss Constitution provides: "Human beings and their environment shall be protected against the misuse of gene technology. The Confederation shall legislate on the use of reproductive and genetic material from animals, plants and other organisms. In doing so, it shall take account of the dignity of living beings as well as the safety of human beings, animals and the environment, and shall protect the genetic diversity of animal and plant species" (English translation available at https://www.fedlex.admin.ch/eli/cc/1999/404/en#fnbck-d7e4606. Accessed 4 April 2025). The data related to the referendum are available at this web address https://www.bk.admin.ch/ch/i/pore/va/19920517/index.html. Accessed 4 April 2025; while the data regarding the voting in the individual cantons can be found here https://www.bk.admin.ch/ch/i/pore/va/19920517/det378.html. Accessed 4 April 2025. See also Michel and Kayasseh (2011); Toomey (2020); Giménez-Candela (2018), p. 22; Kurki (2024), also mentioning other Constitutions or statutes that refer to animal dignity, such as those of Belgium, Luxembourg, Denmark, Finland and the Americas.

have pointed out that the actual application of dignity in relation to animals has not been consistent.[10]

Given the recent nature of some of these reforms (such as the amendment to Article 9 of the Italian Constitution in 2022),[11] jurisprudence—particularly that of the constitutional courts—will play a key role in shaping their interpretation, and the meaning will become clearer with time.[12]

Beyond the specific terminology used to define the object of protection in the Constitutions of different countries, the literature has highlighted certain cross-cutting values that from a comparative perspective, seem to be common across various legal systems when it comes to the constitutional protection of animals: welfare and dignity. The relationship between these two concepts has drawn significant attention from scholars, who identify them as the basis for two distinct approaches—welfarism and dignitarianism—both aimed at guiding animal protection, though from different perspectives.[13] While welfarism focuses primarily on ensuring animal welfare by preventing suffering, pain or distress, dignitarianism emphasizes respecting the animal as an individual, independent of mistreatment or suffering, and recognizing essential aspects of its identity.[14]

Both concepts attract significant scholarly attention and it is precisely in the debates exploring the relationship between welfarism and dignitarianism that insights, valuable for understanding the implications of the constitutionalization of animal protection, arise.[15] One line of reflection focuses on the parallels that can be drawn with human rights: the differences between animal rights and human rights in constitutional frameworks are a broad theme, the contours of which are still being defined. Given the recent inclusion of animals in constitutions, some aspects remain the subject of ongoing debate. These discussions primarily address two issues: the direct holding of rights and their balancing.

From the first perspective, certain organizations have been actively working to extend direct rights to certain animals, such as granting *habeas corpus* to specific species (mainly large mammals such as primates and elephants). Concerning fundamental rights, these activities target various levels, seeking both the introduction of

[10] Blattner (2019) who underlines that while animal dignity has "become a bedrock principle of Swiss animal law", as a value it has "been applied to a few case groups", p. 331.

[11] The constitutional amendment is reported above at note 7. On the reform, see Rescigno (2023).

[12] For example, a case much cited in the literature was decided by the South African Constitutional Court: the Court sought to recognise the intrinsic value of animals, CCSA (8 December 2016) CCT 1/16 *National Society for the Prevention of Cruelty to Animals v Minister of Justice and Constitutional Development and Another*. See also Bilchitz and Wilson (2022).

[13] Kurki (2024).

[14] See for example Perry (2023) and Fasel (2019).

[15] Of course, welfarism and dignitarianism show different possible gradations and the intersections between them can also have different dynamics. A synthesis of these concepts has been proposed in these terms by Kurki (2024): "My overall claim can be put as follows: the notion of dignity is used in animal law discourse to express claims of *more-than-welfarism*. Whereas welfarism is premised on the humane treatment of animals and the prevention of unnecessary suffering, animal dignity is premised on respect for animals".

direct fundamental rights for animals into Constitutions and the initiation of legal actions to have such rights recognized by courts. One example of the first type is a referendum proposed in Switzerland, which aimed to introduce recognition of the rights of non-human primates to life and physical and mental integrity into the Constitution of the Basel-City canton. The referendum was promoted by the opinion group "Sentience Politics" and held in February 2022. Its outcome was negative: participation was 51% with 25.5% of voters in favour.[16]

On the judicial front, there is a notable diversification of outcomes, ranging from rulings in favour (a minority, such as in Pakistan and Argentina),[17] to rejections (currently the majority, especially in the United States, where the *Nonhuman Rights Project* has long been promoting lawsuits aimed at securing such recognition).[18]

It should be clarified that the issue of direct rights of animals must be distinguished from the protection of animals *per se*, which is already ensured by sector-specific legislation. The debate does not therefore concern the legal protection of animals, but rather extension of direct rights to them. The arguments in support of this extension are often presented by associations advocating for it: the evolution of fundamental rights, which views the expansion of rights—from those who had none to those who gain them—as a step forward in the development of civilization. In other words, the categories frequently cited in this process of evolution are women, slaves and minors, who, it is argued, did not have all or some of the rights once, but today these rights are considered inherent to all human beings, grounded in the principles of equality and human dignity, and in line with the post-revolutionary declarations of fundamental rights.[19] The courts, however, hesitate on several fronts regarding this recognition, offering further arguments: from the difficulty of associating responsibilities with rights in the case of animals, to the complexity of the consequences such recognition might bring. This very complexity (a "labyrinth of consequences," as phrased in one ruling) seems to be the general basis for the courts' reluctance, beyond the specific arguments put forth.[20] This reluctance can be inter-

[16] The initiative is reported at https://sentience.ch/en/. Accessed 4 April 2025. See also Blattner and Fasel (2022).

[17] See the cases reported by Jaleel (2024) and Moore (2023).

[18] See the numerous cases reported on the website www.nonhumanrights.org. Accessed 4 April 2025.

[19] See for example the US District Court for the Northern District of California (24 April 2018) in the *Nonhuman Rights Project Statement on Naruto v. Slater*: "The court also claimed implicitly and repeatedly that the word "person" is synonymous with "human being," when that has never been true, is not true today and will never be true. For centuries many humans beings, including slaves, women, children, Jews and fetuses were not 'persons', but 'things', while around the world, courts and legislatures have characterized a chimpanzee, a bear, a national park, a river, a Hindu idol, a mosque, the Amazon rain forest and other nonhumans, as 'persons'." https://www.nonhumanrights.org/blog/nhrp-statement-monkey-selfie-case/. Accessed 4 April 2025.

[20] See Court of Appeals of New York (14 June 2022) Slip Op. 385, *Nonhuman Rights Project, Inc. v. Breheny* in which the Court of Appeals recalls the words of the Appellate Division (which ruled on the same dispute): "The Appellate Division also cautioned that a judicial determination that nonhuman animals are legal "persons" would "lead to a labyrinth of questions that common-law

preted as an awareness of the potential intertwining effects of such recognition, which due to its complexity and partially unpredictable nature, would find a more appropriate place in the legislative framework. Such awareness may thus be primarily linked to the inadequacy of the source (jurisprudence), given the intricate series of intersecting effects that would be better addressed in the domain of legislative authority.

While the issue of direct rights is a relatively recent frontier concept, accompanied by various challenges (such as the representation of animals in court, which remains essential regardless of whether they directly hold rights), another key view is the balancing of competing interests. In this regard, the fundamental characteristics of the current framework of animal rights and the primary differences with human rights become clear.

Firstly, animal rights can be subject to a balancing of interests, even in areas where this would be impossible for humans. One example is the possibility of animals being bought and sold—a practice that, thankfully, is unthinkable for humans but remains permissible for animals, albeit with respect for their sentience. For example, while it is legal to purchase a pet, it is not permissible to harm or kill it, unlike an inanimate object; it is possible to buy a chair and destroy it without legal repercussions, but purchasing an animal comes with the responsibility to treat it humanely, and harming or killing it arbitrarily is prohibited. The right to life, which is unconditional for humans, can, in certain circumstances, be denied to animals, provided a valid justification can be demonstrated. A clear example of this is the use of animals in the food industry, a sector deeply influenced by the cultural traditions and practices of different countries. Social changes can significantly impact this, particularly in the case of individuals whose diets are dictated by religious or other principles (such as vegans or vegetarians). The spread of such dietary preferences can affect consumer demand and also lead to lobbying efforts aimed at securing legal provisions that accommodate these choices (e.g. in schools, prisons, hospitals, etc.), inevitably influencing the market and judicial disputes as well. The cultural changes underlying food choices are and will be fundamental in shaping the supply chain that involves animals, profoundly influencing the meat market and, consequently, the topic of slaughter and cultivated meat.[21]

The legal debate also prompts consideration of another foundational principle of constitutional law: equality. The constitutionally protected rights of individuals are recognized in an egalitarian framework, reinforced by the principle of dignity, which especially since the post-World War II era, has solidified the equality of all human beings, without distinction.[22] This, however, does not apply to animals, whose status is influenced by a variety of factors: their level of development and sentience, as well as the perceived "closeness" to humans, including the role of

processes are ill-equipped to answer," noting that "the decisions of whether and how to integrate other species into legal constructs designed for humans is a matter better suited to the legislative process"".

[21] On the cultural relevance of food choices, see materials and references in Piciocchi (2024).

[22] For all, see the seminal volume by Barak (2015).

emotional attachment to certain animals, such as pets. These criteria are informed by science, but social perceptions also play a key role. For example, social and legal activism often focuses on animals with higher cognitive development. This can be seen in legal actions aimed at granting direct rights, which typically—as seen above—involve large mammals, such as elephants and primates. Similarly, movements like the "Great Apes Project" have sometimes resulted in specific legal consequences: in New Zealand, for instance, a legislative ban was introduced on the scientific use of great apes.[23]

In this regard, direct rights and equality appear to be closely linked, this connection being reflected in the hesitations surrounding the granting of direct rights, as noted by some courts. These concerns can be framed as a "slippery slope" of rights: if, for example, habeas corpus is extended to great apes, would the extension of such rights to cattle or pigs, based on the principle of equality, be a logical next step? A positive answer to this question would undoubtedly have a profound impact on the agri-food sector, with significant economic and labour-related repercussions.

In areas where constitutional protection for animals is applied, there is often a need to balance animal welfare with human needs. For example, their lives are protected, but with the understanding that, in some cases, animals may be sacrificed (e.g. in the food industry); their well-being is ensured, even though they may be used in certain contexts (e.g. circuses or zoos) and so on. It is important to emphasize that even in practices where the use of animals is accepted, sector-specific regulations are in place to safeguard animal welfare. The suffering of sentient beings is therefore regarded as a value to be protected, even when their death is accepted as a necessary outcome due to their intended use.[24] From a general perspective, the balancing of opposing needs, both constitutionally protected, is a well-established and widely used judicial technique, forming an integral part of comparative constitutional jurisprudence. However, in cases involving the constitutional protection of animals, the "core" around which the balancing of interests takes place can often be identified in the concept of necessity. Indeed, the definition of necessity frequently appears in legal acts and jurisprudence, signifying that the use of animals, still permitted in certain areas, is contingent on the existence of a need, the definition of which, like much of the law on this issue, is still being defined.

What exactly does "necessity" mean? Does it refer to the use of animals in the food industry? In entertainment activities? Does the tradition of a people constitute a form of necessity, even when the practices associated with it could be modified?

[23] See the website of the Great Apes Project https://www.projetogap.org/en and the list of countries that have implemented restrictions regarding experimentation with great apes provided on the R&R (Release & Restitution) project website at the following address: https://www.releasechimps.org/laws/international-bans. Both accessed 4 April 2025. See also the New Zealand Animal Welfare Act (1999 No 142) available at https://www.legislation.govt.nz/act/public/1999/0142/latest/whole.html (accessed 4 April 2025) defining "non-human hominid means any non-human member of the family Hominidae, being a gorilla, chimpanzee, bonobo, or orangutan" and providing that "the use of the non-human hominid in the research, testing, or teaching is in the best interests of the non-human hominid."

[24] See for example the consideration made by Nattrass (2004).

And religion? The perception of necessity, in fact, can be tied to complex factors, blending customs and traditions of a country. This is not only in reference to food habits, which are deeply intertwined with the history of the country in which they are practiced, but also cultural manifestations in a broader sense, which reflect rooted traditions, including, though not exclusively, religious practices.[25]

It is not coincidental, in the opinion of the author, that comparative jurisprudence provides several examples of interventions against cultural practices, often deeply rooted in the traditions of countries, where the necessity of such practices does not seem sufficient to justify the sacrifice or suffering of animals. In this regard, several recurring examples in the literature can be highlighted: *Jallikattu* (a practice widespread in certain Indian territories, involving bulls that a crowd of people attempts to grab), which has been deemed contrary to animal welfare, and for this reason, the subject of an unfavourable ruling in 2014. While balancing these practices against animal welfare, the Indian Supreme Court ruled that the latter should take precedence, and that in the absence of any necessity, the welfare of bulls should not be ignored solely for human pleasure.[26] Similar considerations have been made by the Brazilian Federal Supreme Court, which has ruled against certain traditional practices deemed contrary to animal welfare, such as *farra do boi* (which involves the violent killing of cattle),[27] cockfighting[28] and the *vaquejada*, a type of rodeo considered violent.[29] Cultural and religious traditions fall under constitutional protection: from freedom of religion to freedom of expression, as well as the protection of groups and minorities whose identity is often closely tied to rituals and traditions involving animals. In addition to these aspects, there are considerations surrounding freedom of economic initiative: it has been noted that some of the events debated, even in judicial disputes, also have an economic dimension, as they represent shared gatherings of many people with significant impact on the local economy. Thus various interests come into play, all constitutionally protected, but an evolving concept of necessity emerges as the key factor in determining which prevails in each case.

This change in the concept of necessity is likely to have a significant impact on the constitutional value of animal protection, even more than the question of direct rights, in my opinion. From a very general perspective, we are witnessing the development of a concept that represents the "animal condition" at the constitutional level, and appears to be moving toward a synthesis of the two concepts mentioned above (welfarism and dignitarianism). Particular importance seems to be placed on

---

[25] See *Food Justice in Religious Laws: Hypotheses on Cultivated Meat* by Anello in this volume.

[26] Supreme Court of India (7 May 2014) 2014 (7) SCC 547, *Animal Welfare Board of India v. A. Nagaraja and Others*.

[27] See the decision of the Supremo Tribunal Federal of Brasil (3 June 1997) n. 153.531-8/SC, *Recurso Extraordinário*, on which Frota Pestana de Aguiar (2021).

[28] See the two decisions of the Supremo Tribunal Federal of Brasil (26 May 2011) n. 1856, *Briga de Galos, Lei Fluminense no. 2.895/98* and Supremo Tribunal Federal of Brasil (29 June 2005) n. 2.514, *Lei n. 11.366 do Estado de Santa Catarina*, on which Kohler (2019).

[29] See the decision of Tribunal Supremo Federal of Brasil STF, n. 24,165/2016, *Ação direta de inconstitucionalidade*, on which Neto (2018), p. 172.

respecting the ethological characteristics of animals, which is not necessarily limited to merely refraining from behaviours harmful to animal welfare, but also involves respecting the species in question and the animals' inherent traits. The references proposed in literature point, for example, to actions that humiliate animals (e.g. circus performances), or even to respecting their inanimate bodies: a theme that is indeed comparable to the protection of the inanimate human body, where the issue of dignity plays a significant role.[30]

3 Cultivated Meat and Constitutional Protection of Animals

The relationship between the constitutional protection of animals and the prospects that cultivated meat offers for their welfare involve several complex elements. Primarily, both topics are still evolving. The constitutional protection of animals and their welfare has already been discussed: amendments to constitutions to include animals occur at different times, but understanding their full implications requires an analysis of judicial, administrative and private-sector applications. Over time, we will be able to determine how and in what areas these principles have been (or will be) applied. Cultivated meat, too, is an evolving subject: although research in this field dates back some time, its industrial and commercial applications are still in their early stages for several reasons. Firstly, commercialization remains limited; for example, entry into the European market will be contingent on approvals from food safety authorities, as is the case for all Novel Foods.[31] Secondly, different regulations pose challenges: some countries, such as Italy and Hungary, have already imposed or are trying to impose bans or restrictions on the future commercialization of cultivated meat. These regulations are designed for the future, regulating not what exists but what is to come.[32] Finally, the cost factor plays a significant role in the spread and commercialization of cultivated meat, impacting its availability to consumers, alongside the cultural acceptance of this food, which will play an equally important role.[33] Many of the considerations that can be made regarding the

[30] Foster (2011).

[31] See *Cell-Based Meat in the European Union: A Regulatory Crossroads* by Corazza and Formici in this volume.

[32] See the Italian law: *Disposizioni in materia di divieto di produzione e di immissione sul mercato di alimenti e mangimi costituiti, isolati o prodotti a partire da colture cellulari o di tessuti derivanti da animali vertebrati nonché di divieto della denominazione di carne per prodotti trasformati contenenti proteine vegetali* (in OJ no. 281, 1 December 2023), on which Soccol et al. (2024) and Lanzoni et al. (2024). The Hungarian ban on cultivated meat was notified and rejected by the European Union in 2024 (see the Communication from the Commission TRIS/(2024) 1869, Directive 2015/1535 notification 2024/0394). For a comparative analysis of the legal bans regarding cultivated meat, see Fino et al. (2024). See also *Cell-Based Meat in the European Union: A Regulatory Crossroads* by Corazza and Formici; *The Legal Debate on Cultured Meat Around the World: Comparative Insights from the United States, Singapore and Israel* by Pitto in this volume.

[33] See *Cell-Based Meat and Food Policy,* by Mancini in this volume.

impact of cultivated meat on animal welfare, therefore, focus on future and potential scenarios.

In particular, definitions seem to play a very important role: a theme that is traditionally foundational in the legal field, where falling under one definition or another has precise legal consequences. The possibility of relating cultured meat to the different definitions of "food", "meat", etc., presents issues not dissimilar to those of Novel Foods, with the same intertwining of scientific and legal considerations.

The range of terms used for cultivated meat, however, has also a somewhat different relevance, since more than on the legal plane (which must indeed be based on scientific elements to establish a definition), these terms are aimed at consumers and may determine the potential success and circulation of the product. Here, we are therefore more on a cultural and social level than on a legal one.[34] This is an aspect that is naturally not without consequences in the political arena, which inevitably absorbs social demands and influences the laws that Parliament chooses to pass. From this perspective, it is likely that "Frankenstein meat" will have less success than "clean meat", though both terms refer to the same product. In fact, there is a significant body of literature addressing the definitions of this food product and its potential consequences.[35]

While the legal relevance of this issue, as mentioned, is only tangential, it is important to highlight the reference to another concept, "nature", which represents a principle explored in constitutional law. Its broad scope, however, makes it subject to interpretations that can vary significantly.[36] Reference to nature in the legal field is often ambiguous: beyond "nature" referring to the ecosystem, which pertains to an object with a more univocal meaning, the concept of naturalness is invoked in relation to death, procreation, family and so on. According to these arguments, there is a way to die, procreate and form a family that follows the dictates of nature, while behaviours that deviate from these modes are against nature. What goes against these notions of nature is stigmatized, with different aspects of nature being invoked in relation to specific issues. When trying to understand what is meant by this concept, nature ends up taking on different meanings, depending on the boundaries drawn between it and what is presented as contrary to nature. Science and medicine, for example, can be seen as two fields inherently opposed to "nature": the fight against cells that begin to grow abnormally in the human body, for instance, can be viewed as unnatural use of pharmaceutical therapies, just as surgery to remove them is seen as even less natural. Similarly, it is difficult to identify the "naturalness" of the objects that surround us every day, from electric light bulbs to mobile phones and computers. Even in relation to cultivated meat, the concept of nature sometimes

[34] Pasca and Arcese (2025).

[35] On the definitions of cultivated meat and their importance, see Bryant and Barnett (2019) and Dilworth and McGregor (2015).

[36] See *The Regulatory Landscape of Cultured Meat and Fish: Terminology and Consumer Information Provisions* by Rolandi and Sirsi; *Who Hates Cultured Meat? Examining Public Opposition to Cellular Agriculture* by Broad and Glenna in this volume.

emerges in the narrative of this "bio-object" (if it can so be called)[37] and the product that will be (or in some cases already is) introduced to the market. In this case, too, definitions play an important role, because while "clean meat" evokes the idea of greater protection for animal welfare, on the other hand, the concept of "nature" appears to be in contrast with a food product created in a laboratory.

On a more relational level, "nature" has also been linked to the connection with animals, which the disappearance of non-cultivated meat may impact.[38] From this perspective, the appeal to nature seems to imply more of a return to non-intensive farming practices, in line with the idea of a "nature" in which the intersections between food and science are minimized (at least in the aspects that exist in cultivated meat).[39] On the other hand, the reduced contact between humans and animals could, according to some, have a positive effect by lowering exposure to animal-borne diseases: yet another perspective to consider.[40]

By combining the elements considered so far and looking towards the future, two main areas of intersection between animal welfare and the prospects related to cultured meat can be identified.

From an initial perspective, animal welfare is considered within the scope of strict welfarism, meaning the absence of suffering and respect for the animal's ethological characteristics as mentioned above. This debate, in particular, highlights the different connections between the concept of animal welfare—which, as we have seen, is enshrined as a constitutional principle—and the potential developments that cultured meat could bring. The first, most immediately perceptible, issue concerns how concretely these new techniques will allow for the bypassing of animal use in their corporeal dimension. The number of interventions necessary on their bodies (such as biopsies or the use of fetal bovine serum)[41] undoubtedly impacts the reflections on animal welfare, both from the perspective of ethical treatment and the possibility of moving away from practices involving the animal's body.

From this viewpoint, a particularly debated issue concerns the use of fetal bovine serum.[42] The use of fetal bovine serum has been at the centre of criticisms from those who highlight its problematic animal-welfare aspects. The extraction of bovine serum has been analysed both as a scientific issue and as an ethical concern. From the ethical standpoint, particular attention has been given to the methods used to obtain the serum and to potential fetal suffering.[43] Various opinions have been

[37] Regarding the definition of bio-objects, see Webster et al. (2016).

[38] Mancini and Antonioli (2020).

[39] Van der Weele and Driessen (2013).

[40] Mancini and Antonioli (2020). See also contributions in Part "Scientific and Sustainability Challenges" of this volume.

[41] Chriki and Hocquette (2020).

[42] On the use of Fetal Bovine Serum (FBS) for the production of cultivated meat, see *What Is Cell-Based Meat? Food Safety and Sustainability Aspects* by Cheli et al. and *Safety Assessment of Cell-Culture-Derived Food: The Role of the European Food Safety Authority* by Dall'Asta and Gjoni, both in this volume.

[43] See for example Jochems et al. (2002).

raised concerning the methods of fetal euthanasia and the potential for suffering. While this is primarily an ethical issue, it is also based on technical and scientific considerations.[44] Debate on the ethics of using bovine fetal serum focuses on an element that is eminently scientific and central to the field of animal rights: sentience. The debate revolves around whether the bovine fetus can experience pain or other negative consequences and the most suitable methods to avoid these consequences.[45]

From this perspective, the potential impact of progress on the use of this type of material will be crucial.

It should be noted that this kind of evolving relationship is typical of issues involving law and science. Those who study the relationship between legal norms and the life sciences, for example, must always keep in mind that the factual bases on which legal reasoning rests can change, sometimes rapidly, given the pace of scientific advances in certain fields. Once again, this is not exclusive to the scientific or technological realm, as cultural advancement also represents shifting aspects that often influence legal norms (in the examples provided above, cultural changes in food practices could affect legal systems, such as prohibitions on certain foods or their regulation, i.e. the potential commercialization of insect flour). However, in the scientific field, we witness rapid changes, sometimes difficult to reconcile with the timelines of democratic debate in parliaments. There are several examples in this regard, starting with the beginning of life: in very specific cases, science fundamentally alters the factual realities underlying legal regulation, or even our understanding of scientific facts themselves.[46] On a broader scale, some areas have taken a different shape precisely because of scientific advances. A prime case is the emergence of medically assisted procreation techniques and the profound consequences they have had in family law. The definition of death is another classic example, as is the protection of genetic data, or anything related to data in the context of information technology.

As stated, this is a typical aspect of the relationship between law and science, and it is known from the outset that many areas will be subject to change. However, the debate does not become obsolete: it will become so for the subject matter itself, which may change or cease to exist, but the comparison between the various aspects involved scientific, ethical and legal retains its usefulness: a kind of "laboratory" in which the spaces and areas occupied by each aspect are outlined, as well as their possible interactions with the others. The debate on bovine fetal serum, in fact, may be destined to disappear.

If the use of animals were completely excluded, there would be a convergence between welfare and dignity approaches, as the possibility of obtaining meat for consumption without the use of animals. From a welfare perspective, there would

---

[44] Nielsen and Hawkes (2019), Versteegen et al. (2021) and Hahn et al. (2024).

[45] McCann and Treasure (2022).

[46] A well-known and striking example is the drug Levonorgestrel, whose contraceptive and non-abortifacient nature has only been established over time, following a number of studies. See for example Adashi et al. (2023).

be alignment with the focus on reducing animal suffering, as evidenced by current regulations on slaughter, including the recent debate on unconscious slaughter.[47] Similarly, attention to ritual slaughter highlights the search for methods that respect animal welfare as much as possible, with recent calls for reversible stunning.[48]

From this perspective, an "unexpected alliance" mentioned later becomes clear. Cultivated meat marks an unforeseen convergence: while animal rights groups and scientific research are often at odds when it comes to the use of animals for human purposes, in this context, they appear to be aligned. Animal rights organizations themselves, in fact, have highlighted the reduced role of fetal bovine serum (on which see *supra*) in the production of cultivated meat.[49]

From a second perspective, however, some considerations related to the concept mentioned above may arise. These considerations highlight how they connect with environmental concerns, which are of particular interest also in relation to the concept of animal welfare. This reference can also be interpreted from different angles. From a more immediate perspective the intersection between cultivated meat and principles of a constitutional nature concerns the environment, which as mentioned, is often involved in the juridification of animals at constitutional level, typically within the triad of animals, environment and future generations. This perspective also seems to be evolving: much progress can be made, especially regarding energy consumption, which will help reduce not only costs but also criticisms that identify cultivated meat production as unsustainable. The comparison with the resources required for cattle farming is relevant. And this is also an evolving topic, because both monetary and energy costs will be subject to scientific changes.

[47] About the concept of unconscious slaughtering, see Santori (2018), p. 110, who defines it in these terms: "Some solutions for a 'good life' of the animal and for an unconscious slaughter are already available. A free-range or semi-free-range animal - and therefore one in more acceptable ethological conditions - can be stunned and slayed at the breeding place without being transported to the slaughterhouse (legal rules already accept emergency slaughter and the use of a mobile slaughterhouse). Moreover, the butchery can be carried out in unconscious conditions obtained by the use of drugs that can be quickly assimilated and the residues of which are not dangerous for man".

[48] With regard to the debate on reversible stunning and ritual slaughtering, see the European Court of Human Rights decision (4 July 2022), 16760/22 *Executief van de Moslims van België and Others v. Belgium*, and among others, Bottoni (2024). See also the decision of the Court of Justice of the European Union (17 December 2020), C-336/19 C*entraal Israëlitisch Consistorie van België and Others*.

[49] See for example PETA statement *From Lab to Table: PETA Keeps Cultivated Meat Free From Animal Testing*, 3 March 2025 in https://www.peta.org/. Accessed 4 April 2025. Or, in Italy, the Antivivisection League (LAV) comments on the development of food production from cellular agriculture as a solution to the environmental and climate crisis, September 2023, available at https://static.lav.it. Accessed 4 April 2025. See also the Eurogroup for Animals press statement *The correct conversation around cultivated meat: more sustainability and animal welfare, less fear mongering*, 24 January 2024 Available at https://www.eurogroupforanimals.org/news/correct-conversation-around-cultivated-meat-more-sustainability-and-animal-welfare-less-fear. Accessed 17 August 2025.

In this case as well, progress will be crucial in calibrating comparison of the ecological impact (such as water, energy and land use) of meat produced by "traditional" farming with that required for cultivated meat.[50]

The reference to the environment, however, can also be framed in terms of animal welfare, in a second perspective that highlights how the arguments at the intersection of science and law can be quite complex.

In fact, the comparison between animals and the ecosystem reappears, with animals as part of the whole, so that, from one perspective, the replacement of farming could be viewed differently in relation to intensive farming or small-scale farmers, who according to some authors would play a role in relation to the environment.[51] It is clear that in this perspective, animal welfare is only one of the aspects involved, but within the framework where it is linked not to the interests of farmers, but to more sustainable farming practices, we return to the realm of balancing interests. A perspective in which cultivated meat leads to the replacement of animals for food purposes is accompanied by another perspective where respect for animals is placed within a context where biodiversity is combined with other values, looking at how we care for animals, in their interest and in the interest of the environment, but deeming their killing acceptable, from a perspective that views welfare as a quality of life compatible with the use of animals themselves. This is a debate that looks toward the future with current perspectives, where welfarism ends up being characterized by new nuances, similar to those that consider reversible stunning or unconscious slaughter. However, the point of contention remains the acceptability of using animals for food or not. Finally, it is interesting to note how some classic themes of constitutional law are being re-proposed in relation to cultivated meat. Another, in this case classic, theme concerns the relationship between animal dignity and cultivated meat: dignitarianism is indeed evoked not only with reference to animal rights in general but also more specifically in relation to cultivated meat. However, as already observed, and one could say, as always in reference to dignity, interpretations of this principle and its applications in this area may not be unanimous.

On one hand, it has been noted that cultivated meat separates food from animals.[52] This is another recurring theme in science, which not only alters realities assumed by legal norms (as seen in the examples above) but also causes dissociations. Just as science separates procreation from the sexual relationship between a man and a woman, or the genetic relationship with parents, or genetics dissociates the ownership of data from the possible social family group, so too does cultivated meat, upon closer inspection, create a separation between food and animals. It is precisely in this separation that the concept of dignity arises in debates about the acceptability of cultivated meat. The degree of dissociation between food and

[50] See *What Is Cell-Based Meat? Food Safety and Sustainability Aspects* by Cheli et al. in this volume.

[51] Chriki and Hocquette (2020).

[52] Chauvet (2018).

animals seems to be a determining factor, for those who accept the use of animals for food, in evaluating the acceptability of cultivated meat. In this perspective, it appears that even in relation to animals, we see a range of nuanced meanings, making dignity a broad concept that can be interpreted differently depending on the context in which it is considered.[53] With reference to the degree of separation between meat and animals (and without intending to delve into the realm of philosophy of law), one can discern a concept of dignity in this case linked to the use of animals for food. This concept of dignity seems incompatible for those who believe that the use of animals for such purposes is not acceptable.[54]

More generally, a problem can be identified in the connection between cultivated meat and the use of animal-derived cells: even in small quantities, this link may be unacceptable for those who reject animal-derived products. Naturally, the connection to animals presents a higher level of unacceptability when linked to concerns about their suffering, as in the debate surrounding fetal bovine serum. From this perspective, overcoming the need for serum does not change the situation for those who consider not the well-being of animals or the dissociation between animals and food, but the very possibility of using animals as a means. An additional argument thus emerges, that of the "illusion" of "victimless consumption", where the dissociation between meat and slaughter takes on a particular meaning. In this context, the concept of the failure to respect animal dignity seems to appear.[55]

This argument is of great interest because it goes beyond the connection with practices associated with animals (even in the absence of suffering) and goes somehow to the heart of the concept of dignity. This issue concerns the view that consuming meat still represents an acceptance of the possibility of using animals, the existence of a dichotomy between "us" and "them", which implies a failure to recognize dignity as an intrinsic value of the animal itself.[56] From this perspective, the two approaches in the field of constitutional protection of animals—focused respectively on welfare and dignity—seem to be evoked.

However, the specific issue addressed here allows us to observe an interesting aspect: the potential areas of conflict between the two approaches, with differing views on the significance to be attributed to the actual use of animals for food and the symbolic value of their use. This can lead to a sometimes problematic confrontation between the two, in which cultivated meat could play an important role.

[53] Obviously, this is a subject we are only mentioning here in relation to the much broader and more complex topic, for which see for example Nussbaum (2006a, b).

[54] Chauvet (2018).

[55] See for example the debates and opinion reported by Dilworth and McGregor (2015).

[56] Chauvet (2018).

4 Conclusions

The themes of constitutional protection of animals and cultivated meat offer a valuable opportunity for comparison. Both represent an ongoing process, the analysis of which necessarily presents the undefined contours of what is currently unfolding rather than something explored ex post. The constitutional protection of animals is a widespread and in many countries, recent phenomenon that lends itself to different interpretations, particularly jurisprudential ones, on which the actual impact of such provisions will depend. This will also relate to the evaluation of the constitutionality of various sector-specific regulations that embody the protection of animal welfare in the different areas where it is at stake. How the development of the cultivated meat sector intersects with this ongoing process and the constitutional values involved will be determined by the spread of this product in the food market. This will depend on many factors: competitiveness, consumer acceptance and technological advancement. Above all, cultural aspects will be crucial, as they will determine the success of this and other changes in food habits.

I think the same can be said, for example, of insect flours, where consumer acceptance will largely depend on cultural acceptance of what are defined as Novel Foods[57]—a designation that reflects not only their legal framework but also their essence and the issues that arise from it.

Cultural aspects are also significant in the intersections between constitutional animal protection and cultivated meat. Compared to other themes, these intersections present very different refracted and nuanced facets, which like in a kaleidoscope, take on different shapes. The very concepts of animal welfare and animal dignity can indeed align with the possibilities opened by cultivated meat, especially if the focus is on elimination of animal slaughter for food. However, the aspects previously mentioned and discussed in the literature focus on other potential considerations, such as departure from a "natural" dimension and the preference for sustainable farming. Cultural factors may also require total dissociation of cultivated meat from any animal cell—an aspect that is currently not possible but could potentially be impacted by scientific progress.

Only scientific and technological development will indicate the paths that this new food product will take, but a complex set of factors will determine its spread, particularly in terms of how it is deemed compatible with the values and principles that constitutionally protect animals. The protection afforded to animals could better define the contours of this product, which are not yet clear but may depend on varying sensitivities that attribute different degrees of importance to the protection of nature, environment, animal welfare and the dignity of sentient beings. Definitions in progress not only seem to be evolving but also to be situated differently according to the cultural standpoint of the observer. As mentioned at the beginning, it is a very complex matter.

[57] See *Socio-Demographic and Behavioural Determinants of Consumer Attitudes to Cell-Based Meat and Insect-Flour-Based Products: An Empirical Study of Italians' Propensity to Accept New Foods for Human Nutrition and Pet Food* by Degli Antoni and Faillo in this volume.

References

Adashi EY, Cohen IG, Wilcox AJ (2023) The FDA declares levonorgestrel a nonabortifacient—a 50-year saga takes a decisive turn. JAMA Health Forum 4(8):e232257

Barak A (2015) Human dignity: the constitutional value and the constitutional right. Cambridge University Press, Cambridge

Bilchitz DI, Wilson AP (2022) Key animal law in South Africa. In: Andrew Knight A, Clive Phillips C, Sparks P (eds) Routledge handbook of animal welfare, 1st edn. Routledge, London, pp 426–435

Blattner CE (2019) Protecting animals within and across borders: extraterritorial jurisdiction and the challenges of globalization. Oxford University Press, New York

Blattner E, Fasel R (2022) The Swiss primate case: how courts have paved the way for the first direct democratic vote on animal rights. Transnatl Environ Law 11(1):201–214

Bolliger G (2016) Legal protection of animal dignity in Switzerland: status quo and future perspectives. Schulthess, Zurich

Bottoni R (2024) Religious slaughter and the conflict between secular and religious interests: dispelling the myth. Diritto Pubblico Comparato ed Europeo Online 65(3):1643–1656

Bryant CJ, Barnett JC (2019) What's in a name? Consumer perceptions of in vitro meat under different names. Appetite 137:104–113

Cao D, White S (2016) Animal law and welfare - international perspectives. Springer, Cham

Cerda-Guzman C, Insunza X (2024) Constitutional status of animals in Chile's draft Constitutions of 2022 and 2023: two contrasting approaches. https://blog-iacl-aidc.org/2024-posts/2024/3/7/constitutional-status-of-animals-in-chiles-draft-constitutions-of-2022-and-2023-two-contrasting-approaches. Accessed 4 Apr 2025

Chapaux V, Mégret F, Natarajan U (eds) (2023) The Routledge handbook of international law and anthropocentrism. Routledge, New York

Chauvet DJ (2018) Should cultured meat be refused in the name of animal dignity? Ethic Theory Moral Prac 21(2):387–411

Chriki S, Hocquette JF (2020) The myth of cultured meat: a review. Front Nutr 7(7):1–9

Dilworth T, McGregor A (2015) Moral steaks? Ethical discourses of in vitro meat in academia and Australia. J Agric Environ Ethics 28:85–107

Eisen J (2017) Animals in the constitutional state. Int J Constitutional Law 15(4):909–954

Fasel RN (2019) The old 'new' dignitarianism. Res Publica 25:531–552

Fino M, Anzà B, Bairati L, Bertini I, Biolatti B, Biressi S, Cannizzo FT, Cavallarin L, Conti L, Deriu M, Gargioli C, Loera B, Lo Sapio L, Marchisio D, Pallante L, Stano S, Torri L, Bertero A, Massai D (2024) Cultivated meat beyond bans: ten remarks from the Italian case toward a reasoned decision-making process. One Earth Commentary 7(12):2108–2111

Foster C (2011) Human dignity in bioethics and law. Hart Publishing, Oxford

Frota Pestana de Aguiar L (2021) Farra do boi, olhos vazados da Justiça: 24 anos desrespeitando o STF. https://www.conjur.com.br/2021-mai-04/lucia-frota-farra-boi-olhos-vazados-justica/. Accessed 4 Apr 2025

Giménez-Candela M (2018) Dignity, sentience, personality: the legal relationship between animals and humans. Derecho Anim (Forum Anim Law Stud) 9(2):17–28

Hahn O, Peters K, Hartmann A, Dannenberger D, Kalbe C (2024) Potential of animal-welfare compliant and sustainably sourced serum from pig slaughter blood. Cell Tissue Res 397(3):205–214

Jaleel H (2024) The judicial recognition of animal sentience: developments in Pakistan and India. In: Kotzmann J, Rodriguez F (eds) The legal recognition of animal sentience: principles, approaches and applications. Hart Publishing, Oxford, pp 135–154

Jochems CE, van der Valk JB, Stafleu FR, Baumans V (2002) The use of fetal bovine serum: ethical or scientific problem? Altern Lab Anim 30(2):219–227

Kelch TG (2017) Globalization and animal law: comparative law, international law and international trade, 2nd edn. Kluwer Law International, Alphen aan den Rijn

Kohler NS (2019) A inconstitucionalidade da "Rinha do Galo" sob o aspecto da vedação de tratamento cruel aos animais e o conflito aparente com a proteção à cultura. https://jus.com.br/. Accessed 4 Apr 2025
Kurki V (2024) What is animal dignity in law? J Anim Law Ethics One Health:9–23
Lanzoni D, Rebucci R, Formici G, Cheli F, Ragone G, Baldi A, Violini L, Sundaram TS, Giromini C (2024) Cultured meat in the European Union: legislative context and food safety issues. Curr Res Food Sci 8:100722
Le Bot O (2023) Constitutional animal law, independently published, Paris
Le Bot O (2024) Constitutional animal law: trends and impacts. https://blog-iacl-aidc.org/2024-posts/2024/2/27/constitutional-animal-law-trends-and-impacts. Accessed 4 Apr 2025
Mancini MC, Antonioli F (2020) To what extent are consumers' perception and acceptance of alternative meat production systems affected by information? The case of cultured meat. Animals 10(4):656
McCann TJ, Treasure C (2022) Addressing animal welfare issues in fetal blood collection for fetal bovine serum production. Altern Lab Anim 50(5):365–368
Michel M, Kayasseh ES (2011) The legal situation of animals in Switzerland: two steps forward, one step back—many steps to go. J Anim Law 7(1):1–33
Moore KC (2023) The case for the legal protection of animals humanity's shared destiny with the animal kingdom. Macmillan Palgrave, Cham
Nattrass KM (2004) "Und Die Tiere": constitutional protection for Germany's animals. Anim Law 283(10):292–294
Neto JA (2018) Borrowing justification for proportionality. Springer, Cham
Nielsen OB, Hawkes PW (2019) Fetal bovine serum and the slaughter of pregnant cows: animal welfare and ethics. BioProcessing J 18. https://bioprocessingjournal.com/afp/J18OA-Hawkes.pdf. Accessed 4 Apr 2025
Nussbaum MC (2006a) Frontiers of justice: disability, nationality, species membership. Harvard University Press, Boston
Nussbaum MC (2006b) The moral status of animals. https://www.arcusfoundation.org/wp-content/uploads/2014/09/The_Moral_Staus_of_Animals_by_Martha_Nussbaum_2006.pdf. Accessed 4 Apr 2025
Oylan MS, Rosenblum JD (2021) Animals and the law in antiquity. Brown University, Providence, Rhode Island
Pasca MG, Arcese G (2025) Sustainability and quality of cultured meat: consumer perceptions. Sustainability 17:1633
Perry MW (2023) The dignitarian return. Eur J Polit Theory. https://journals.sagepub.com/doi/full/10.1177/14748851231216878. Accessed 4 Apr 2025
Piciocchi C (2024) Courts, pluralism and law in the everyday: food, clothing and days of rest. Routledge, Abingdon
Rescigno F (2023) The entrance of animal beings in the Italian Constitution. The still incomplete path of animal subjectivity. J Ethics Leg Technol 5(1):51–71
Santori P (2018) Unconscious slaughtering. Derecho Anim. Forum Anim Law Stud 9(2):109–111
Soccol CR, Forte Maiolino Molento C, Glufke Reis G, Grace Karp S (eds) (2024) Cultivated meat. Springer, Cham
Toomey J (2020) Constitutionalizing nature's law: dignity and the regulation of biotechnology in Switzerland. Law Biosci 7:1–33
Van der Weele C, Driessen C (2013) Emerging profiles for cultured meat; ethics through and as design. Animals 3(3):647–662
Van Hoorick G, Veriners E (2024) Some explorations about extending the right to a healthy environment to future generations and animals. In: Schoukens H, Bouquelle F (eds) The right to a healthy environment in and beyond the anthropocene: a European perspective. Edward Elgar, Cheltenham, pp 227–245
Versteegen RJ, Murray J, Doelger S (2021) Animal welfare and ethics in the collection of fetal blood for the production of fetal bovine serum. Altern Anim Exp 38(2):319–323
Webster A, Vermeulen N, Tamminen S (2016) Bio-objects life in the 21st century. Routledge, Abingdon

Socio-Economic and Cultural Issues

Food Justice in Religious Laws: Hypotheses on Cultivated Meat

Giancarlo Anello

Abstract Religious laws significantly influence meat consumption patterns across various cultures, primarily through dietary laws and prohibitions. The comprehensive laws of religion regarding meat consumption are complex, but they all seem to share a basic principle of balance: meat is a lawful food that can be eaten within certain boundaries. These boundaries are determined by thorough assessment from two perspectives. On one hand, various religious customs and legal regulations acknowledge that meat production is resource-intensive and involves the utilization of elements such as soil, pasture, cropland and livestock. On the other, religious laws view animals primarily as valuable assets that should only be killed under specific circumstances, recognizing their role in agricultural society and the benefits they provide. This chapter first examines the significance of religious laws in global food systems before delving into a comparative analysis of religious regulations on meat production and consumption. It then discusses the concept of "meat security" according to religious laws. Finally, the analysis shifts to the contemporary topic of cultivated meat, which raises new issues and questions rooted in old cultural, ethical and economic dilemmas surrounding the consumption of meat.

Keywords Religious laws · Meat consumption · Food security · Food justice · Rituality

---

G. Anello (✉)
University of Parma, Department of Humanities, Social Sciences, and Cultural Industries, Parma, Italy
e-mail: giancarlo.anello@unipr.it

G. Formici et al. (eds.), *Cell-Based Meat in the European Union and Beyond*,
https://doi.org/10.1007/978-3-032-14624-3_9

1 Introduction

In recent decades, technical and scientific progress has paved the way for an innovative method of cultivated meat production, starting with stem cells, often obtained by biopsies from live animals. These cells are then cultured and differentiated in bioreactors by cell culture and tissue engineering techniques, using culture medium containing growth factors from various sources. The goal is to produce fibres and tissues. The resulting product—often referred to as "synthetic", "artificial" or "lab-grown" meat, or "clean", "slaughter-free", "cultured", "cultivated" or "cell-based" meat[1]—is intended to be an alternative source of animal protein.[2]

In a world facing urgent challenges such as food insecurity,[3] population growth and limited natural resources, innovation in the agri-food sector and Novel Food products, including cell-based meat, are at the centre of lively debate in the political, legislative,[4] economic,[5] scientific and ethical realms. Disinformation and ideological considerations could hinder these discussions.

The overarching laws of religion regarding meat consumption are complex,[6] yet they share a fundamental principle of balance: the consumption of meat is permissible and at times even blessed, but within certain limits. These limitations are based on a twofold assessment. On one hand, different religious customs and legal regulations highlight the resource-intensive nature of meat production, which involves the exploitation of various elements such as soil, pasture, cropland, and livestock itself. On the other hand, religious laws often view animals as valuable and only to be slaughtered under specific circumstances, acknowledging their role in agriculture and the benefits they provide. This eco-religious understanding of the value of animals is rooted in two key principles: firstly, that taking a life is a significant moral issue, justifiable only when there is a genuine need for human sustenance; and secondly, that misappropriation occurs when humans fail to recognize that an animal's life belongs to God, not to them.[7] These principles are based on practical observations: much of the meat consumed today is a product of intensive and competitive

---

[1] Bryant (2020) and Bondi (2022). See also Part "Scientific and Sustainability Challenges" of this volume.

[2] Leroy and Praet (2015).

[3] See the Report *Collective Action for Ending a Collective Problem* (2023).

[4] See *Cell-Based Meat in the European Union: A Regulatory Crossroads* by Corazza and Formici; *The Legal Debate on Cultured Meat Around the World: Comparative Insights from the United States, Singapore and Israel* by Pitto; *Animal Welfare as a Constitutional Value: The Debate in Comparative Law and the Emerging Role of Cultivated Meat* by Piciocchi, all in this volume.

[5] See *Cell-Based Meat and Food Policy* by Mancini in this volume; see also Part "Socio-Economic and Cultural Issues" of this volume.

[6] More in general, religious food codes are not only expressed as negative duties but also influence the behaviours of economic agents and become food market factors, see Alicino (2023), p. 92.

[7] Linzey (2004), p. 190.

farming that monopolises vast tracts of cropland to cultivate fodder, making one person's food another person's hunger, dispossession and poverty.[8]

Since many religious norms are focused on environmental protection and the promotion of ecological well-being,[9] this chapter seeks to explore the concept of balance as it relates to religious laws and their ecological implications. By examining the relevance of religious laws in global food systems and conducting a comparative analysis of religious rules on meat production and consumption, the chapter delves into the idea of "meat security" from a religious perspective. Finally, the research returns to the specific and contemporary field of cultivated meat, addressing new issues and questions that resonate with the historical, cultural, ethical and economic challenges surrounding meat consumption and its deeper meanings.

2 Food, the Sacred and the Law: Towards a Holistic Perspective on Food Security

At a global level, religious institutions influence changes in the law in a variety of ways. Many of the world's most influential religions, such as Judaism, Islam and Hinduism, contain their own legal systems that interact with the secular law of states. In addition to these larger systems that drive religious legal change, individuals inspired by religious principles and norms routinely appear before national and international legislative and judicial bodies to advocate recognition of their religiously-inspired versions of human and civil rights as alternatives to the more secular conceptualizations often enshrined in national and international law.

Taking the characteristics of global law into account, this chapter offers some insights into important differences between "religion" and "the religious": the former perhaps superficially or at least incompletely understood at institutional or state level, the latter understood in a more anthropologically informed and nuanced way that nevertheless impacts institutions and states. Religiosity can be considered the universalizing force that guides individuals in choosing the best instruments and remedies to achieve their religious goals when they act as private players on the global scene.

Globalization is not limited to markets and trade; it has important implications for law as well. To understand both religion and law on a global level, we must recognize that global law is the result of economically driven globalization. Processes that originated in the economic sphere led to emergence of a global community of actors and practitioners. For example, the process of globalization has impacted international trade, giving more and more relevance to practices and values of multinational economic players and the international community at large, so that in many respects it resembles the historical notion of *Lex Mercatoria*.

[8] Linzey (2004), p. 191.

[9] Jenkins and Chapple (2011).

Moreover, international law gradually displaced many of the legal mainstays of the pre-globalized world and grew to emphasize international rather than local economic governance. New international norms have allowed individuals, firms, corporations and non-governmental organizations to bring claims before international jurisdictions. As a result, global legal rules are not the rules of particular legal systems but rules that arise from the convergent behaviour of institutional and economic players of different legal systems. Today, the objective of global law is to identify a uniform set of rules, principles and procedures to manage global human rights, interests, goods, groups and cultures. Identifying these approaches involves drawing from a variety of international practices in political and jurisprudential contexts.

In such a scenario, religious laws can intersect the food regulatory system in many ways, making a comprehensive perspective on food security relevant in many respects.

The link between food and religion encompasses concepts and values such as celebration, sharing, community, solidarity, interconnectedness, dignity, self-restraint, appreciation and a long-term vision of purpose and good deeds that extend beyond an individual's lifetime. These concepts and values are also present in secular contexts to some extent. However, religious laws can provide a rich foundation for envisioning a better world where generations of humans can flourish. Religious rules are based in community life and offer a perspective that considers human and non-human needs organically. In times of insecurity and existential threat, these narratives can help bridge the gap between the familiar and the unknown, offering different interpretations to counteract dominant opposing narratives.

Many religious rituals involve food, such as the blessing and sharing of specific food items. Religious communities often observe dietary rules, like kosher food and slaughter for Jewish communities, as a matter of religious freedom. In Islam during Ramadan and in other religions, food is associated with fasting as a way to exercise patience, introspection, discipline and compassion. Wasting food is considered wrong in many religious traditions due to the belief that food is a gift to be grateful for, as often expressed by prayer before meals.

Recent crises, like COVID-19, showcased the active role of religious groups in addressing food security challenges. The World Council of Churches has emphasized the urgent need to protect a sustainable future in the face of climate change-related crises.[10] Religious and faith-inspired organizations are working to ensure the flourishing of future generations of humans, based on their religious teachings and principles.

---

[10] At the General Assembly held in Karlsruhe in 2022 WCC addressed the climate emergency as a priority for churches. See the document "The Living Planet: Seeking a Just and Sustainable Global Community", https://www.oikoumene.org/sites/default/files/2022-10/ADOPTED-PIC01.2rev-The-Living-Planet-Seeking-a-Just-and-Sustainable-Global-Community.pdf. Accessed 8 April 2025.

Since a significant percentage of the global population identifies with a religious group,[11] religions deserve consideration in discussions on food security for several reasons:

1. Religious traditions and communities are often able to reach people who may be reluctant to accept a secular message: religious actors and the narratives of religious traditions can reach people who cannot be reached by secular narratives and appeals. Religious actors can be multipliers who bring important information and calls to action to their communities. In these processes, religious actors should not be viewed as passive recipients. Rather, there are processes of translation and mutual influence at work between religious and secular actors from which both sides can benefit.
2. Religious institutions have experience of transnational collaboration and solidarity: they often extend solidarity beyond their own circle. Faith-inspired organizations operate at local, regional, national and global levels. Faith-inspired organizations are supported in their efforts by funds raised among religious communities, where people are motivated by the values of their faith to donate money towards fighting food insecurity. In some contexts, religious actors may be judged more trustworthy and altruistic by donors and recipients than secular actors.[12]
3. Religious institutions and communities of faithful have established networks that can be relied on in times of food crisis: they have many years of experience in alleviating hunger and malnutrition, also in areas with political instability and conflicts. During the COVID-19 pandemic, many religious actors extended their initiatives to alleviate food insecurity. For example, Sikh communities, who always welcome guests and offer free meals in their houses of worship, stepped up their effort and engaged in the large-scale distribution of free meals. In the words of a member of the Irish Sikh community: "Langar [i.e. the distribution of free meals] was started by Guru Nanak to feed the visiting Sangat [i.e. community] who were coming to listen to his sermons and discussion. If you are hungry you cannot concentrate on listening. The same vegetarian food was served to everyone irrespective of caste, creed, status, king, saint or pauper. The principle of equality and giving respect to all: love all hate none. It is service of mankind, whether it is peace or disaster or natural calamities. Even during the Coronavirus pandemic, Sikh charities and Gurdwaras opened their door to community for food, medicines and oxygen cylinder supply."[13]

---

[11] *The Global Religious Landscape* (2012), https://www.pewresearch.org/religion/2012/12/18/global-religious-landscape-exec/. Accessed 8 April 2025; as to the global population growth projections, see *The Future of World Religions: Population Growth Projections, 2010-2050* (2015) https://www.pewresearch.org/religion/2015/04/02/religious-projections-2010-2050/. Accessed 8 April 2025.

[12] Gill (2013).

[13] Chabba (2021).

4. Religious traditions provide values that connect individuals and groups: human relationships rely on values that sustain them, both in every-day life and in times of crisis. Community and solidarity are not created by bureaucratic acts alone but require a glue that binds people together on an interpersonal level. While solidarity is a value also frequently referred to in secular contexts, actual solidarity may often not be based on philosophical beliefs but on religious practice. An example from the Jewish tradition is found in a quote of Maimonides: "To eat and drink at a festival in the company of your family without providing for the poor and distressed is not 'the joy of the commandment' but the joy of your stomach. It is a disgrace."[14]
5. Religious literature offers holistic models of human flourishing and interconnectedness: to be effective, discussions on health and food have to take the whole person into account. They offer anthropological models that include the physical, mental, spiritual, psychological and social aspects of what it means to be human. Religious models tend to view the human being not in isolation but interconnected with plant and animal life and often even the cosmos as a whole. The following lines are part of the Zen Buddhist Oryoki eating ritual:

First, this is for the Three Treasures;
Second, it is for our teachers, parents, planet and all sentient beings;
Third, it is for every being in all realms of existence.
Thus we eat this food with everyone.[15]

These resources deserve to be rediscovered in the context of the environmental and climate crises, which are major threats to global food security.

6. Religious traditions provide metaphors and narratives that connect food security with other issues: language about food security should be inclusive and do justice to the diversity of human experience. They may provide stimulating narratives for rethinking the way we talk about food, hunger and food security. Faith-inspired actors are able to "turn" the narratives when the debate tends to polarize. They can provide metaphors and narratives, such as the various creation stories, that can motivate people to engage with humans, animals and the environment. Challenging food waste, for example, can be linked to Jewish and Muslim narratives: "If the Torah warns us not to destroy fruit trees (Deut. 20, 19–20), then we should be even more careful about not destroying the fruit itself" (Rabbi Yishmael); "The Messenger of Allah said, 'Satan appears at everything done by you; he appears even at one's dinner. When a morsel of any of you falls, he should pick it up and remove any dirt on it, and then eat it. He should not leave it for Satan, nor should he wipe his hand with the towel until he has licked his fingers, for he does not know in what portion of the food the blessing lies'" (Riyad as-Salihin 751).

---

[14] Maimonides, Mishneh Torah, Rest on a Holiday 6:18, https://www.sefaria.org/sheets/290888?lang=bi. Accessed 8 April 2025; see also Segal and Blazer (2020).

[15] Daido Loori (2012), p. 75, https://terebess.hu/zen/mesterek/bring.pdf. Accessed 8 April 2025.

7. Religious traditions can provide principles for good agriculture and use of natural resources: they offer overarching principles, such as good stewardship of creation, and specific regulations for agriculture and natural resource use, such as for wood and water in specific contexts. In the Talmud, for example, it is stated that "a man should not eat before he feeds his livestock." Many religious narratives are about the place of humans in an ordered cosmos, where there is enough for everyone. Equally, there are narratives about the dire consequences of recklessly exploiting the earth and not respecting the gift of food and other natural resources. Such narratives also provide a long-term perspective (including the "eternal") that links up with consideration for the environment and future generations; elsewhere they emphasise a cyclic sense of time and the importance of the cycles of life and the growth of plants in agriculture.

Food security is a complex issue because of the global interconnectedness of food systems at different levels, and because of the behavioural shifts necessary to improve food security under climate change, shrinking resources and world population growth. The 2021 Intergovernmental Panel on Climate Change (IPCC) report highlighted the interconnectedness of energy systems and food systems, and the significant climate change-related impacts that can be expected in many areas, from malnutrition to ocean ecosystems, including water availability and food production.[16] Addressing food security requires a holistic approach that takes into account the varied fundamental needs of human beings. Such a broad vision may be provided by the rules, values and practices of religious traditions. All major religious traditions emphasize the importance of feeding the hungry and helping those in need, whether or not they belong to one's community. Mahatma Gandhi considered self-determination of the poor to be the touchstone of political progress. According to Gandhi, the only acceptable form in which God can appear to a starving and idle people is through work and the promise of food as wages.[17] Pope Benedict XVI wrote in his encyclical on the Eucharist, *Deus Caritas Est*, that worship and Eucharistic communion include being loved and loving others in return, and that a Eucharist devoid of tangible love is inherently fragmented.[18] Religious traditions bridge the gap between the long-standing experience of communities and the new challenges of today's world. In India, the Ganges River has been recognized as a person to protect it from exploitation and pollution, in the traditional Indian view of the river as a goddess. This is an example of the intersection of religious, social, environmental and legal activism, which is likely to become more prevalent with the mounting pressures of ecological crises. Religious traditions appeal to the heart and the mind, inspiring deep reflection and collective action. The role of religion in addressing food insecurity includes issues such as combatting food waste, reflecting theologically on the benefits of community gardening in urban areas, promoting

---

[16] https://www.ipcc.ch/assessment-report/ar6/. Accessed 8 April 2025.

[17] Gandhi (2006), p. 257.

[18] https://www.vatican.va/content/benedict-xvi/en/encyclicals/documents/hf_ben-xvi_enc_20051225_deus-caritas-est.html. Accessed 8 April 2025.

dietary transitions, and advocating subsidiarity as a principle for navigating the different levels and power dynamics of food systems. Religious perspectives bring a strong focus on providing the basic necessities of life for all, emphasizing that this goal should not be overshadowed by concern for profit or market logic. Religious perspectives offer a different set of priorities for engaging in the public sphere, serving as a source of constructive criticism in policy discussions.

Such discussions on food security require expert knowledge and precise focus, but also call for a preamble or inclusive vision that ties these aspects together. Religious traditions can offer a horizon of meaning and a perspective towards the greater good for all. The sacred texts, narratives, values and practices of religious communities can unite people in a vision of shared humanity, promoting just and sustainable agricultural principles, and advocating a wholesome diet that upholds the dignity of people, animals and the environment.

3 Religions and Meat Security

As mentioned, religious laws typically view animals as valuable resources that can be killed under specific circumstances, recognizing their importance in an agricultural society and the benefits they provide. In contrast, today's goals focus on establishing a sustainable meat production system and improving food security.[19] This section aims to explore the potential connection between these two goals to determine whether religious and global perspectives can complement each other. One common element between the two sets of rules is the practice of sharing meat with members of the community. It is important to note that this practice is rooted in anthropology rather than ritual, stemming from the collective nature of hunting, which requires cooperation among hunters and the sharing of any prey obtained. Since Homer's Odyssey, hunting, particularly for large and dangerous animals, has historically played a significant role in building relationships within communities.[20] The perishable nature of meat and the excess produced from hunting large prey led to sharing behaviour with minimal costs, reducing risks and promoting stability within the community. Throughout history, hunting and the sharing of meat have been instrumental in shaping intra-community relationships. The anthropologist William Robertson Smith believed that sharing food and drink creates bonds of friendship and mutual obligation among individuals.[21]

Over time, animals have been utilized for various purposes such as labour, egg and milk production, wool, and soil fertilization, in addition to their meat.[22] This

[19] Leroy and Praet (2015), p. 200.

[20] See Homer, The Odyssey, Book 3, verses 416–463, and Book 19, verses 418–454.

[21] This quotation can be found in Fiddes (1989), p. 55. Some examples are in the masterpiece of Homer, The Odyssey, Book 3, verses 416–463 and especially Book 19, verses 418–454 that describe preparation of the ox for communal meals and boar hunting.

[22] Chiles and Fitzgerald (2018), p. 6.

human experience has underscored the communal significance of animal killings, as the excess meat was often shared among community members, fostering gift exchange and social cohesion.[23] Livestock came to symbolize wealth, with meat consumption serving as a means to demonstrate social status and strengthen group unity. This interpretation aligns with the long-term environmental aspects of religious ideologies.

The economic role of meat has been intertwined with cultural norms throughout history. While grain was essential for survival, regulations prevented sellers from profiting excessively from it.[24] Meat, however, was considered a luxury and did not face the same restrictions, livestock production thus becoming a source of wealth creation. This distinction resulted in two agricultural systems: one focused on sustenance and the other on luxury and celebration.

Traditional and contemporary teachings on meat consumption in various religious laws have portrayed meat as a strong symbol, both in positive and negative contexts. Meat taboos have been prevalent in human culture, despite its nutritional benefits and esteemed value, leading to theories that suggest functionalist or symbolic reasoning or that stem from evolutionary psychology.[25]

In the Jewish tradition, dietary laws known as kashrut outline what is kosher, including prohibition of certain animals and the requirement for specific slaughtering techniques.[26] In Ancient Judaism, sacrifices were only made in one temple, limiting access to meat consumption for most Jews.[27] These regulations serve to reinforce community identity and adherence to religious principles, as they dictate not only individual behaviour but also communal practices surrounding food.[28] In the Jewish tradition, "Reverence for Life… is the constant lesson of the laws of Kashrut",[29] even with respect to animals. More in depth, vegetarianism in Judaism is associated with an ideal world, symbolized by Eden, where meat consumption is only permitted after Noah's ark came to ground. This conflict between dietary commandments in an ideal world and permission to eat meat in a fallen world is highlighted in the passages Gen 1:28–30 and Gen 9:1–4.

In the Catholic tradition, abstaining from meat consumption has been a form of spiritual fasting and a response to gluttony.[30] Reasons for abstinence in Catholicism include health concerns, spiritual or philosophical ascetic practices,[31] and

[23] *Ibid*, p. 4.

[24] Moreno (2008).

[25] Leroy and Praet (2015), p. 206.

[26] Regenstein et al. (2003).

[27] Leviticus 17:3–4; Deuteronomy 12:20; see a comment in Schwartz (2025).

[28] Bonne and Verbeke (2007).

[29] Kalechofsky (2004), p. 172.

[30] Berkman (2004).

[31] *Ibid*, p. 201: "From the beginnings of the Church, abstinence from animal flesh seems to be identified as part of the virtuous or ascetic life, largely as a key element in the Christian's response to the capital vice of gluttony. For gluttony—at least according to many of the church fathers—is *the* primal vice of the Christian faith. Conversely, abstinence (e.g. fasting) is a virtue by which one

eschatological beliefs related to the Eucharist.[32] While eschatological abstinence is suggested, it is not a strict command.[33]

Likewise in Islam, the concept of halal defines permissible food, which includes specific slaughtering practices that must be adhered to in order for meat to be considered acceptable. The Quran and Hadith provide guidelines that govern not only the types of animals that can be consumed but also the methods of slaughter.[34]

Furthermore, in the Islamic tradition, the issue of meat consumption can be linked to social justice and the dietary differences between affluent and less affluent members of society.[35] The main sources for the rules on this matter are Q. 5:1 and Q. 6:145. According to this perspective, the majority of meat eaters are seen as contributing to serious ecological and social harm, particularly in comparison to a balanced vegetarian diet. This viewpoint is still accepted today, as demonstrated by a *fatwa* (legal opinion) issued by the Imam of Paris in 2001, stating that animal sacrifice on Eid al-Adha is not necessary and can be replaced by donating a third of the price of a sheep to the poor.[36] In an article, Muslim scholar Shaid Muttaqi argues that animal sacrifice in Islam is mostly a custom derived from pre-Islamic Arab society.[37]

Principles and guidelines regarding meat consumption can also be found in other Eastern religious traditions. Hindu dietary practices are influenced by a combination of religious beliefs and cultural customs, where the veneration of cows leads to a widespread avoidance of beef among Hindus.[38] The caste system further complicates meat consumption, as certain castes may abstain from meat altogether, while others may have restrictions on specific types of meat.[39] This reflects a broader trend where religious affiliation and cultural identity shape dietary habits, often leading to significant variations in meat consumption within a single country, as seen in India. Thus, Hinduism supports vegetarianism based on principles such as non-killing of cows, non-violence, monism from the Upanishads (which reject sharp distinctions

overcomes the vice of gluttony. Early Christian references to fasting typically refer not to periods of total abstinence from food, but to diets that restrict the amount of food eaten (certainly no more than necessary for nutrition) and to diets that avoid certain types of food (i.e. meat). Fasting, along with the Eucharist, are the most important food practices in the Catholic tradition. In the Scriptures and the Catholic tradition, fasting is often aligned with prayer and has a number of spiritual purposes. For Aquinas, fasting is a means not only of atoning for and preventing sins but also, more generally, of raising the mind to spiritual matters. Whereas gluttony puts one to sleep, fasting draws one to prayer".

[32] The healthful effects of moderate fasting are received by the contemporary church, which, for example, discusses it in the CEI document (1999). See the remarks of Alicino (2023), pp. 103–105.

[33] Berkman (2004), p. 203.

[34] Regenstein et al. (2003).

[35] Foltz (2004), p. 209.

[36] *Ibid*, p. 216.

[37] Muttaqi (2016).

[38] Ferry (2020). See also Devi et al. (2014).

[39] Ramachandran et al. (2017).

between beings, including humans and animals), and the concept of reincarnation[40] involving all living creatures.

Buddhism likewise promotes vegetarianism, moderation, and compassion and non-violence towards all sentient beings. Buddhist values have a significant impact on food consumption habits by encouraging vegetarianism, moderation and compassion. The promotion of vegetarianism within Buddhism has important implications for sustainability, as it leads to decreased meat consumption and a shift towards plant-based diets, thereby helping to address environmental challenges associated with meat production.[41] Conversely, Confucianism takes a humanistic approach and allows consumption of meat.[42]

A final remark should be made on the symbolic meaning of meat in relation to blood. Fiddes pointed out: "Blood's symbolic significance is widespread. Traditionally, we have viewed blood as the essence of life itself: when too much is shed, life comes to an end. We become faint at the sight of it. We suggest guilt by saying we have blood on our hands, a stain that Lady Macbeth could not remove and led to her madness. Blood signifies kinship, such as blood brotherhood, noble blood, or blood feud. It is used to determine inheritance, like blood lineage, or to describe inherited attributes. The upper classes in British society are referred to as blue-blooded, suggesting they are so civilized that they are spiritually elevated above mortality, and no longer characterized by red blood. Red meat may be suitable for their consumption, but apparently does not align with their own physiology."[43]

4 Legal Hypotheses on Cultivated Meat

Innovations in cultured meat production have multiplied the number of problems facing contemporary interpreters of religious laws. Moreover, regulation of cultivated meat consumption in religious laws varies significantly across faiths and cultures:

1. Judaism: The consumption of cultivated meat is subject to *kashrut*, the Jewish dietary laws. In general, if cultivated meat derives from *kosher* animals and is produced in a way that observes these laws,[44] it may also be considered *kosher*. But many legal authorities will likely await further development of these practices before asserting definitive rulings. Many elements of the production of cultivated meat are under scrutiny by the rabbis. Although cultivated meat appears to comply with some of the guiding principles of Judaism, such as respect for the environment or animal welfare, its nature as a *kosher* food is controversial.

[40] Gaffney (2004).

[41] Berzin (2014).

[42] Huang (2017).

[43] Fiddes (1989), pp. 93–94.

[44] Regenstein et al. (2003).

Traditionally, *kosher* meat derives only from permitted animals according to the Torah.[45] Moreover, the animals must be ritually slaughtered, so the preliminary legal question about cultivated meat is whether this food can be considered to be derived from an animal or not. Secondarily, if the stem cells used for cultivation are regarded as meat, or meat derivatives, their extraction from permitted animals slaughtered in accordance with religious precepts is not in itself evidence of conformity: for example, current methods of cellular agriculture often opt for (non-blood) extraction of stem cells from live animals. This process is an insurmountable obstacle to the qualification of cultivated meat as *kosher*, due to lack of the indefectible prerequisite of ritual slaughter (not to mention that the Noachian precepts of the Jewish religion prohibit the eating of live animals).[46] Qualifying stem cells as entities in their own right—i.e. distinct from the tissue from which they are derived—likewise does not guarantee that cultivated meat is *kosher* when it comes from prohibited animals, such as pigs. This hypothesis can be excluded because it violates precise prohibitions. The Talmud contains the principle that all food that comes from an impure species is considered impure (Chullin 117b:6-118a:4). In application of this principle, rabbinic literature considers foods such as eggs and milk—which are derived from animals, but are not meat—and *kosher* animals born of non-kosher parents to be forbidden. Thus, the prevailing opinion is to consider cultivated meat *kosher* only when it is derived from stem cells extracted from licit animals.[47]

2. Islam: Similarly, Islamic *halal* dietary laws, which regulate what kinds of food are religiously lawful, are under discussion to determine the regulatory framework of cultivated meat. Like in Judaism, in Islam the main point is whether this new food can be counted as admissible meat. Traditionally speaking, the meat must derive from *halal* animals and they must be slaughtered according to Islamic requirements. According to the sources, some of the revelations to the Prophet Muhammad concerned the origin and significance of food, which is not only indispensable for physical well-being, but also necessary for worshipping Allah; hence the provision of permitted and forbidden foods and the meticulous regulation of meat in accordance with the rules of the Quran.[48]

   Minority opinion has it that the process of producing cultured meat is in conflict with natural laws, in particular with the precept that by tampering with human and animal creation, Muslims act in the vest of Allah.[49] From this perspective, cultivated meat is an experiment with nature from which no advantage is derived, in the same way as activities such as human cloning or cosmetic surgery not based on medical prescriptions. However, the current prevailing opinion is that meat derived from the cells of permitted animals, slaughtered according

---

[45] *Ibid.*

[46] Astori (2010).

[47] Saporito (2024), p. 7.

[48] Regenstein et al. (2003), p. 112.

[49] Saporito (2024), p. 10.

to religious law and certified as *halal* by religious authorities, is permissible. More precisely, cultured meat is *halal* if the stem cells are derived from a permitted animal, slaughtered without prior stunning until completely bled, and if no animal blood or serum has been used in the production process. However, the technological nature of the process of production poses several problems: for example, the culture-medium normally used in the cell proliferation process is serum extracted from animal blood by centrifuging. This process separates blood cells, platelets and clotting factors but if the serum itself does not meet the dictates of the Quran it generates a cultivated product that can be considered impure. By contrast, if the culture medium is plant-based or consists of a mixture of artificial chemicals, the cultivated meat will be considered *halal* because the plant-based or chemical medium is not considered an impure ingredient, as long as it is not toxic or harmful to human health. Other details are still under debate.[50] Of course, meat derived from pork or other prohibited animals is completely forbidden.

3. Hinduism: In Hinduism, *ahimsa* (non-violence) is a core principle, and many Hindus are consequently vegetarians. The acceptability of cultivated meat may therefore depend on the interpretation of how the meat is produced and whether it aligns with the values of non-violence. In the dharmic tradition, food plays a strong symbolic and material role. This religious tradition accepts the concepts of *dharma*, *karma* and reincarnation, common to Hinduism, Buddhism, Sikhism and Jainism. In Hinduism (*Sanatana dharma*), the scriptures hold food responsible not only for the individual's birth and bodily development, but also for his or her psychological temperament. On one side, the "three qualities" (*triguna*) on which man's physiognomic traits and character depend are in fact closely related to the food that is consumed. Thus, foods of the *sattvic* type—milk and foods derived from plants—produce calmness and mental serenity; those of the *rajasic* type—spicy, sour and bitter foods—generate energy; while those of the *tamsic* type—such as meat and alcohol—dull the mind and lead to the degeneration of human nature.[51] Eating meat, for example, means absorbing negative vibrations that alter the nervous system and amplify feelings such as fear, anger and jealousy which are harmful to health. On the other side, the killing of innocent animals in order to fill one's stomach is contrary to karma and has harmful consequences for man due to the evil he has inflicted on the animal.

   Based on these findings, the principle of non-violence (*ahimsā*), one of the pillars of dharmic religions, is generally interpreted as synonymous with vegetarianism, even though Hindu scriptures are not explicit on this point. As a matter of fact, vegetarian regimes best exemplify the rule of "compassion" towards animals and respect for all life forms; consequently, most Hindus abstain from the consumption of beef (moreover the cow is considered a sacred animal), meat derivatives, and in the case of lacto-vegetarians, even eggs. As a result, the doc-

[50] Hamdan et al. (2024), p. 8.

[51] Agte and Chiplonkar (2007).

trine tends to be against the consumption of cultured meat, as it sees the use of animal stem cells as an example of human arrogance, or in any case as a product not suitable for a vegetarian diet.[52]

Similar considerations apply to other Dharmic religions, such as Sikhism, Buddhism and Jainism, which are ostensibly more respectful, especially in the Jain version, of the dictates of a vegetarian diet. However, there is no legal prohibition in these religions against a Novel Food such as cultured meat. For example, Sikhs do not disdain the consumption of meat derived from animals that have been killed instantaneously; thus, cultivated meat derived from stem cells extracted from such meat seem permissible.[53]

4. Buddhism: many Buddhists advocate non-harm and are often vegetarians. Since the first precept of Buddhism is abstention from killing cultivated meat may not violate the precept.[54] However, views can differ significantly among Buddhist communities regarding the ethical implications of cultivated meat.[55] Although food is a privileged channel of communion with God, there is currently a lack of interest in the literature on the subject of cultivated meat. According to a few articles, there is a conviction that cultivated meat may not violate the principles of the *Shojin Ryori* diet, which envisages the total absence of meat, fish and animal derivatives in order to attain enlightenment and live in harmony with the world. More in general, there are currently no dogmatic or practical obstacles to entry of cultivated meat into the diets of Buddhists of south-east Asia.[56]
5. Christianity: There is a wide range of beliefs among Christians regarding meat consumption. Christianity has relative freedom of dietary rules, so there are no specific restrictions on the consumption of meat or related products. The basic assumption is that God rules over men, while men rule over animals. There are no rules or specific prohibitions, except for general calls for moderation in the form of suggestions, not impositions. Dietary prescriptions only appear in canon law concerning fasting and abstinence:[57] for the Orthodox Church, fasting is preparatory to communion with God; during special liturgical periods, such as Lent, meat consumption is discouraged; among Christian texts, we find a list of forbidden foods considered impious in Leviticus.[58] Thus, most Christian denominations do not have specific laws governing meat, including cultivated varieties, but ethical considerations may influence individual choices.

To conclude, as cultivated meat technology advances and becomes more widespread, religious leaders and scholars are likely to continue exploring its implications in the context of their respective teachings and values.

---

[52] Hamdan et al. (2021), p. 16.

[53] Saporito (2024), p. 11.

[54] Kwon and Park (2023), p. 346.

[55] *Ibid*, p. 347.

[56] Saporito (2024), p. 12.

[57] Anello (2024).

[58] Saporito (2024), pp. 6–7.

5 Conclusions

"Food is important to people on a very fundamental level."[59] The emerging field of cultivated meat, which is primarily of interest to technologists and the food industry, should therefore be brought to the public's attention through information and debate on a larger scale. This notion is supported by the historical significance of food and meat in religious traditions, where animals were considered valuable and were subject to strict regulations.

Ensuring a stable meat supply for communities with religious dietary requirements poses a significant social and commercial challenge, but also offers potential opportunities. The sources of meat are limited and its production is resource-intensive. Over the centuries, meat production has had a notable environmental impact, leading to the conversion of an increasing amount of forest and other land into pasture. Despite this, meat was not always a primary food source and was often consumed for reasons of taste and social status, rather than nutritional necessity.[60]

Meat production and consumption were viewed as essential human pursuits, tied closely to the relationship between humans and the natural world. Meat was a symbol of human dominance over animals.[61] In religious laws, the ecological consequences of meat consumption were understood as a trade-off, since meat may provide valuable nutrients but also promotes a predatory instinct in humans that needs to be controlled. Meat contains some essential amino acids that are found less abundantly in plants, so that vegan or vegetarian diets without meat or fish are not fully equivalent to diets with meat.[62]

The negative impacts of the current agribusiness system are seen as both a health concern and an ethical issue, as the profligate production of meat depletes valuable resources and harms the environment.[63] This has led some to argue that meat consumption should be restricted. In an eco-holistic perspective, such restrictions would be in the interests of Earth's biodiversity and the web of life. They would not be a reflection of human weakness, but rather recognition of the need for balance and sustainability.

To address these challenges, large-scale production of cultured meat could offer a more sustainable alternative to traditional meat production methods. It could help meet increasing world demand for meat while also aligning with religious dietary requirements and broader sustainability goals.

[59] Fiddes (1989), p. 278.

[60] Chiles and Fitzgerald (2018), p. 8.

[61] Fiddes (1989), p. 10; Fiddes (1994), p. 277.

[62] Mariotti and Gardner (2019).

[63] Kalechofsky (2004), p. 174.

References

Agte VV, Chiplonkar SA (2007) Linkage of concepts of good nutrition in yoga and modern science. Curr Sci 92(7):956–961

Alicino F (2023) Alimentazione sostenibile fra menù teologali e teodicee secolarizzate. In: VV AA (ed) L'etica ambientale e lo sviluppo sostenibile nel diritto delle religioni. Giuffrè, Milano, pp 81–132

Anello G (2024) Persone capitali. Storie di entità giuridiche. Marietti 1820, Bologna

Astori D (2010) The seven Noachian precepts between monotheistic religions and human liberties. Dionysiana 4:356–389

Berkman J (2004) The consumption of animals and the Catholic tradition. In: Sapontzis SF (ed) Food for thought: the debate over eating meat. Prometheus Books, Amherst, pp 198–208

Berzin A (2014) Improving the environment through vegetarianism https://studybuddhism.com/en/advanced-studies/history-culture/interreligious-dialogue/improving-the-environment-through-vegetarianism. Accessed 8 Apr 2025

Bondi D (2022) Cultivating meat: ethical and ontological reflections about cultured meat. Teoria. Rivista di Filosofia 42(1):149–159

Bonne K, Verbeke W (2007) Religious values informing halal meat production and the control and delivery of halal credence quality. Agric Hum Values 25(1):35–47

Bryant CJ (2020) Culture, meat, and cultured meat. J Anim Sci 98(8):1–7

CEI document (1999) The Christian meaning of fasting and abstinence. Pastoral note of the Italian episcopate. Notiziario della Conferenza Episcopale Italiana

Centre for Religion, Human Values, and International Relations, Dublin City University (2023) Collective action for ending a collective problem: a multi-stakeholder project on global food security report. Centre for Religion, Human Values, and International Relations, Dublin City University, 25 July 2023. https://www.dcu.ie/sites/default/files/inline-files/Report%20-%20final%20%283%29.pdf. Accessed 8 Apr 2025

Chabba S (2021) COVID: India's Sikh community helps amid pandemic. https://www.dw.com/en/india-covid-sikhs/a-57756219. Accessed 8 Apr 2025

Chiles RM, Fitzgerald AJ (2018) Why is meat so important in western history and culture? A genealogical critique of biophysical and political-economic explanations. Agric Hum Values 35(1):1–17

Daido Loori J (2012) Bringing the sacred to life – the daily practice of Zen ritual. Shambhala, Boston. https://terebess.hu/zen/mesterek/bring.pdf. Accessed 8 Apr 2025

Devi SM, Balachandar V, Lee SI, Kim IH (2014) An outline of meat consumption in the Indian population – a pilot review. Korean J Food Sci Anim Resour 34(4):507–515

Ferry M (2020) What's India's beef with meat? Hindu orthopraxis and food transition in India since the 1980s. Sociol Forum 35(2):511–534

Fiddes N (1989) Meat: a natural symbol. PhD thesis, University of Edinburgh, Edinburgh

Fiddes N (1994) Social aspects of meat eating. Proc Nutr Soc 53(2):277–279

Foltz RC (2004) Is vegetarianism un-Islamic? In: Sapontzis SF (ed) Food for thought: the debate over eating meat. Prometheus Books, Amherst, pp 209–222

Gaffney J (2004) Eastern religions and the eating of meat. In: Sapontzis SF (ed) Food for thought: the debate over eating meat. Prometheus Books, Amherst, pp 223–235

Gandhi R (2006) Mohandas: a true story of a man, his people, and an Empire. Penguin Books India, New Delhi

Gill A (2013) Religious liberty and economic development. Rev Faith Int Aff 11(4):5–23

Hamdan MN, Post M, Ramli MA, Kamarudin MK, Md Ariffin MF, Zaman Huri NMF (2021) Cultured meat: Islamic and other religious perspectives. Int J Islamic Civiliz Stud 8(2):11–19

Hamdan MN, Jalil RA, Ramli MA, Ramli N, Ibrahim MNA, Ab Rahman MF, Thaidi HAA, Rahman NNHA (2024) A review of the discussions on cultivated meat from the Islamic perspective. Heliyon 10(7):e28491

Huang CC (2017) Food ethics in ancient China: a Confucian perspective. J Cult Interact East Asia 8:27–34
Jenkins W, Chapple CK (2011) Religion and environment. Annu Rev Environ Resour 36:441–463
Kalechofsky R (2004) The Jewish diet and vegetarianism. In: Sapontzis SF (ed) Food for thought: the debate over eating meat. Prometheus Books, Amherst, pp 168–176
Kwon J, Park Y (2023) Buddhist perspectives on cultivated meat. Nat Food 4(5):346–347
Leroy F, Praet I (2015) Meat traditions. The co-evolution of humans and meat. Appetite 90:200–211
Linzey A (2004) The theological debate about meat eating. In: Sapontzis SF (ed) Food for thought: the debate over eating meat. Prometheus Books, Amherst, pp 186–197
Mariotti F, Gardner CD (2019) Dietary protein and amino acids in vegetarian diets—a review. Nutrients 11(11):2661
Moreno A (2008) Feeding the democracy: the Athenian grain supply in the fifth and fourth centuries BC. Oxford University Press, Oxford
Muttaqi SA (2016) The sacrifice of 'Eid al-Adha': an Islamic perspective against animal sacrifice. https://www.animalsinislam.com/islam-animal-rights/sacrifice-eid-al-adha/. Accessed 8 Apr 2025
Ramachandran R, Kumar A, Gopi Sundar KS, Singh Bhalla R (2017) Hunting or habitat? Drivers of waterbird abundance and community structure in agricultural wetlands of southern India. Ambio 46(5):613–620
Regenstein J, Chaudry M, Regenstein C (2003) The kosher and halal food laws. Compr Rev Food Sci Food Saf 2(3):111–127
Saporito L (2024) Global meat complex, nuove tecnologie alimentari, sostenibilità: La prospettiva delle religioni sulla carne coltivata. Diritto Pubblico Europeo Rassegna online 22(2):1–15
Schwartz RH (2025) When eating meat was a sacrifice. https://www.myjewishlearning.com/article/when-eating-meat-was-a-sacrifice/. Accessed 8 Apr 2025
Segal I, Blazer S (2020) The Maimonides model for a regimen of health: a comparison with the contemporary scenario. Rambam Maimonides Med J 11(4):e0029
The future of world religions: Population growth projections, 2010-2050 (2015). https://www.pewresearch.org/religion/2015/04/02/religious-projections-2010-2050/. Accessed 8 Apr 2025
The Global Religious Landscape (2012). https://www.pewresearch.org/religion/2012/12/18/global-religious-landscape-exec/. Accessed 8 Apr 2025

Who Hates Cultured Meat? Examining Public Opposition to Cellular Agriculture

Garrett M. Broad and Leland Glenna

Abstract Despite its limited availability as a consumer product, recent years have witnessed significant scepticism, concern, and outright rejection regarding the emerging science and industry of cultured meat. Far from a cohesive group, opposition comes from a diverse assemblage of stakeholders, motivated by a heterogeneous set of sometimes contrasting ideological principles and personal incentives. The aim of this chapter is to catalogue the networked actors, arguments and broader socio-technical ideologies that are reflected in this diffuse opposition to cultured meat. We suggest that methods from consumer psychology and marketing, which have been used extensively to understand public opposition, are valuable but incomplete for understanding this terrain; instead, the task calls for greater engagement with conceptual principles highlighted in critically-oriented domains such as the social studies of science, communication and media studies, and the anthropology of food. Through this lens, issues related to culture, ideology, power, politics and our unique digital communication environment become vital to understanding the cultured meat debates. From this foundation, the chapter explores the perspectives of vocal advocates against the technology, describing three broad but intersecting perspectives, based in cultural tradition, techno-economic critique, and conspiracy, that have featured prominently in the oppositional coalition.

G. M. Broad (✉)
Rowan University, Ric Edelman College of Communication, Humanities and Social Sciences, Glassboro, NJ, USA
e-mail: broad@rowan.edu

L. Glenna
Pennsylvania State University, Department of Agricultural Economics, Sociology, and Education, University Park, PA, USA
e-mail: llg13@psu.edu

G. Formici et al. (eds.), *Cell-Based Meat in the European Union and Beyond*,
https://doi.org/10.1007/978-3-032-14624-3_10

Keywords Meat · Cultured meat · Consumer acceptance · Socio-technical imaginaries · Conspirituality

> Our results indicate that cultured meat is not necessarily a less resource-intensive protein product than conventional meat and, in fact, may lead to significantly greater environmental impact if the industry is unable to fully transition from pharmaceutical-grade ingredients to food/feed-grade inputs.—Derrick Risner and colleagues, Food Science Researchers.[1]

> We are the first nation to ban it, to the chagrin of multinational companies that hoped to make monstrous profits, jeopardizing the jobs and health of citizens.—Francesco Lollobrigida, Italian Minister of Agriculture, Food Sovereignty and Forestry.[2]

> We cannot advocate a food made from animals and at the same time claim to want to end the exploitation and use of animals. It is impossible.—Lori Girshick, Feminist Vegan Sociologist.[3]

> Study: Bill Gates' Lab Grown Meat Causes Cancer in Humans[4]—Sean Adl-Tabatabai, Author at the Digital Media Outlet *People's Voice*.

1 Introduction

For a product that has been consumed by only hundreds, perhaps a few thousand, individuals at the time of writing, cultured meat has already elicited an impressive range of strongly-held opinions. Within the cultured meat industry and in much of the social science literature on the topic, a central question has been to explore what factors might predict support or acceptance for the product, and from there to leverage product development and strategic communication to achieve those goals.[5] Often remarked upon, but less often made the central focus of inquiry, have been questions about scepticism, concern, or outright rejection and opposition to cultured meat. Understanding this aspect, however, offers a variety of useful insights, not only about the potential barriers facing cultured meat as it aims to become a viable consumer product, but also about the broader nature of contemporary society's relationship to meat, cultural perceptions of food technology, and the digital information ecosystem that shapes modern debates about the politics of food.

Far from a cohesive group, opposition to cultured meat comes from a diverse assemblage of individuals, communities of interest and professional organizations. These varied stakeholders are motivated by a heterogeneous set of sometimes contrasting ideological principles and personal incentives. The strength of opponents' attitudes ranges from firmly held to potentially malleable, while their claims-making

[1] Risner et al. (2025).

[2] Quoted in Hogg (2023).

[3] See Girshik (2019).

[4] Adl-Tabatabai (2023).

[5] See Tsvakirai et al. (2024).

differs substantially in terms of the types and quality of evidence put forth. Some draw from empirically sound technical arguments, others from conspiratorial misinformation; in other contexts, concern emerges from implicit negative associations, while for some their opposition is a product of deeply-held social or economic value systems. The aim of this chapter is to catalogue the networked actors, arguments and broader socio-technical ideologies that are reflected in the diffuse opposition to cultured meat. A fully exhaustive compendium is beyond the scope or intention of this chapter, of course. Rather, the goal of this work is to provide those interested in the future of cultured meat with frameworks for understanding the nature of debate and provide examples of illustrative perspectives, as well as to offer some brief recommendations for how to best engage with this landscape moving forward.

Methodologically, this chapter is informed by years of multi-methodological engagement by members of the research team, both individually and collaboratively, in the social science of meat, alternative proteins and cellular agriculture, including as part of a USDA National Institute of Food and Agriculture AFRI Competitive Grant project (PENW-2018-08967). The themes explored in the pages ahead are grounded in a review of interdisciplinary scholarly literature and critical analysis of media and policy communications; they are further informed by our own findings from surveys and focus groups, although this latter subset of empirical data is not featured directly. Parts "Scientific and Sustainability Challenges" and "Legal and Policy Debate" of this volume can be referred to in order to understand the technical and regulatory dynamics of cultured meat. From here, the work begins with a discussion of key assumptions in consumer research and marketing psychology, as it outlines key findings with respect to consumer opposition to cultured meat, while also noting the limitations of this approach. It then follows with a discussion of how concepts from the critical studies of food and technology—including disciplines such as Science and Technology Studies (STS), anthropology, sociology, and communication and media studies—aid in understanding the cultural debates that surround cultured meat. From there, the chapter uses those concepts to explore the perspectives of vocal advocates against the technology, describing three broad but intersecting critical perspectives—based in cultural tradition, techno-economic critique, and conspiracy—that have featured prominently in the oppositional coalition. It concludes with suggestions for future research and public engagement practices for both supporters and critics alike.

2 Theoretical Underpinnings of Cultured Meat Perceptions

2.1 Consumer Acceptance Approaches

The vast majority of research into public perceptions of cultured meat has been conducted using frameworks and methods widely employed in consumer research, marketing psychology, and related areas of scholarship. Research in this domain has long studied the factors that shape consumer perceptions of meat and meat products, noting that consumer preferences derive from a combination of intrinsic and extrinsic attributes. Sensory qualities, nutrition composition, and food safety are generally regarded as the most important intrinsic characteristics, with price the dominant extrinsic characteristic; additional considerations, for some consumers, include concerns about chemical additives, animal welfare, and environmental impact, among others. Other key personal factors—such as lifestyle and marketing—are intertwined with these intrinsic and extrinsic factors in shaping consumer preferences and purchasing behaviour.[6]

A subset of food-related consumer-preference research has explored consumer perceptions of novel foods and new food technologies, with a particular focus on theories that explain consumer aversion. A mix of models from traditional economics, behavioural economics, psychology, and risk sociology has been used to explain these dynamics. As Jayson Lusk and colleagues summarized, over time, "Research has evolved from a rationalist discourse about the lay-expert divide in risk assessment to an understanding that much of human decision making is driven by heuristics and feelings."[7] Key heuristics as identified in this literature include those shaped by general affective associations, levels of trust in producers and the food industry, as well as the common notion that "natural-is-better." There are also a variety of framing effects (shaped by media discourse and information provision such as food labels), preferences based in cultural differences, and individual personality differences (including concepts such as "disgust sensitivity" and "food technology neophobia").[8]

Much of the research in the study of consumer preferences and novel food technology has been driven by what can be termed the "consumer acceptance" paradigm.[9] That is to say, the primary goal of the research has been to catalogue the factors that are associated with a willingness to try or purchase the food product, as well as those associated with resistance or rejection, and from there devise practical strategies for promoting further openness and uptake, both among those already interested and those who are reluctant. This general paradigm has been adopted by many scholars in the burgeoning literature on consumer perceptions of cultured meat,[10] expounding in detail on the challenges to consumer acceptance; the key personal, product-related, regulatory and external factors that shape consumer perceptions; and the strategic communication approaches that might be taken to bolster levels of consumer acceptance globally.

---

[6] See de Araújo et al. (2022).

[7] Lusk et al. (2014), p. 399.

[8] See Siegrist and Hartmann (2020).

[9] See Broad and Biltekoff (2023) and Biltekoff and Guthman (2023).

[10] See *Navigating Consumer Acceptance of Cultured Meat: Insights from Social Science Research* by Burdorf and Lin-Hi; *Socio-Demographic and Behavioural Determinants of Consumer Attitudes to Cell-Based Meat and Insect-Flour-Based Products: An Empirical Study of Italians' Propensity to Accept New Foods for Human Nutrition and Pet Food* by Degli Antoni and Faillo, both in this volume.

The assumptions of the consumer research and marketing psychology literature, combined with the specific empirical findings of work on novel foods and cultured meat, have helped craft a paradigm for understanding consumer perceptions and developing strategies for public communication. The field recognizes that eating behaviour is shaped by a mix of both rational and automatic heuristic thinking; that demographics and social environments are key predictors of decision-making; and that improvements in the product, price, place, and promotion should be central to marketing activities. Researchers and the emerging cultured meat industry recognize, therefore, that certain identifiable groups will be harder to reach as potential consumers, and that resources for outreach should be devoted accordingly, with those segmented as potential early adopters the focus of initial targeting. Among those who resist, targeted information that comes from trusted sources—combined with improvements in product taste, price, and convenience—may allow broader market expansion in the years to come.

To summarize some of the key takeaways of this scholarship to date, acceptance ratings of cultured meat have been quite low across a wide variety of cultural contexts, even when compared to most other meat alternatives, and especially when compared to conventional meat counterparts. Some research has suggested that younger, more highly educated people, and individuals with stronger levels of meat attachment express relatively higher acceptance ratings.[11] To return to the guiding question that serves as the title of this chapter, flipping the frame of consumer acceptance to that of rejection or scepticism, it can be said that the demographics associated with opposition to cultured meat include being older, from a rural community, politically conservative, and possessing lower levels of formal education. There is also some evidence that women are less interested in the product than men, and that some national cultures—notably the French—may be particularly against the idea of cultured meat. With respect to attitudes and personality variables, concerns about naturalness, high levels of disgust sensitivity, high levels of food technology neophobia, a general distrust of the food and agricultural industries, and particular opposition to Genetically Modified Organisms (GMOs) in food have been identified as predictors of rejection. From an informational perspective, those who are less aware of cultured meat products and processes are less likely to believe in the technology's potential benefits, are generally indifferent to values-based food and meat purchasing, and are less likely to endorse it. In terms of willingness to eat the product, some research has also shown that vegetarians and vegans are less likely than omnivores, possibly due to negative associations with the sensory characteristics of meat and their existing satisfaction with plant-based offerings and alternative protein options. Some of the biggest barriers to the embrace of cultured meat among consumers are nearly impossible to know at this stage given its novelty and highly limited distribution—that is to say, consumers remain unsure about the sensory qualities of cultured meat, unclear on whether it will actually be available

[11] See Onwezen and Dagevos (2024).

at an affordable price, as well as concerned about potentially unknown long-term safety and health risks because products are not yet widely available.[12]

As useful as research in this empirical and application-oriented tradition is, however, there are significant limitations that narrow the field's understanding of the landscape of debate about cultured meat, in general, and the emerging coalition of steadfast opponents, specifically. The heavy reliance on survey research methods, even when occasionally combined with interviews or other qualitative research approaches, relies on responses from individual consumers, separated from the social dynamics that are central to understanding real-world eating behaviour. Through this process, difficult to measure cultural elements are left underexplored or simplified via the use of imperfect if validated survey measures.[13] The low levels of consumer awareness of the topic, alongside the near impossibility of doing real-world choice experiments, create significant risks of forced-response bias, as respondents consider their potential behaviour only in hypothetical scenarios.[14] In addition, the work tends to overlook the potential of highly engaged opponents, even if a small-but-vocal minority, to dominate conversation in the public sphere, stalling progress for cultured meat across both social norm and policy levels. To better understand these dynamics, and to further investigate the question of who hates cultured meat and why, it is therefore important to delve into literature from the critical studies of food and technology.

2.2 Critical Theories of Food, Meat and Tradition

Rather than any formal discipline, the critical study of food is best understood as existing at the intersection of diverse scholarly approaches that are interested in the cultural practices and implications of food production and consumption. This includes traditional disciplines such as sociology and anthropology, but also other interdisciplinary fields such as Science and Technology Studies (STS) or communication and media studies. It is well beyond the scope of this chapter to cover all concepts relevant to debates about cultured meat, but a few frameworks that are particularly connected to the case are worth covering at this stage.

The first relates to the uniquely important role of meat in human society. The procurement and eating of meat, of course, is closely connected to the history of human evolution and the nature of nutrition security over time. Yet, its biological implications are only part of the story, as meat is also central to fundamental social and economic structures, understandings of community and human-environment relations, and individual cultural identities.[15] In his work *Meat: A natural symbol,*

---

[12] See Bryant and Barnett (2020), Hartmann and Siegrist (2023), Mancini and Antonioli (2020) and Pakseresht et al. (2022).

[13] For an overview, see Harris et al. (2009).

[14] See To et al. (2024).

[15] See Smil (2013). For a focus on eco-religions and cultured meat, see also *Food Justice in Religious Laws: Hypotheses on Cultivated Meat* by Anello in this volume.

the anthropologist Nick Fiddes argued that eating meat represents human control of the natural world and serves as a statement of humanity's supreme power. In response to the question, "Why do we value animal flesh so highly, in spite of the consequences for the creatures involved?," his analysis concluded that we do not esteem meat *in spite* of the domination of sentient beings, but rather partly *because* of that power: "It is not that we each consciously exult in our mastery of nature whenever we bite into a piece of flesh, but we are brought up within a culture which has regarded environmental conquest as a laudable goal, and which has deployed meat as a primary means to demonstrate it."[16]

Elsewhere, the critical feminist scholar Carol J. Adams noted that people with power have always eaten meat, going further to outline the intersecting economic and gender-based power dynamics inherent to the phenomenon: "Dietary habits proclaim class distinctions, but they proclaim patriarchal distinctions as well. Women, second-class citizens, are more likely to eat what are considered to be second-class foods in a patriarchal culture: vegetables, fruits, and grains rather than meat. The sexism in meat eating recapitulates the class distinctions with an added twist: a mythology permeates all classes that meat is a masculine food and meat eating a male activity."[17]

The environmental ethic of Fiddes and the feminist vegetarian perspective of Adams are surely not shared by all. Still, even among strong proponents of the belief that meat should remain a primary component of contemporary foodways, there is little debate about the fundamental point that "meat traditions" are fundamentally intertwined with "elements of gender differentiation, cooperation and reciprocity, social stratification and power, religion, cultural expression, and identity."[18] And while these traditions have deep historical roots, as with all traditions, they are also in constant flux, being shaped and reshaped based on social, economic, and environmental contingencies. As the sociologist Anthony Giddens once explained, all traditions are "invented traditions" not just in the modern period, but throughout human history. Giddens added that such traditions "always incorporate power, whether they are constructed in a deliberate way or not." Importantly, in the contemporary age of globalization, at a time when tradition has retreated in the face of cosmopolitanism, it is *fundamentalism* that emerges as a counter-response: "Fundamentalism is beleaguered tradition. It is tradition defended in the traditional way—by reference to ritual truth—in a globalizing world that asks for reasons. Fundamentalism, therefore, has nothing to do with the context of beliefs, religious or otherwise. What matters is how the truth of beliefs is defended or asserted."[19] As subsequent sections of this chapter will demonstrate, some of the core opposition to cultured meat emerges from those who see the enterprise as an affront to the ritual truth of "traditional" meat.

[16] Fiddes (2004), p. 228.

[17] Adams (2020), p. 4.

[18] Leroy and Praet (2015), p. 200.

[19] Giddens (2003), pp. 40–49.

2.3 Socio-Technical Imaginaries

Intersecting with these discussions of culture, tradition, and power are considerations of the role of technological innovation in society. The concept of technology, in and of itself, is often equated with specific artifacts—machines, tools, inventions. But as the scholar Sheila Jasanoff pointed out, technological objects are "thoroughly enmeshed in society, as integral components of social order" adding that it has been a central aim of the field of science and technology studies to bring "social thickness and complexity back into the appreciation of technological systems." She further defined the notion of "socio-technical imaginaries" as key to understanding debates about technology and society, describing the concept as, "collectively held, institutionally stabilized, and publicly performed visions of desirable futures, animated by shared understandings of forms of social life and social order attainable through, and supportive of, advances in science and technology." Such visions can emerge from the minds of single individuals or collectives, "gaining traction through blatant exercises of power or sustained acts of coalition building." Fundamentally, socio-technical imaginaries "encode not only visions of what is attainable through science and technology, but also of how life ought, or ought not, to be lived; in this respect they express a society's shared understandings of good and evil."[20]

In Charles Mann's book *The Wizard and the Prophet,* he outlined two archetypal approaches to advancing sustainability that help to illustrate socio-technical imaginaries in action. As he described, "Prophets look at the world as finite, and people as constrained by their environment. Wizards see possibilities as inexhaustible, and humans as wily managers of the planet." In this framework, wizards are champions of industrial-scale growth and development, supportive of technologically-driven and capital-intensive approaches to solving environmental challenges; undergirding this belief is a clear pessimism about the potential for humanity to voluntarily alter its behaviour in a way that would make a substantive pro-environmental difference. Prophets, by contrast, champion "stability and preservation as our future and our goal." They are deeply critical not only of large-scale technological innovations, but also of the coalition of corporate and governmental interests that are often at the helm of those technologies. Instead, they push forward a vision of democratically-led and locally-oriented social change, voluntary simplicity or degrowth, and nature-based solutions as the key to sustainability. Indeed, the assumed goodness of nature is often central to the prophet's worldview, building common cause with many traditionalists who demonstrate, in the words of religious scholar Alan Levinovitz, that "the appeal to natural goodness is among the most influential arguments in the history of human thought."[21]

To be clear, the wizard and prophet framework represents ideal types, with many individuals possessing elements of both—however, especially among vocal advocates within sustainability debates, one or the other pole tends to be dominant. This

---

[20] Jasanoff (2015), pp. 2–4.

[21] Levinovitz (2020), p. 3.

is much to the chagrin of scholars like Levinovitz, who believes that an "oppositional binary between "natural" and "unnatural" inhibits constructive dialogue about humanity's most pressing problems."[22] In reality, the wizard and prophet perspectives both come with strengths and weaknesses. Wizardly thinkers point to a host of local and global data points—related to extreme poverty, life expectancy, childhood mortality, hunger, malnutrition, pollution, and many more—to make the case that our long-term developmental trends have led to a "silent miracle of human progress."[23] They also lament that things could be even better if technological innovation were not frequently undermined by the public's risk aversion, by widespread misinformation, and by a failure of leadership to better communicate the benefits of innovation to the public *en route* to gaining acceptance.[24]

This perspective, however, is countered by those who insist that, in the words of Jasanoff, "institutional deficiencies, unequal resources, and complacent storytelling continue to hamper profound reflection on the intersections and mutual influences of technology and human values." Jasanoff further argued that, "the power to set the rules of the game for governing technology rests with capital and industry, not with the political representatives of the working, consuming, and too often suffering masses."[25] From a prophet-oriented perspective, there are plenty of examples of human flourishing that are driven not by technology but by nature-based solutions and collective action instead, and which are often erased from major policy debates. There are also many examples of destructive technologies, as well as many technologies better understood as "non-disruptive disruptions" those that make only incremental (at best) environmental change while accruing significant benefits to capital interests and foreclosing other more potentially transformative alternatives.[26] Advocates in this vein acknowledge that technology has a role to play in social change, but contend that its historical interconnection with capitalist logics actually constrains its ability to deliver on large-scale well-being. As degrowth advocate Jason Hickel put it, once we are "liberated from the growth imperative, we will be free to focus on different *kinds* of innovations—innovations designed to improve human and ecological welfare, rather than innovations designed to speed up the rate of extraction and production."[27] Distinctions between wizardly and prophetic sociotechnical imaginaries play an important role in shaping the landscape of debate on cultured meat, as specific examples later in this chapter will show.

[22] *Ibid.*

[23] Rosling et al. (2018), p. 51.

[24] See Juma (2016).

[25] Jasanoff (2016), pp. 265–266.

[26] See Goldstein (2018).

[27] Hickel (2020), p. 157.

2.4 Digital Media, Conspirituality and Diagonalism

The concept of socio-technical imaginaries helps demonstrate how debates about future food system pathways are often grounded in good-faith differences in worldview, empirical interpretation, or the combination of both. However, to understand the nature of debates in the modern public sphere—related to health, technology, and many other vital concerns—analysts must also reckon with the widespread prevalence of misinformation (false or inaccurate information, whether intentional or not), disinformation (misinformation actively intended to mislead), and conspiratorial thinking. Reflecting on the contentious debates about genetically modified organisms (GMOs), for instance, the author Mark Lynas carefully recounted how opponents of the technology relied on misinformation, pseudoscience, and fear-mongering rhetoric to impede its development and deployment. This is not to say that proponents of GMOs were always in the right—as Lynas conceded, the industry too often bulldozed over broader moral, economic, political, and ecological concerns that were worthy of deliberation, sometimes slandering opponents in the process. What was needed in the GMO debates, and in future food technology deliberations in general, are middle way critiques that balance concerns about unequal power relations with an insistence on evidence-based public sphere debate. As Lynas argued, "Let's use science as the wonderful tool that it is, but let's also respect people's feelings and moral intuitions about the proper extent of human intrusion into the biosphere."[28]

Especially in the age of digital and social media, misinformation at the intersection of public health, technology, and "wellness" culture has run rampant, particularly supercharged during the COVID-19 pandemic and vaccination campaign. The emergence of what has been termed a "conspirituality" movement—which blends New Age alternative health precepts with a conspiratorial worldview that assumes the problems of the world result from the actions of diffuse, nefarious cabals of power-brokers who must be stopped at all costs—has been closely catalogued by co-authors and podcasters Derek Beres, Matthew Remski, and Julian Walker. Their work chronicles the "explosion of new conspirituality themes through the interplay of COVID-denialism, the reactionary politics of Trump, and a cultish surge of prophetic spiritual dynamics." They added that "Conspiritualist viral messaging slipped into an already seamless e-commerce delivery system that leverages social media, email marketing, and websites designed to sell books, video courses, and online conferences to passionate followers with shared fears and longings."[29] Importantly, they echoed Lynas's message that there are real reasons to be sceptical of dominant power structures, but criticized the way these concerns are channelled by the conspiritualists: "True believers of conspirituality turn legitimate critiques that can be addressed through democratic policymaking into apocalyptic crises that can only be

---

[28] Lynas (2018), p. 270.

[29] Walker et al. (2023), p. 20.

solved by miracles. They've found a compelling storytelling machine that amplifies their passions."[30]

The conspirituality concept overlaps significantly with another astute observation in contemporary critical theory—the emergence of "diagonalism." As Callison and Slobodian outlined:

> Born in part from transformations in technology and communication, diagonalists tend to contest conventional monikers of left and right (while generally arcing toward far-right beliefs), to express ambivalence if not cynicism toward parliamentary politics, and to blend convictions about holism and even spirituality with a dogged discourse of individual liberties. At the extreme end, diagonal movements share a conviction that all power is conspiracy… Diagonal movements trade in both familiar and novel fantasies about elite control. They attack allegedly "totalitarian" authorities, including the state, Big Tech, Big Pharma, big banks, climate science, mainstream media, and political correctness.[31]

Further documenting the rise of these diagonalist coalitions, author Naomi Klein detailed how, "These formations bring together many disparate political and cultural strains: the traditional right; the QAnon conspiratorial hard right; alternative health subcultures usually associated with the green left; a smattering of neo-Nazis; parents (mainly white mothers) angry about a range of things happening and not happening in schools (masks, jabs, all-gender bathrooms, anti-racist books); small-business owners enraged by the often-devastating impacts of COVID controls on their bottom lines…".[32] Combining a set of traditionalist cultural arguments and a prophet-aligned socio-technical imaginary with a heavy dose of conspirituality, some of the key players in these diagonalist communities have seized upon meat as a central symbol. The work of Marek Muller and colleagues, for instance, chronicled popular "primal" social media influencers and their broader rhetorical networks in their advocacy of extreme levels of meat and animal product consumption. Drawing on conspirituality and diagonalism as analytical concepts, the researchers showed how "seemingly unrelated ideas—spirituality, meat consumption, masculinity, and race—as well as groups—alternative belief systems, white nationalists and health and diet influencers—can be closely linked within primal rhetorical networks."[33] The discussion that follows in this chapter, highlighting the perspectives of some of cultured meat's staunchest and most media savvy opponents, will further show the dynamics of conspirituality and diagonalism in action.

---

[30] *Ibid*, p. 56.

[31] Callison and Slobodian (2021), para. 2–3.

[32] Klein (2024), p. 100.

[33] Muller et al. (2024), p. 5.

3 Cultured Meat's Opponents

The review of theories offered in the previous sections of this chapter was wide-ranging, offering a set of conceptual frameworks that can serve as a primer to understand the diverse influences that shape public concern, in general, and steadfast opposition, in particular, with respect to the future of cultured meat. The remainder of this chapter will further illustrate key arguments that have been made by those sceptical or firmly opposed to the enterprise, pointing throughout to those frameworks outlined above. It will proceed from here with a discussion of critiques based primarily in considerations of cultural tradition, then explore those offered on technical or economic grounds, before it concludes with critiques of a conspiritualist or diagonalist nature. In so doing, it will point to both overlaps and distinctions between key arguments and participants across these different camps, showing the heterogeneous complexity of the coalition that has emerged as a counter to cultured meat's hopeful proponents.

3.1 The Traditionalist Critique

The traditionalist critique that has emerged in response to the development of cultured meat includes at least two dominant and intersecting strains—one concerned about the implications of the novel food for public health; the other concerned about its impact on local communities, culture, and occupational ways of life. This first objection aligns closely with a common finding in the consumer perceptions literature, with opposition grounded in a fundamental aversion to the "unnaturalness" of the technology, motivated by a belief that the product is simply gross from a gustatory perspective or too novel to be trusted in terms of its health and safety impacts. Here, some of the most prominent voices of opposition have been those with connections to movements for so-called "good food"—that is, those who advocate for the value of local, organic, whole food cuisines—as well as those with long-standing opposition to GMOs in the food system.[34]

In an article framed as a set of questions for the science and industry of cultured meat, Errol Schweizer—a long-time grocery and natural products merchandising professional, as well as Board Member for the non-profit Non-GMO Project—drew direct links between the novelty of the technology and its potential harms to human and public health. Schweizer offered comparisons to the introduction of trans fats into foods, which were later phased out due to its potential contributions to heart disease and cholesterol; he further lamented the rise of GMOs and the use of glyphosate in GMO crop production, which he noted have been the subject of health and environmental concerns, cancer lawsuits, and hefty settlements. From there, he offered a set of critical questions for the cultured meat industry—focused on an

[34] See also Broad (2023b).

intersecting set of concerns related to health and safety, nutritional composition, environmental impact, and social equity—posed as key to "learning from some of the unforeseen consequences of previous large scale food tech innovation."[35]

Schweizer and many other good food movement critics of cultured meat have concurrently advocated reductions in meat consumption overall, arguing against the current system of industrialized animal food production. In an editorial that pushed back against enthusiasm for both plant-based and cultured animal product alternatives, for instance, journalist and advocate Mark Bittman drew from the natural argument in a pithy rejoinder to alternative protein's boosters: "It's funny, in a way, because the perfect meat alternative exists, and has always existed. It's natural, delicious, sustainable—even soil boosting! It takes no research dollars and is the world's most important protein source. It's called the legume. Sure, it's not smoked brisket or a juicy burger, but it has fed cultures for centuries."[36]

Others who rely on appeals to nature push back not only against alternative proteins as a solution, but also forcefully reject calls for eating less meat at all. Seth Itzkan is the co-founder of Soil4Climate, a non-profit that advocates for expansions of "regenerative agriculture" and "holistic planned grazing" in the production of grass-fed beef. "Meat, as we know, is the most healthful food on earth and there would have been no human evolution nor higher order thinking without it," Itzkan wrote in an online post. Criticizing both alternative proteins and large-scale industrialized Confined Animal Feeding Operations (CAFOs), he asserted "The insidious flaw with the "less meat" argument is that it implies that meat is bad (when, of course, it isn't) while looking the other way as it advances soil-depleting GMO soy, faux meat products at the expense of nutritionally superior, regenerative beef and dairy alternatives that are essential for enhancing soil carbon, reviving pasture ecosystems, and just now gaining a foothold in supermarkets."[37]

Itzkan's remarks speak to a valorisation of "natural" approaches to food production, but also underline another key element of the traditionalist critique, that which integrates broader concerns about the impacts of cultured meat on food cultures, local economies, and specific interest groups in the animal agricultural professions. It is this coalition of concerns that has been leading the helm to outright ban the development and sale of cultured meat in specific jurisdictions. Notably, in 2023, Italy became the first nation to ban cultured meat via legislation advanced by the Minister of Agriculture, Food Sovereignty and Forestry, Francesco Lollobrigida, part of the right-wing government of Prime Minister Giorgia Meloni.[38] In a Facebook post, Lollobrigida celebrated the ban: "In defence of health, of the Italian production

[35] Schweizer (2021), para 4. A response from cultured meat researcher Stout (2021), published on the website of the non-profit New Harvest, provided responses to many of the questions posed, while also pushing back on what he saw as several pieces of incorrect information, misunderstandings, and an overly fearmongering tone.

[36] Bittman et al. (2021), para 5.

[37] Itzkan (2020), para 1.

[38] On this topic, see *Cell-Based Meat in the European Union: A Regulatory Crossroads* by Corazza and Formici, in this volume.

system, of thousands of jobs, of our culture and tradition, with the law approved today, Italy is the first nation in the world to be safe from the social and economic risks of synthetic food."[39]

Similar initiatives have been put forth in several other European nations,[40] including Hungary, France, Austria, and Romania, often as part of broader packages that also place labelling restrictions on plant-based products. A joint note from the Austrian, French, and Italian delegations to the Council of the European Union—supported by ten other European delegations—argued that the production of "lab-grown artificial cell-based food…represents a threat to primary farm-based approaches and genuine food production methods that are at the very heart of the European farming model." Criticizing the economies of scale required to create cultured meat, they called for "pre-emptive action against the monopolization of food production and towards the diversification of primary farm-based food production guaranteed by European farmers."[41] At the time of writing, the European Commission has rejected these national-level bans, calling for a unified European approach to pre-market safety testing and authorization, but pressure to stop the nascent industry before it can begin remains.[42]

The United States has also been the site of similar efforts[43]—the States of Florida and Alabama have officially passed pre-emptive bans, while lawmakers in a host of other states have offered up various versions of bans or partial restrictions, with additional momentum picking up in the early days of the second Trump administration.[44] Florida Commissioner of Agriculture Wilton Simpson was proud to champion his state as the first-in-the-nation to ban what he referred to as lab-grown meat: "We must protect our incredible farmers and the integrity of American agriculture. Lab-grown meat is a disgraceful attempt to undermine our proud traditions and prosperity, and is in direct opposition to authentic agriculture."[45] To be clear, not all agricultural interests have been supportive of these efforts—several large meat industry corporations have a financial and scientific interest in the cultured meat industry, while others simply reject restrictions on consumer choice or oppose the heavy-handed regulatory approach on libertarian grounds.[46] Still, central to all of these oppositional efforts—in the United States (US), across the European Union (EU), and in other contexts around the world—has been an organized campaign of resistance by key interest groups in the agricultural community.

---

[39] Quoted in Barbuti (2023), para. 4.

[40] See *Cell-Based Meat in the European Union: A Regulatory Crossroads* by Corazza and Formici, in this volume.

[41] General Secretariat of the Council of the European Union (2024), p. 2.

[42] See European Commission (2024) and Mridul (2024); see also *Cell-Based Meat in the European Union: A Regulatory Crossroads* by Corazza and Formici, in this volume.

[43] See *The Legal Debate on Cultured Meat Around the World: Comparative Insights from the United States, Singapore and Israel* by Pitto in this volume.

[44] *Ibid.* See also Stevens (2025) and Grunwald (2025).

[45] Quoted in Executive Office of the Governor Ron DeSantis (2024), para 3.

[46] See Ogles (2024).

Specifically, Cattlemen's Associations from the US, Canada, Mexico, and Australia have signed a series of joint statements and engaged in national and international lobbying in an attempt to limit the expansion of cultured meat, all while trying to promote the purported benefits of cattle production for wildlife habitat, natural water filtration, and greenspace protection.[47] In Italy, the agriculture organization Coldiretti, which claims to represent 1.6 million farmers, was a key player in the advocacy effort against what they termed "Frankenstein meat."[48] Coldiretti also worked with the think-tank Farm Europe and other aligned European groups to launch Eat Europe in 2022, a multistakeholder Brussels-based organization with a stated mission to "develop with its members and partners data-driven, science-based, pragmatic solutions to drive a resilient and sustainable food systems transformation."[49] In practice, Eat Europe, whose President has previously worked as an executive in the beef industry, has been actively opposed to EU proposals to reduce pollution from large livestock farms, limit pesticide usage, and revive damaged ecosystems.[50] Documents from the Eat Europe coalition provided a list of talking points to be used by its members and others opposed to the "synthetic imitation of meat." The materials criticized cultured meat in terms of its environmental impact, uncertain nutritional value, monopoly risk, health risks, impacts on biodiversity, and disconnection from nature. Appeals to nature and tradition were at the heart of these claims, as they asserted that a world that includes cultured meat "supposes a vision of society where food is a corollary, and not the pivotal building stone, of culture and traditions."[51]

The individuals and organizations featured in this section vary widely in terms of their visions of optimal food systems, incorporating supporters of mostly whole food plant-based diets, advocates for pasture-based livestock rearing, and defenders of large-scale animal food production systems, They are united, however, by a deep scepticism about cultured meat as a product and an enterprise, in large part because of a semi-religious conceptualization that it is contrary to their understanding of what nature and tradition intend. As the insights of Giddens would suggest, especially when read alongside the work of historians of modern food systems and meat production, the cultural traditions they defend from the perceived attack of cultured meat are themselves "invented traditions," constructed across recent generations on account of a variety of economic, agricultural, and technological developments.[52] Yet, appeals to tradition and the valorisation of natural practices remain powerfully persuasive tools, both for those aligned with meat industry powers and those positioned as alternative food movement actors, in their shared critique of cultured meat.

---

[47] See National Cattleman's Beef Association (2024); AGDAILY Contributors (2024).

[48] See Coldiretti Giovani Impresa (2021).

[49] Eat Europe, https://www.eat-europe.eu/areas-of-work/. Accessed 21 April 2025.

[50] See Boren (2024).

[51] Farm Europe (2022), p. 6.

[52] See Smil (2013).

3.2 The Techno-Economic Critique

Some of the strongest opposition to the cultured meat enterprise has come from those who are sceptical of its fundamental technical viability, as well as of the economic and market-based relations that underpin its development and industrial structure. There are overlapping themes between this techno-economic critique and many of the traditionalist perspectives discussed in the previous section, and many traditionalist opponents draw from this techno-economic opposition in presenting their arguments. With that said, the techno-economic opposition positions itself as more empirically grounded than driven by ideology, even though their empirical assessments are often undergirded by deeper value-based commitments, reflecting a socio-technical imaginary that deems other types of investment as more worthwhile than cultured meat in an effort to promote food system sustainability.

From a purely technical perspective, the field of cultured meat has been subject to a variety of criticisms, from internal and external critics alike. The pessimistic takeaway of this research—conducted through Life Cycle Analysis (LCA) and Techno-Economic Assessment (TEA)—is that the potential for scaling up cultured meat to the point of legitimate market availability remains extremely limited, or nearly impossible, due to a host of challenges. These include, but are not limited to, energy requirements and associated carbon emissions, biosecurity and food safety, bioreactor design, cell line and growth media optimization, and scaffolding materials.[53] Taken together, the costs of the enterprise are understood to be so large and the negative environmental impacts so significant that cultured meat will have little chance of making it to market at all—and then, of course, there is the problem of whether there will be any market demand, a topic thoroughly explored in the consumer perception literature.

Research by independent analyst David Humbird, funded by Open Philanthropy (an effective altruist-aligned organization with interests in animal protection and alternative protein development), was one of the first to provide a comprehensive TEA that called into question the technology's fundamental assumptions via a data-driven analysis. A 2021 publication referred to efficiency enhancements and low-cost input developments as "necessary but insufficient conditions for displacement of conventional meat by cultured meat," and the work ultimately concluded that "Capital- and operating-cost analyses of conceptual cell-mass production facilities indicate economics that would likely preclude the affordability of their products as food."[54] A 2025 publication by Risner and colleagues offered an LCA with a particularly pessimistic perspective on cultured meat, concluding that "Our findings indicate that cultured meat may not outperform traditional meat production in environmental terms."[55]

[53] See Goswami et al. (2023).

[54] Humbird (2021), p. 3239.

[55] Risner et al. (2025), p. 72.

These technical assessments provided confirmation for many researchers and observers who had long been sceptical of the scientific foundations of the field on the basis of environmental, market-based, regulatory, and ethical considerations.[56] Researchers supportive of the endeavour have offered retorts to these critiques, often acknowledging as legitimate many of the technical and environmental challenges, but offering up a variety of pathways through which innovation could provide solutions.[57] As discussed by New Harvest's Executive Director Isha Datar, one of the biggest challenges faced by the industry is perhaps self-inflicted—that is to say, the "hype cycle" that led to significant investment from venture capitalists and the proliferation of start-ups also inflated expectations about the inevitability of technological advancement and the acceleration of product timelines.[58] As these promises were not met and other market dynamics shifted, a "cell ag winter" characterized by start-up closures, pivots, and funding shortfalls ensued.[59] As a coalition of researchers put it in a commentary for *Nature Food,* in such a moment of crisis, the field would be better suited to reorient away from a "race to market" and toward a "race to mission," grounded in a set of key mission-oriented objectives: create slaughter-free solutions, centre health and safety, pursue sustainability proactively, scale for impact, and practice open science.[60]

To a number of other critics, however, the technical challenges of cultured meat are too inextricably linked to its market-based model of innovation and food system change to be redeemed. Here, the target is not solely the technology of cultured meat, but rather the ideology of "solutionism," a Silicon-Valley inspired worldview that "presumes rather than investigates the problems that it is trying to solve" and from there proposes "technological fixes" that are projected to deliver relatively easy and often profitable answers to complex problems.[61] Writing about cultured meat, the critical food system scholar Julie Guthman argued that the capital-intensive push for simulacra of animal products "takes off the table other solutions, even techno-fixes, which might address the underlying cause of the problem—namely the human practices and political economic exigencies that have made animal agriculture so detrimental."[62] Advocates in this domain argue that cultured meat proponents fundamentally misunderstand the problem when they frame their objective as creating a solution that will keep up with global demand for animal products; instead, the real problem is the intersecting social, political, economic, and ecological crises caused by a capital-driven industrial food system that creates demand for cheap meat to begin with. This all occurs, those critics note, at the same time as hundreds of millions of individuals go hungry, food and agricultural corporations

[56] See Chriki and Hocquette (2020).

[57] See Specht et al. (n.d.) and Tuomisto and Ryynänen (2024).

[58] Datar (2021).

[59] Datar (2023).

[60] Holmes et al. (2022).

[61] As defined by Morozov (2013), p. 6.

[62] Guthman (2024), p. 102.

make billions of dollars, and climate change and other environmental crises run rampant. As Philip Howard, a researcher focused on market consolidation in the food system, pointed out, cellular agriculture firms are increasingly controlled by the same firms that dominate industrial meat, fish, and animal feed processing, and therefore "do not challenge the inequities in global food systems." As an alternative, he and others have called for "increasing support for more demonstrated and less centralized strategies, such as agroecology, integrated multi-trophic aquaculture, increasing diversity in food systems and greater self-sufficiency."[63]

An additional critical perspective grounded in socio-technical ideology includes a perhaps unlikely opponent of a product that aims to replace meat from animals, namely elements of the animal rights and vegan communities, particularly those attached to radically-oriented animal liberation philosophies rather than incrementalist visions of social change. One key concern relates to the reliance on animals in the cellular agriculture process, either as sources of cells or growth mediums. While advocates of the technology insist that the long-term need for animals in the process will be limited, and ultimately humane, ethical concern about the potential for persistent animal use or exploitation in the process leads to objections among many hard-line animal rights proponents.[64] Their other driving consideration is more aligned with the techno-economic concerns expressed earlier in this section, with a belief that support for cellular agriculture distracts and detracts from more fundamental efforts to change industrial agriculture, in general, and the relationship between humans and non-human animals, specifically. As animal studies scholar Elan Abrell argued, "the strategy of focusing on investment in this technology as the main solution to the problem of animal agriculture has amounted to a disastrous hijacking of both financial and activist resources to make a small group of venture capitalists richer in the short term—indeed, one might be hard-pressed to distinguish this strategy from other manifestations of disaster capitalism."[65]

Once again, the critical voices highlighted in this section are not univocal in their opposition to cultured meat. They are clearly linked, however, by scepticism about the technology's fundamental necessity and viability, as well as doubt about the industry's social, economic and ecological potential. Through a mix of empirical investigations and value-oriented assessments, cultured meat is rejected as a poor solution for promoting food system sustainability, especially when compared to other potential options that are otherwise available for financial and cultural investment. Which of those alternative options they prefer ultimately depends on the socio-technical imaginaries that animate their analysis and advocacy.

[63] Howard (2022), p. 799.

[64] See Poirier (2022) and Jiménez Rodríguez (2023).

[65] Abrell (2023), p. 156.

3.3 The Conspiritualist Critique

To this point, the opposing perspectives outlined in this work have been treated as good faith critiques, grounded in defensible, if contestable, empirical or ideological commitments. However, no treatment of opposition to cultured meat would be complete without a discussion of the significant segment that is wholly unempirical in its claims about public health risk and/or driven by conspiratorial ideation. Importantly, this "conspiritualist" perspective is not necessarily a category unto itself—rather, it is part of a diagonalist coalition that includes elements of both the traditionalist and techno-economic critical perspectives. This worldview is supercharged by a digital media environment that is rife with misinformation and disinformation, as well as by a global political moment characterized by deep cynicism towards authoritative institutions (government, mainstream media, scientific establishments and technology industries) and democratic governance in general. To be clear, there are plenty of good reasons to be sceptical about dominant institutions, but rather than offering evidence-based and democratically-informed paths forward, the conspiritualists and diagonalists offer their own set of self-interested solutions, miracles, or calls for anti-democratic revolution.

Given the diffuse nature of this network, it would be impossible to offer a full accounting of the diagonalist attack on cultured meat, so a few operative examples will have to suffice. One place to look is in the spread of internet memes that have proven to be a major force in contemporary political opinion-making and social movements.[66] Commonly displayed as text and images, often without clear attribution or sourcing, memes go viral across social media platforms and can make their way into other discussions on podcasts and news commentary shows, and eventually into mainstream political discourse. Widely-distributed memes about cultured meat have claimed, for instance, that the products approved for sale in the U.S. are actually grown from human cells, or alternately from cancer cells—two claims that had to be met with official denials from regulatory authorities.[67] A number of memes revolve around powerful global figures such as Bill Gates, whose advocacy and financial support for alternative proteins has led to accusations that he is part of a cabal of figures enforcing a "Great Reset" that will eventually lead to a ban on conventional meat-eating and forced consumption of insects and cultured meat instead.[68]

Not simply the domain of anonymous internet memes, these types of claims have been adopted, remixed, and amplified by well-known and influential personalities across the conspirituality and diagonalist landscape. Indeed, an in-depth social media listening analysis by the Changing Markets Foundation tracked how reporting about a University of California, Davis pre-print study critical of the environmental impacts of cultured meat was seized upon and transformed in ways that went well beyond the study's actual claims. Several waves of amplification took place,

[66] See Mina (2019).

[67] See Reuters Fact Check (2023) and Frank (2023).

[68] See Scott-Reid (2024).

boosted by established researchers aligned with the meat industry, popular science news outlets, and high-profile conservative political figures (including Donald Trump Jr.). From there, the narrative was further fuelled by a mix of "attention grabbing misinfluencers" who consistently disparage alternative proteins while attempting to enhance the image of conventional meat and dairy production, many of whom are "affiliated with alt-right and extreme right ideologies, often pushing anti-vax, anti-immigration, Islamophobic and/or antisemitic content."[69] As the report concluded, "This complex and multifaceted evolution of the narrative illustrates the inter-twining of science, politics, sensationalism, conspiracy theories and misinformation in the digital age."[70]

In a discussion of what they termed the "meat culture war," researchers Jan Dutkiewicz and Gabriel Rosenberg tracked how extremist voices were increasingly becoming part of establishment conservative talking points on climate and environmental policy in the US: "Actual members of Congress now customarily accuse any policy that might reduce meat consumption, and some policies that simply touch on greenhouse emissions, of ripping meat from Americans' jaws."[71] Another study by the European Fact-Checking Standards Network described how farmers' protests in the Netherlands and other parts of the EU in the 2020s, initially sparked by a diverse set of local issues and demands, were "weaponized by a wide array of actors to spread climate misinformation and narratives undermining climate action." Far-right politicians, including some of those discussed earlier in this chapter, proved to be influential amplifiers in these contexts, such that, "Many false claims revolved around perceived European Union control over various aspects of life: What people eat (lab-grown meat, insects), what they grow (pesticides, home-grown production), and even weather manipulation (through chemtrails or destroying dams)."[72]

The case of Vandana Shiva, a well-known Indian activist and scholar long involved in environmentalism, food sovereignty, anti-globalization, and anti-GMO efforts, shows that these developments are not simply a right-wing phenomenon. Her perspective blends elements of the alternative food movement's preference for traditionalist approaches to food and agriculture, a techno-economic critique of corporate industrialism, and ecofeminist spirituality. But her work has also been criticized as propagandistic, misinformed, self-aggrandizing, and harmful to the communities that she claims to represent, particularly due to her advocacy against agri-food biotechnology and vaccines.[73] In recent years, she has been outspoken in her opposition to what she calls the "corporate push for synthetic foods," with cultured meat among the products she sees as "false solutions that endanger our health

[69] Changing Markets Foundation (2023), p. 56.

[70] *Ibid*, p. 52.

[71] Dutkiewicz and Rosenberg (2023), para 4.

[72] European Fact-Checking Standards Network (2024), p. 35.

[73] See Tagliabue (2023).

and damage the planet."[74] As part of coalitions like Regeneration International,[75] Shiva and her Navdanya International organization work in close collaboration with the Organic Consumers Association, Joseph Mercola and others with long and documented histories as purveyors of pseudoscience in food and agricultural debates, sellers of unproven alternative health and supplement products, and COVID-19 conspiracies and anti-vaccine misinformation.[76] Shiva mused, for instance, about the "Great Reset" and the perils of Bill Gates on the podcast of noted conspirituality influencer Russell Brand, calling on the audience to, "not allow the totalitarian takeover of food to make it fake food and push it as the next liberation."[77]

This left-coded opposition to cultured meat aligns in many ways with the "Make America Healthy Again (MAHA) Movement," the brainchild of former environmental lawyer turned anti-vaccine conspiracist Robert F. Kennedy Jr (RFK Jr.). In the 2024 US Presidential Election, RFK Jr. initially planned to run as a Democrat before switching to an independent candidacy, then eventually dropping out and aligning with Donald Trump. The stated aim of MAHA is to lead the charge to "transform America's approach to public health, environmental sustainability, and government accountability," with key focus areas that include tackling chronic disease, advancing regenerative agriculture, protecting natural habitats, exposing corporate corruption, and eliminating environmental toxins.[78] But as Tom Frieden, former director of the Center for Disease Control and Prevention put it, "(T)o make America healthier, we need an accurate diagnosis and effective prospection. Although the focus on combating chronic disease is appropriate, the MAHA combination of sound science, pseudo-science, and profiteering by so-called "wellness" companies isn't the answer."[79] As an investigation by the Green Queen media outlet explained, RFK Jr's public statements about cultured meat have generally come in the form of reposting articles from the anti-vaccine organization Children's Health Defense, for which he served as chairman up until he resigned to run for president. In one instance, he shared an article titled "The Fake Meat Scam" editorializing that the "industry's fake meat is just another name for ultra-processed food, full of genetic engineering and pesticide-laden ingredients to look as much like meat as possible." In another instance, he shared an article published on the website for Children's Health Defense that was actually a republication of work by Shiva's Navdanya International. As RFK Jr. put it in his post: "Biotech + agribusiness giants are tapping into the "fake food" market in hope of profiting off consumers who are

---

[74] Navdanya International (2022).

[75] See the website of Regeneration International at https://regenerationinternational.org/. Accessed 21 April 2025.

[76] See Frenkel (2021).

[77] See the August 1, 2023 episode recording and transcript of the "Stay Free with Russel Brand" podcast, titled "Vandana Shiva - The Truth About Food!" at https://app.podscribe.ai/episode/87033333. Accessed 21 April 2025.

[78] See the website of Make America Healthy Again at https://www.maha.vote/. Accessed 21 April 2025.

[79] Frieden (2024), para 2.

critical of industrial food production—but fake food doesn't address root problems of industrial food + its eco + health consequences."[80] At the time of writing, RFK Jr. had taken the helm as the Secretary of Health and Human Services in Donald Trump's administration. While an outright ban on cultured meat is unlikely, and the first Trump administration was generally supportive of its development, early indications suggest the diagonalist perspective could further limit the product's emergence this time around.[81]

Ultimately, the diagonalist coalition that has emerged in opposition to cultured meat takes aspects of both the traditionalist and techno-economic critiques, remixing those arguments with a heavy dose of conspirituality and meme culture to maximize attention in the digital media ecosystem. Often overlooked as fringe or inconsequential to real policymaking, recent years have seen ideas from these movements take on more central roles in public opinion formation and public policy. In so doing, they link opposition to cultured meat with a host of other culture war battles, many with connection to right-wing reactionary politics, others associated with left-coded wellness and spirituality themes. Leveraging the global public's legitimate frustration with authoritative public health institutions, key voices use misinformation and disinformation to undermine evidence-based deliberation, all while claiming they provide access to truths that are being withheld by dangerous cabals of global leaders.

4 Conclusion

In contrast to a burgeoning academic literature on the factors that might promote consumer acceptance of cultured meat, this chapter instead examined factors that are associated with rejection. In so doing, it argued that although research from consumer psychology and marketing perspectives can be useful, other more critically-oriented theoretical frameworks could provide additional insights to better understand the dynamics of opposition. Peoples' attitudes to new food products involve more than narrowly defined individual preferences. Rather, attitudes to cultured meat emerge from collective forms of communication across physical and digital contexts, driven by intersecting cultural attachments to particular food and meat traditions, by socio-technical imaginaries that may be critical or supportive of dominant food system structures, and in some instances, by an affinity for conspirituality in seeking answers to complex modern public health challenges. In providing a thematic overview of key oppositional arguments and advocates, we outlined the ideologically diverse collection of individuals, communities and professional organizations that have taken issue with the science and industry of cultured meat.

[80] See Mridul (2025) for overview of the topic and the link shared by RFK Jr. at https://childrenshealthdefense.org/defender/false-solutions-who-benefits-from-fake-food/. Accessed 21 April 2025.

[81] See Grunwald (2025).

Scholars and practitioners interested in a world where accessible cultured meat becomes a reality would be well advised to take this disparate coalition seriously, as well as the merits and limits of its contentions.

Although it was not the primary intention of this chapter to offer strategic recommendations, a few brief notes in this vein are warranted. Perhaps the most important message for cultured meat advocates is that concern or opposition is not the product of a single attribute that can be remedied by a single experience, new fact or persuasive message. Perceptions are the product of a complex combination of sensory engagement, rational consideration and implicit affective associations. To build broader support for the enterprise, different publics will need to be engaged across each of these domains: they will need the opportunity to actually taste, smell, enjoy and integrate the products into their personal and cultural foodways; to learn more about the safety, nutritional quality and environmental impacts of the process; and to feel comfortable with the political and economic institutions that are driving its development and expansion. The best way to advance such an initiative is to engage not only in "downstream" persuasion efforts once the path of cultured meat has been decided upon by innovators and regulators, but also to engage in "upstream" practices of participatory dialogue in order to build shared understandings of its potential risks and benefits.[82]

Through such a process, it should also become clear that some individuals and communities will never be interested in cultured meat at all, and that is perfectly reasonable. The goal should not be to persuade all individuals to eat it, but rather to help everyone grasp its fundamental technical aspects; to understand its value proposition, comparative advantages, and ongoing limitations; to respectfully address their concerns, including those that may arise from misinformation; and to invite those interested into a process of envisioning how cultured meat could be part of a sustainable food future.[83] Similarly, researchers have a useful role to play in tracking the development of cultured meat technology, holding the industry accountable with respect to its claims and promises, documenting the contours of public support and opposition, and advancing platforms for transparent engagement to inform evidence-based public policy formation. By facilitating discussions that support not just one-way technical communication from experts to publics, but also foreground conversations that explore the importance of cultural tradition and desired sociotechnical imaginaries, the industry will be better suited to address its own limitations while providing a check on the growing power of diagonalism in food system debates.

[82] See Broad and Biltekoff (2023).

[83] See Broad (2023a) on improving agri-food biotechnology deliberations.

References

Abrell E (2023) The empty promises of cultured meat. In: Adams CJ, Crary A, Gruen L (eds) The good it promises, the harm it does: critical essays on effective altruism. Oxford University Press, Oxford, pp 149–159

Adams CJ (2020) The sexual politics of meat: a feminist-vegetarian critical theory. Bloomsbury Academic, London

Adl-Tabatabai S (2023) Study: Bill Gates' lab grown meat causes cancer in humans. The People's Voice. https://thepeoplesvoice.tv/study-bill-gates-lab-grown-meat-causes-cancer-in-humans/. Accessed 21 Apr 2025

AGDAILY Contributors (2024) Multi-national cattlemen join in lab-grown beef statement. AGDAILY. https://www.agdaily.com/livestock/multi-national-cattlemen-join-in-lab-grown-beef-statement/. Accessed 21 Apr 2025

Barbuti A (2023) Italy becomes the first country in the world to ban lab-grown meat. New York Post. https://nypost.com/2023/11/18/news/italy-becomes-the-first-country-in-the-world-to-ban-lab-grown-meat/. Accessed 21 Apr 2025

Biltekoff C, Guthman J (2023) Conscious, complacent, fearful: agri-food tech's market-making public imaginaries. Sci Cult 32(1):58–82

Bittman M, Hope-Danieri C, McCart M (2021) We don't need fake 'meat'; there are better solutions. The Bittman Project. https://bittmanproject.com/how-to-replace-meat/. Accessed 21 Apr 2025

Boren Z (2024) How a livestock industry lobbying campaign is turning Europe against lab-grown meat. Unearthed. https://unearthed.greenpeace.org/2024/07/30/cultivated-backlash-livestock-industry-lobbying-europe-lab-grown-meat/. Accessed 21 Apr 2025

Broad GM (2023a) Improving the agri-food biotechnology conversation: bridging science communication with science and technology studies. Agric Hum Values 40(3):929–938

Broad GM (2023b) Understanding the (fake) meat debates: the alternative protein ideological circle. Nutr Today 58(4):181–188

Broad GM, Biltekoff C (2023) Food system innovations, science communication, and deficit model 2.0: implications for cellular agriculture. Environ Commun 17(8):868–874

Bryant C, Barnett J (2020) Consumer acceptance of cultured meat: an updated review (2018–2020). Appl Sci 10(15):5201

Callison W, Slobodian Q (2021) Coronapolitics from the Reichstag to the Capitol. Boston Review. https://www.bostonreview.net/articles/quinn-slobodian-toxic-politics-coronakspeticism. Accessed 21 Apr 2025

Changing Markets Foundation (2023) Truth, lies and culture wars. Social listening analysis of meat and dairy persuasion narratives. https://changingmarkets.org/report/truth-lies-and-culture-wars-social-listening-analysis-of-meat-and-dairy-persuasion-narratives/. Accessed 21 Apr 2025

Chriki S, Hocquette JF (2020) The myth of cultured meat: a review. Front Nutr 7:507645

Coldiretti Giovani Impresa (2021) Consumi: Le 5 bugie della carne Frankenstein. https://giovanimpresa.coldiretti.it/notizie/attualita/pub/consumi-le-5-bugie-della-carne-frankenstein/. Accessed 21 Apr 2025

Datar I (2021) Where are we on the hype cycle? https://new-harvest.org/where-are-we-on-the-hype-cycle/. Accessed 21 Apr 2025

Datar I (2023) Where are we on the hype cycle? Part II. https://new-harvest.org/where-are-we-on-the-hype-cycle-part-ii/. Accessed 21 Apr 2025

de Araújo PD, Araújo WMC, Patarata L, Fraqueza MJ (2022) Understanding the main factors that influence consumer quality perception and attitude towards meat and processed meat products. Meat Sci 193:108952

Dutkiewicz J, Rosenberg GN (2023) Why right-wingers are so afraid of men eating vegetables. New Republic. https://newrepublic.com/article/171781/meat-culture-war-crickets. Accessed 21 Apr 2025

European Commission (2024) Classical detailed opinion response to Draft act prohibiting the production and placing on the market of laboratory-grown meat. https://technical-regulation-information-system.ec.europa.eu/en/notification/26066. Accessed 22 Aug 2025

European Fact-Checking Standards Network (2024) A fertile ground for disinformation. https://climatefacts.efcsn.com/reports/3760d048-299e-4ee9-9b0b-bedb524584d8. Accessed 21 Apr 2025

Executive Office of the Governor Ron DeSantis (2024) Governor DeSantis signs legislation to keep lab-grown meat out of Florida. https://www.flgov.com/eog/news/press/2024/governor-desantis-signs-legislation-keep-lab-grown-meat-out-florida. Accessed 21 Apr 2025

Farm Europe (2022) Synthetic imitation of food: pros and cons. https://www.farm-europe.eu/wp-content/uploads/2022/12/EN-Synthetic-food-PROS-CONS.docx.pdf. Accessed 21 Apr 2025

Fiddes N (2004) Meat: a natural symbol. Routledge, London

Frank BJ (2023) Stem cells, not cancer cells, used to make lab-grown meat. USA Today. https://www.usatoday.com/story/news/factcheck/2023/07/13/no-lab-grown-meat-is-not-cultivated-from-cancer-cells-fact-check/70410454007/. Accessed 21 Apr 2025

Frenkel S (2021) The most influential spreader of coronavirus misinformation online. New York Times. https://www.nytimes.com/2021/07/24/technology/joseph-mercola-coronavirus-misinformation-online.html. Accessed 21 Apr 2025

Frieden T (2024) What Trump and RFK Jr.'s 'Make America Healthy Again' gets right—and very wrong. STAT News. https://www.statnews.com/2024/10/31/maha-make-america-healthy-again-trump-rfk-jr. Accessed 21 Apr 2025

General Secretariat of the Council of the European Union (2024) The CAP's role on safeguarding high-quality and primary farm-based food production. https://data.consilium.europa.eu/doc/document/ST-5469-2024-REV-1/en/pdf. Accessed 21 Apr 2025

Giddens A (2003) Runaway world: how globalization is reshaping our lives. Routledge, London

Girshik L (2019) Why I am against "clean meat." https://loribgirshick.weebly.com/against-clean-meat.html. Accessed 21 Apr 2025

Goldstein J (2018) Planetary improvement: Cleantech entrepreneurship and the contradictions of green capitalism. MIT Press, Cambridge

Goswami AB, Charlesworth J, Biazik JM, Rybchyn MS, le Coutre J (2023) Cultured meat technology: an overview. In: Post M, Connon C, Bryant C (eds) Advances in cultured meat technology. Burleigh Dodds, Cambridge, pp 3–25

Grunwald M (2025) Why lab-grown meat could be cooked under Trump. https://heatmap.news/lifestyle/lab-grown-meat-trump-rfk. Accessed 21 Apr 2025

Guthman J (2024) The problem with solutions: why Silicon Valley can't hack the future of food. University of California Press, Berkeley

Harris JE, Gleason PM, Sheean PM, Boushey C, Beto JA, Bruemmer B (2009) An introduction to qualitative research for food and nutrition professionals. J Am Diet Assoc 109(1):80–90

Hartmann C, Siegrist M (2023) Consumer attitudes to cultured meat products: improving understanding and acceptance. In: Post M, Connon C, Bryant C (eds) Advances in cultured meat technology. Burleigh Dodds, Cambridge, pp 341–356

Hickel J (2020) Less is more: how degrowth will save the world. Random House, New York

Hogg R (2023) Italy bans lab-grown meat and 'tofu steak' in a bid to protect prosciutto and the people who make it. https://www.aol.com/finance/italy-bans-lab-grown-meat-121223916.html?guccounter=1. Accessed 21 Apr 2025

Holmes D, Humbird D, Dutkiewicz J, Tejeda-Saldana Y, Duffy B, Datar I (2022) Cultured meat needs a race to mission not a race to market. Nat Food 3(10):785–787

Howard PH (2022) Cellular agriculture will reinforce power asymmetries in food systems. Nat Food 3(10):798–800

Humbird D (2021) Scale-up economics for cultured meat. Biotechnol Bioeng 118(8):3239–3250

Itzkan S (2020) The insidious flaw in the "less meat" argument—we need soil, not soy. https://sethitzkan.medium.com/the-insidious-flaw-in-the-less-meat-argument-we-need-soil-not-soy-500742b7d550. Accessed 21 Apr 2025

Jasanoff S (2015) Future imperfect: science, technology, and the imaginations of modernity. In: Jasanoff S, Kim SH (eds) Dreamscapes of modernity: sociotechnical imaginaries and the fabrication of power. University of Chicago Press, Chicago, pp 1–33

Jasanoff S (2016) The ethics of invention: technology and the human future. WW Norton and Company, New York

Jiménez Rodríguez A (2023) Cultured meat, clean meat… queer meat? A vegan queer ecofeminist perspective on the implications of cellular agriculture. Front Sustain Food Syst 7:1104731

Juma C (2016) Innovation and its enemies: why people resist new technologies. Oxford University Press, Oxford

Klein N (2024) Doppelganger: a trip into the mirror world. Random House, New York

Leroy F, Praet I (2015) Meat traditions. The co-evolution of humans and meat. Appetite 90:200–211

Levinovitz A (2020) Natural: how faith in nature's goodness leads to harmful fads, unjust laws, and flawed science. Beacon Press, Boston

Lusk JL, Roosen J, Bieberstein A (2014) Consumer acceptance of new food technologies: causes and roots of controversies. Annu Rev Resour Econ 6(1):381–405

Lynas M (2018) Seeds of science: why we got it so wrong on GMOs. Bloomsbury Publishing, London

Mancini MC, Antonioli F (2020) To what extent are consumers' perception and acceptance of alternative meat production systems affected by information? The case of cultured meat. Animals 10(4):656

Mina AX (2019) Memes to movements: how the world's most viral media is changing social protest and power. Beacon Press, Boston

Morozov E (2013) To save everything, click here: the folly of technological solutionism. PublicAffairs, New York

Mridul A (2024) EU Commission and Member States call Hungary's proposed cultivated meat ban 'unjustified'. Green Queen. https://www.greenqueen.com.hk/eu-commission-hungary-lab-grown-meat-cultivated-ban-tris-notification/. Accessed 21 Apr 2025

Mridul A (2025) With Trump incoming, what does RFK Jr really think about cultivated meat? Green Queen. https://www.greenqueen.com.hk/rfk-jr-robert-f-kennedy-lab-grown-cultivated-meat-donald-trump/. Accessed 21 Apr 2025

Muller SM, Rooney D, Cerja C (2024) Long live the liver king: right-wing carnivorism and the digital dissemination of primal rhetoric. Front Commun 9:1338653

National Cattleman's Beef Association (2024) American and Australian cattle producers commit to partnership through joint statement signing. https://www.ncba.org/ncba-news/news-releases/news/details/37941/american-and-australian-cattle-producers-commit-to-partnership-through-joint-statement-signing. Accessed 21 Apr 2025

Navdanya International (2022) The corporate push for synthetic foods. https://navdanyainternational.org/the-corporate-push-for-synthetic-foods-false-solutions-that-endanger-our-health-and-damage-the-planet/. Accessed 21 Apr 2025

Ogles J (2024) Meat institute argues cellular agriculture ban will hurt Florida's food economy. Florida Politics. https://floridapolitics.com/archives/662944-meat-institute-argues-cellular-agriculture-ban-will-hurt-floridas-food-economy/. Accessed 21 Apr 2025

Onwezen MC, Dagevos H (2024) A meta-review of consumer behaviour studies on meat reduction and alternative protein acceptance. Food Qual Prefer 114:105067

Pakseresht A, Kaliji SA, Canavari M (2022) Review of factors affecting consumer acceptance of cultured meat. Appetite 170:105829

Poirier N (2022) On the intertwining of cellular agriculture and animal agriculture: history, materiality, ideology, and collaboration. Front Sustain Food Syst 6:907621

Reuters Fact Check (2023) Fact check: no evidence lab-grown meat authorized in the US contains human cells. https://www.reuters.com/article/fact-check/no-evidence-lab-grown-meat-authorized-in-the-us-contains-human-cells-idUSL1N39Y0I2/. Accessed 21 Apr 2025

Risner D, Negulescu P, Kim Y, Nguyen C, Siegel JB, Spang ES (2025) Environmental impacts of cultured meat: a cradle-to-gate life cycle assessment. ACS Food Sci Technol 5(1):61–74. https://pubs.acs.org/doi/pdf/10.1021/acsfoodscitech.4c00281?ref=article_openPDF. Accessed 30 Dec 2024

Rosling H, Rosling O, Rönnlund AR (2018) Factfulness: ten reasons we're wrong about the world—and why things are better than you think. Flatiron Books, New York

Schweizer E (2021) What questions should we be asking about cell-based meats? Forbes. https://www.forbes.com/sites/errolschweizer/2021/05/06/what-questions-should-we-be-asking-about-cell-based-meats/. Accessed 21 Apr 2025

Scott-Reid J (2024) The Bill Gates bug-eating conspiracy, explained. Sentient. https://sentientmedia.org/bill-gates-bug-eating-conspiracy/. Accessed 21 Apr 2025

Siegrist M, Hartmann C (2020) Consumer acceptance of Novel Food technologies. Nat Food 1(6):343–350

Smil V (2013) Should we eat meat? Evolution and consequences of modern carnivory. Wiley, New York

Specht L, Swartz E, Bomkamp C (n.d.) Preliminary review of technical assumptions within the Humbird analysis. https://gfi.org/cultivated/preliminary-review-of-humbird-report/. Accessed 21 Apr 2025

Stevens R (2025) Lab-grown meat bans in 2025: which states are restricting cultivated meat? statecapitallobbyist.com/food/lab-grown-meat-bans-in-2025-which-states-are-restricting-cultivated-meat/. Accessed 22 Apr 2025

Stout A (2021) "What questions should we be asking about cell-based meats?" A fellow responds. https://new-harvest.org/forbes-response-cultured-meat/. Accessed 21 Apr 2025

Tagliabue GM (2023) Scientists sign another letter regarding Dr Vandana Shiva's anti-scientific and unethical stances. European Scientist. https://www.europeanscientist.com/en/features/scientists-sign-another-letter-regarding-dr-vandana-shivas-anti-scientific-and-unethical-stances/. Accessed 21 Apr 2025

To KV, Comer CC, O'Keefe SF, Lahne J (2024) A taste of cell-cultured meat: a scoping review. Front Nutr 11:1332765

Tsvakirai CZ, Nalley LL, Tshehla M (2024) What do we know about consumers' attitudes towards cultured meat? A scoping review. Future Foods 9:100279

Tuomisto HL, Ryynänen T (2024) Environmental impacts of cultivated meat. In: Soccol CR, Molento CFM, Reis GG, Karp SG (eds) Cultivated meat: technologies, commercialization and challenges. Springer Nature, Cham, pp 277–297

Walker J, Remski M, Beres D (2023) Conspirituality: how new age conspiracy theories became a public health threat. Random House, Toronto

Navigating Consumer Acceptance of Cultured Meat: Insights from Social Science Research

Nick Lin-Hi and Katharina Burdorf

Abstract This chapter explores the growing body of literature on consumer acceptance of cultured meat, offering a comprehensive overview of the factors that influence acceptance. By synthesizing existing research, the chapter categorizes these factors into three key antecedents: personal, product-related and external factors. Personal factors, such as demographic traits, psychological characteristics and individual awareness, significantly shape consumer perceptions of cultured meat. Product-related factors, including health perceptions, taste expectations and the framing of the product (e.g. labelling and terminology), further impact the willingness of consumers to adopt this innovation. Additionally, external factors—ranging from societal and environmental concerns to trust in the food industry and social influences—play a critical role in determining broader acceptance. The chapter also provides a critical reflection on the limitations of current research on consumer acceptance, highlighting the challenges of relying on intention-based measures and the need for more robust studies that account for actual consumer behaviour. In light of these complexities, businesses in the cultured meat sector are encouraged to take a strategic approach in addressing consumer concerns, actively shaping public perceptions and fostering trust. Practical implications for businesses are discussed, emphasizing the importance of tailored communication strategies and the need to consider consumer expectations and cultural sensitivities when crafting market strategies. The chapter concludes by emphasizing the ongoing need for interdisciplinary research and dialogue to navigate the complexities of consumer acceptance and support the successful market introduction of cultured meat.

Keywords Attitudes · Consumers · Perceptions · Psychology · Radical innovation

N. Lin-Hi (✉)
University of Vechta, Faculty I: Management of Social Services, Vechta, Germany
e-mail: nick.lin-hi@uni-vechta.de

K. Burdorf
Hydrosol GmbH and Co. KG, Ahrensburg, Germany
e-mail: kburdorf@hydrosol.de

G. Formici et al. (eds.), *Cell-Based Meat in the European Union and Beyond*,
https://doi.org/10.1007/978-3-032-14624-3_11

1 Introduction

Ensuring a liveable planet for future generations requires clear and practical responses to the pressing question of what can concretely be done today. This urgency is amplified by the limited effectiveness of current strategies aimed at achieving sustainable development. Voluntary reductions in consumption, frequently advocated as a solution by academics and policymakers, remain largely ineffective, as most individuals are unwilling to continuously restrict their lifestyles for the sake of sustainability.[1] Political interventions promoting sustainability also face substantial resistance as they are often perceived by the public as restrictive measures rather than valuable investments in a better future.[2]

Radical innovations for sustainable development have genuine potential to overcome these limitations. Such innovations induce transformative market shifts, significantly reducing the social and ecological footprints of products, services and processes without sacrificing essential consumer attributes such as price, quality or convenience.[3] The logic underlying such innovations is that technological advancement can simultaneously improve environmental/social outcomes and enhance consumer value. In an ideal case, such innovations redefine sustainability by shifting its meaning from a compromise or sacrifice to a source of added value and improved quality of life. As a result, sustainable products, services and processes can naturally outcompete traditional alternatives in the market by offering superior benefits with respect to the status quo. A prime example of such a radical innovation for sustainable development is cultured meat, which could significantly improve the sustainability of meat production while simultaneously providing added consumer value compared to conventional meat.[4]

The diffusion of radical innovations such as cultured meat is, however, associated with a variety of technical and social challenges. From a technical perspective, introducing innovations to the market requires scaling production methods from laboratory settings to industrial-scale manufacturing. From a social perspective, the challenges relate to ensuring favourable public perceptions and acceptance. In particular, to successfully establish themselves on the market, radical innovations have to gain social legitimacy, which is most clearly reflected in consumer acceptance—the crucial foundation underlying purchasing decisions.

In general terms, consumer acceptance refers to the willingness of individuals to embrace a new product or innovation and incorporate it into their consumption habits. Widespread consumer acceptance of cultured meat cannot be taken for granted since innovations that diverge substantially from established consumption patterns

[1] El Haffar et al. (2020).

[2] Gonzalez-Arcos et al. (2021).

[3] Shang et al. (2020).

[4] Lin-Hi et al. (2022). On this topic, see also *What Is Cell-Based Meat? Food Safety and Sustainability Aspects* by Cheli et al. in this volume.

and familiar practices often encounter strong consumer resistance.[5] Such resistance is especially pronounced in the context of radical food innovations because consumers often associate the consumption of novel and unfamiliar food products with health risks.[6] Since consumer acceptance is ultimately a fundamental precondition for cultured meat to establish a robust market presence, it is a significant challenge for the success of this radical innovation.[7]

This contribution addresses consumer acceptance of cultured meat, providing a comprehensive overview of the literature. The next section examines three primary antecedent categories that influence consumer acceptance: personal factors, product factors and external factors. These categories serve as a framework for understanding the different influences that guide consumer perceptions and behaviour. This is followed by a critical look at the limitations of existing research on consumer acceptance and the need for further research. We then elaborate on the business-related implications of consumer acceptance research, offering companies in the cultured meat sector general recommendations for building trust and developing effective market strategies. The chapter concludes with a final remark.

2 Research on Consumer Acceptance of Cultured Meat

In recent years, the social science perspective on cultured meat has gained considerable attention as illustrated by an increasing number of comprehensive reviews and meta-analyses addressing this topic.[8] This heightened focus underscores the crucial role consumer acceptance plays in determining the commercial success and societal integration of cultured meat in the future. Reflecting this trend, social science research in this area has expanded significantly, employing a wide range of methods, including surveys,[9] experimental studies,[10] qualitative approaches[11] and mixed-method designs.[12] Studies have also been conducted in different geographic contexts and cultural settings, highlighting the complexity and context-dependent nature of consumer acceptance of cultured meat. Such diversity of research approaches and settings allows a nuanced understanding of the factors influencing acceptance, recognizing that consumer responses may vary significantly across cultural, social and

[5] Heidenreich and Kraemer (2015).

[6] Pliner and Salvy (2006).

[7] Post et al. (2020).

[8] See Bryant and Barnett (2018, 2020), Deliza et al. (2023), Kouarfaté and Durif (2023), Laureati et al. (2024), Lewisch and Riefler (2023a), Pakseresht et al. (2022), Pivoraite et al. (2024), Siddiqui et al. (2022) and Tsvakirai (2024).

[9] See Lazou et al. (2024) and Pilařová et al. (2023).

[10] See Arango et al. (2023a) and Leite et al. (2024).

[11] See Ho et al. (2023) and van Der Weele and Driessen (2019).

[12] See Rehman et al. (2024) and Ruzgys and Pickering (2020).

individual backgrounds.[13] This emerging body of research therefore provides essential insights into the multifaceted dynamics involved in consumer acceptance, helping researchers, policymakers and practitioners to better address consumer concerns and facilitate the successful market introduction of cultured meat.

We now present a structured overview of the literature on the factors that influence consumer acceptance of cultured meat. This overview draws on several meta-studies and systematic reviews, as well as individual peer-reviewed papers. Three categories of factors consistently emerge: personal factors, product-related factors and external factors. We use these categories as a framework for organizing the literature; they provide a clear and accessible introduction to the field and support a more comprehensive understanding of the different drivers of consumer acceptance. As illustrated in Fig. 1, each of these categories comprises a set of specific sub-factors that shape consumer perceptions and intentions in distinct ways. Note that this overview does not aim to be exhaustive or to capture every factor discussed in the literature. Rather, it offers a high-level synthesis of the most frequently identified and thematically grouped antecedents found across meta-analytical work to date.

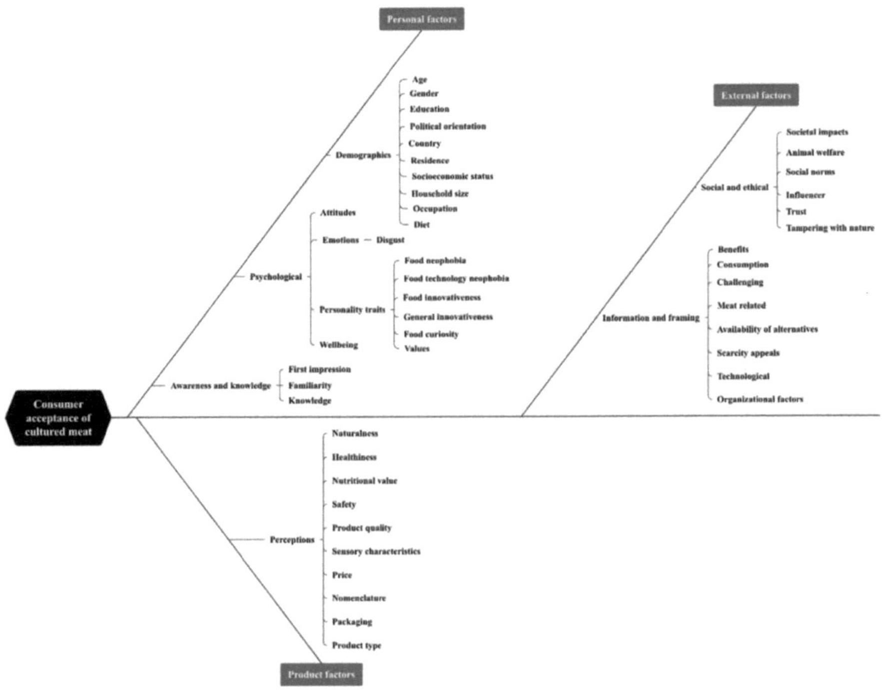

Fig. 1 Typical antecedents for consumer acceptance of cultured meat

[13] For example, see *Food Justice in Religious Laws: Hypotheses on Cultivated Meat* by Anello in this volume.

2.1 Personal Factors

Personal factors refer to individual attributes—both demographic and psychological—that shape how people perceive, evaluate and respond to cultured meat. Research into demographic traits offers valuable insights into the types of consumers who may be more inclined to accept this innovation in the food sector. While findings vary, a profile of a "typical" cultured meat consumer emerges: those more inclined to accept cultured meat tend to be younger,[14] male,[15] highly educated individuals[16] who hold liberal political views,[17] live in urban environments,[18] and enjoy eating meat.[19] However, these patterns are not universally applicable. For example, in some contexts or after some interventions—such as after providing information of the health-related benefits of cultured meat—women have demonstrated higher acceptance levels than men.[20] Furthermore, consumer acceptance of cultured meat varies across cultures, reflecting differences in food traditions, societal values and technological openness.[21]

In addition to demographic traits, psychological factors also play a crucial role in shaping consumer acceptance. For example, positive attitudes to cultured meat as a food product were identified as drivers of acceptance.[22] Emotions also play a crucial role in shaping consumer acceptance, negative emotions, particularly feelings of disgust, emerging as significant barriers to adoption.[23] Many consumers find the concept of cultured meat unappealing,[24] and reducing these negative emotional reactions is essential for fostering greater openness to this product. Indeed, a reduction in disgust has been associated with increased willingness to try cultured meat.[25] Personality traits, such as food neophobia, i.e. avoidance of novel foods,[26] and food technology neophobia, i.e. fear or distrust of new food production technologies,[27] also play a significant role. These traits have been consistently linked to lower

---

[14] See Baum et al. (2022) and Lazou et al. (2024).

[15] See Chia et al. (2024) and Cornelissen and Piqueras-Fiszman (2023).

[16] See Espinosa and Treich (2023) and Liu et al. (2023).

[17] See Baum et al. (2023) and Bryant et al. (2019b).

[18] See Fernandes et al. (2022) and Shaw and Mac Con Iomaire (2019).

[19] See Rombach et al. (2022) and Sikora and Rzymski (2023).

[20] See Bryant et al. (2019b) and Piochi et al. (2022).

[21] See Lewisch and Riefler (2023a) and Siddiqui et al. (2022).

[22] See Dupont and Fiebelkorn (2020).

[23] See Pakseresht et al. (2022). For analysis of the Italian scenario, see *Socio-Demographic and Behavioural Determinants of Consumer Attitudes to Cell-Based Meat and Insect-Flour-Based Products: An Empirical Study of Italians' Propensity to Accept New Foods for Human Nutrition and Pet Food* by Degli Antoni and Faillo in this volume.

[24] See Rosenfeld and Tomiyama (2022).

[25] See Arango et al. (2023b).

[26] Pliner and Hobden (1992).

[27] Cox and Evans (2008).

acceptance.[28] In contrast, greater curiosity and openness to new food experiences are associated with greater acceptance of cultured meat.[29]

Consumer awareness and knowledge about cultured meat are other important factors influencing acceptance.[30] A more positive first impression of cultured meat is strongly correlated with greater willingness to consume it.[31] Prior familiarity with the concept of cultured meat is a strong predictor of acceptance,[32] although this relationship is not consistent across all studies.[33] Thus, the findings in the context of cultured meat align with innovation diffusion theory,[34] according to which knowledge about an innovation is a crucial first step towards fostering its acceptance.

Overall, existing research demonstrates that consumer acceptance of cultured meat is shaped by a complex interplay of demographic and psychological factors. A nuanced understanding of these factors is essential for businesses aiming to design effective marketing and communication strategies. By considering individual characteristics and broader societal trends, companies can better tailor their communication approaches to foster acceptance of this radical innovation.

2.2 Product Factors

Product factors refer to the characteristics of cultured meat itself, and a substantial body of research highlights the significant role these product-related factors play in shaping consumer acceptance.[35] Notably, the objective characteristics of the product are not decisive in shaping acceptance, but rather consumers' subjective perceptions and interpretations of these attributes. Many of these perceptions tend to function more as barriers than as drivers of acceptance. One of the most prominent barriers identified is the perceived lack of naturalness of cultured meat.[36] Consumers often associate cultured meat with artificiality, which can strongly undermine its acceptance, and attempts to reframe cultured meat as "natural" have largely proven ineffective.[37]

Perceptions of naturalness are closely linked to concerns about health. Consumer fears regarding the long-term health effects of consuming cultured meat have been

[28] See Boereboom et al. (2022) and Pivoraite et al. (2024).

[29] See Rombach et al. (2022) and Sikora and Rzymski (2023).

[30] See Onwezen and Dagevos (2024).

[31] See Engel et al. (2024).

[32] See Chia et al. (2024) and Rolland et al. (2020).

[33] See Baum et al. (2022) and Dupont et al. (2022).

[34] Rogers (1983).

[35] See To et al. (2024) and Tsvakirai (2024).

[36] See Chia et al. (2024) and Hibino et al. (2023).

[37] See Bryant et al. (2019a).

repeatedly documented by qualitative and quantitative studies.[38] Such concerns act as a significant barrier to acceptance. However, perceived health benefits—particularly those related to improved food safety, such as reduced risk of foodborne illnesses due to sterile production conditions—can partly mitigate these concerns and foster greater willingness to consume cultured meat.[39] The option of enhancing the nutritional profile of cultured meat, for example, by enriching it with omega-3 fatty acids or other beneficial compounds, is a driver of acceptance.[40]

In addition to health-related factors, sensory perceptions, particularly taste, play a decisive role in shaping consumer acceptance. Research has consistently shown that anticipated taste is a major predictor of purchase intentions.[41] The opportunity to taste cultured meat, even in experimental settings such as product labelling or sampling, has been shown to significantly increase consumer acceptance.[42] This highlights the importance of sensory appeal—taste and texture—in influencing consumer attitudes and behaviours toward cultured meat.

The terminology used to describe cultured meat also significantly impacts consumer perceptions. For instance, names such as "cultivated meat" generally evoke more positive attitudes than names like "lab-grown meat," which can amplify associations with artificiality and technological manipulation.[43] Similarly, packaging design, particularly eco-friendly visuals, such as green colour schemes that symbolize sustainability, has been found to enhance acceptance.[44] These visual cues align the product with broader societal trends toward environmental responsibility and sustainability.

Finally, the type of the cultured meat product can also influence consumer acceptance. Products that closely resemble familiar food formats, such as nuggets or burgers, tend to be more readily accepted. Research indicates that consumer acceptance varies by product category: cultured chicken nuggets, for example, are often perceived more favourably than cultured beef burgers, presumably due to differences in perceived taste, health considerations and emotional responses such as disgust.[45]

Overall, the interplay of product factors, ranging from perceived naturalness and healthiness to sensory qualities, nomenclature, packaging and product type, illustrates the complex ways in which consumer perceptions can either constrain or facilitate the acceptance of cultured meat.

[38] See Pilařová et al. (2023) and Shaw and Mac Con Iomaire (2019).

[39] See Fu et al. (2023).

[40] See Bryant et al. (2019b).

[41] See Espinosa and Treich (2023) and Wilks and Phillips (2017).

[42] See Rolland et al. (2020).

[43] See Asioli et al. (2021) and Hallman et al. (2023).

[44] See Califano et al. (2023).

[45] See Vural et al. (2023).

2.3 External Factors

External factors play a crucial role in shaping consumer perceptions and the acceptance of cultured meat.[46] These factors encompass a wide range of societal, environmental and social influences, as well as how information about the product is communicated. Potential societal and environmental benefits associated with cultured meat are an external factor influencing consumer acceptance. Research has consistently shown that perceived environmental impact and animal welfare concerns[47] are primary motivators for many consumers to accept cultured meat. Specifically, the claim of reduced environmental footprint of cultured meat production and its potential to reduce harm to animals compared to conventional meat production have been identified as key drivers of acceptance.[48] In fact, many consumers perceive more societal benefits of cultured meat—such as reduced harm to animals and the environment—than personal benefits from consuming it.[49] However, these perceptions are not universally shared, and in some regions concerns regarding the broader societal implications of cultured meat persist. Critics have raised concerns about the potential disruption cultured meat could cause to local animal husbandry practices and traditional food cultures.[50] Some studies also highlight fears that cultured meat may undermine established cultural eating traditions and impact local agricultural industries.[51]

Social influences also play a critical role in shaping consumer acceptance.[52] Research has demonstrated that individuals are more likely to accept cultured meat if they are surrounded by positive attitudes and intentions in their social networks.[53] When consumers are embedded in supportive social environments, whether family, friends or broader community groups, they are more likely to view cultured meat favourably and be open to accepting it. This highlights the importance of societal norms and social validation in the acceptance of new food innovations. Conversely, in cultures where traditional food practices are deeply rooted, social norms can create resistance to new food products. For instance, in Turkey, the consumption of red meat is not only a dietary preference but also a social norm, which negatively affects the acceptance of cultured meat.[54]

Social media have also emerged as a powerful tool in shaping public perceptions and influencing consumer acceptance of cultured meat. Research indicates that

---

[46] See Pivoraite et al. (2024).

[47] On this topic, see *Animal Welfare as a Constitutional Value: The Debate in Comparative Law and the Emerging Role of Cultivated Meat* by Piciocchi in this volume.

[48] See de Oliveira et al. (2021) and Sikora and Rzymski (2023).

[49] See Bryant and Barnett (2020).

[50] See Shaw and Mac Con Iomaire (2019).

[51] See Wilks and Phillips (2017).

[52] See Pivoraite et al. (2024).

[53] See Engel et al. (2024).

[54] See Baybars et al. (2023).

recommendations from micro-influencers, individuals with smaller niche followings, tend to have a greater impact on consumers' willingness to purchase cultured meat than recommendations from mega-influencers.[55] The perceived credibility and trustworthiness of micro-influencers, particularly when experts on the subject, significantly enhance the effectiveness of social media campaigns promoting cultured meat. Trust, in a broad sense, has been identified as a major determinant of consumer acceptance. For example, high levels of trust in the food industry generally go with greater acceptance,[56] while distrust in biotechnology[57] and scientific institutions[58] can impede it.

Another critical external factor is the framing of information about cultured meat. How the product is presented and communicated can have a profound impact on consumer perceptions and acceptance.[59] For example, counter-messaging strategies that highlight the negative aspects of conventional meat production—such as its environmental impact—can transform these weaknesses into potential strengths for cultured meat.[60] By framing cultured meat as a solution to the problems associated with traditional meat production, marketers can capitalize on these concerns to increase consumer acceptance. Moreover, presenting cultured meat as "real meat" has been found to foster more positive attitudes toward it, while more technical framings, such as detailed descriptions of the production process, tend to evoke negative perceptions related to artificiality and unnaturalness, acting as barriers to acceptance.[61]

In conclusion, external factors such as societal benefits, social influences and the framing of information play a pivotal role in the acceptance of cultured meat. These factors highlight the importance of not only understanding the individual product characteristics that shape consumer attitudes but also recognizing the broader social, cultural and environmental contexts in which acceptance takes place.

3 Critical Reflections on Consumer-Acceptance Research on Cultured Meat

Research on consumer acceptance of cultured meat has developed considerably in recent years, providing important initial insights into the factors that may drive or hinder acceptance. Existing studies have contributed to a better understanding of

---

[55] See Leite et al. (2024). On this point, see *Who Hates Cultured Meat? Examining Public Opposition to Cellular Agriculture* by Broad and Glenna in this volume.

[56] See Siegrist and Hartmann (2020).

[57] See Hwang et al. (2020).

[58] See Lewisch and Riefler (2023b).

[59] See Pivoraite et al. (2024).

[60] See Baum et al. (2022).

[61] See Bryant and Dillard (2019).

key antecedents, including personal characteristics such as values, attitudes and emotional responses, as well as product-related perceptions like taste, healthiness and naturalness. Research has also highlighted the importance of external influences, including media framing, trust in institutions and broader societal narratives about food innovation. Taken together, these findings offer valuable indications of the opportunities and challenges associated with the market introduction of cultured meat.

However, despite these contributions, a closer examination reveals several limitations that complicate the interpretation and generalization of existing findings. The complexity of consumer acceptance processes, the provisional character of the information provided in studies, methodological inconsistencies, cultural variability across study contexts, and the frequent reliance on intention-based measures point to important challenges that current research has yet to address. These limitations suggest that while the existing body of research forms an important foundation, it must be interpreted with caution and further refined in future work. Against this background, this section critically reflects on central challenges indicated in the literature. In so doing, we aim to promote a more nuanced understanding of consumer acceptance of cultured meat and to identify key areas for future research and practical action.

Consumer acceptance of cultured meat is a highly complex and dynamic phenomenon. Rather than being driven by a single factor, it results from the interaction of a wide range of influences that span individual attitudes and values, product-specific characteristics, and broader social and cultural contexts. These factors do not operate in isolation but often interact in nuanced and sometimes unpredictable ways. For instance, a consumer's openness to innovation may be moderated by perceptions of naturalness or trust in the food industry. Moreover, acceptance is not static and can evolve with changes in knowledge, familiarity and public discourse on cultured meat.[62] This complexity makes it challenging to draw simple conclusions about what drives or hinders acceptance, and it underscores the need for multidimensional approaches in research and practice alike. Understanding consumer acceptance therefore not only requires identifying relevant factors but also appreciating the ways they combine and influence one another.

Studies on consumer acceptance of cultured meat must provide participants with sufficient information about the product to ensure that their responses are based on a sound understanding of what cultured meat is. This includes essential details about its pricing, taste, production process, healthiness, safety, environmental impact and ethical considerations. Such information is particularly crucial when studying radical innovations, since the inherent novelty of these products often means that participants otherwise lack the necessary background knowledge to develop an informed opinion.[63] However, since cultured meat is still in the development phase, much of the information provided in studies is provisional and may

[62] Bryant and Barnett (2018).

[63] Schoormans et al. (1995).

change with technological progress. Thus the information given to participants may later prove to be outdated and/or inaccurate, casting doubt on the validity of earlier findings. Studies may also differ in the type and amount of information they provide, which can lead to substantial variation in consumer responses and make it difficult to compare results across studies. Existing findings should therefore be interpreted with caution, and it is important to acknowledge that consumer perceptions of cultured meat are likely to evolve as new and more accurate information becomes available as the product itself continues to mature. Continuous research is therefore essential to monitor how emerging knowledge and technological advances influence consumer acceptance over time.

Methodological differences across studies further complicate the interpretation of findings on consumer acceptance of cultured meat. Studies vary considerably in how acceptance is measured, approaches ranging from structured surveys with closed-ended questions to qualitative interviews and experimental designs. Each method captures different aspects of consumer attitudes and provides varying levels of depth. The specific framing and wording of study materials and survey questions can also significantly influence participants' responses. For instance, emphasizing the technological aspects of cultured meat may evoke greater scepticism, whereas highlighting the environmental or ethical benefits may foster more positive reactions. These methodological inconsistencies make it difficult to directly compare findings across studies and emphasize the need for greater standardization. Developing consistent research designs and comparable measures of acceptance would enhance the robustness of the evidence and support a clearer understanding of consumer perceptions and intentions regarding cultured meat.

Besides methodological variations, the heterogeneity of sample populations and cultural contexts is an additional challenge for interpreting consumer acceptance research. Acceptance drivers identified in one geographic or cultural setting cannot necessarily be transferred to others. For example, consumers in societies with a high degree of technological openness and strong environmental awareness may react more favourably to cultured meat than those in contexts where traditional food practices and scepticism about food innovations are more prevalent. Such cultural and regional differences significantly limit the generalizability of existing findings. Future research should therefore adopt context-sensitive approaches that explicitly account for socio-cultural, economic and political factors. Only by incorporating these contextual variables can a more nuanced and comprehensive understanding of consumer acceptance dynamics across different markets be developed.

Due to the very limited availability of cultured meat and existing legal restrictions, nearly all current studies on consumer acceptance rely on intention-based measures, such as willingness to try or willingness to buy the product. While these indicators provide valuable insights into consumers' expressed openness toward cultured meat, they do not capture actual purchasing or consumption behaviour. This reliance on intention data leads to the well-documented "attitude-behaviour gap"[64] which is the discrepancy between what individuals state they are willing to do and their actual actions in real-world contexts. In other words, consumers may behave very differently from how they originally expressed themselves, which for example is often the case in the area of sustainable consumption.[65] As a result, findings based solely on intention measures only offer a limited and preliminary understanding of consumer acceptance. In addition, current research on consumer acceptance of cultured meat provides little basis for reliably estimating the potential market size of this innovation. This underscores a need for studies that move beyond self-reported intentions and investigate actual consumer behaviour, for instance through experimental purchasing environments, longitudinal field studies or real-market pilot projects. Only by bridging the gap between intentions and behaviour can a more accurate and practically relevant understanding of consumer acceptance of cultured meat be achieved.

4 Implications for Business

The question of how consumers will respond to cultured meat is central to companies operating in this emerging sector. In order to realistically assess the market opportunities and develop effective strategies for successful market introduction, businesses in the cultured meat industry need to understand consumer-acceptance drivers and barriers. As a radical innovation that challenges established consumption habits and perceptions of naturalness, cultured meat is unlikely to be adopted automatically. Companies are therefore well advised to proactively deal with the factors influencing consumer reactions and to engage in public communication. This is especially important given the mixed opinions surrounding cultured meat and the various arguments that continue to be raised against it. Companies in this sector cannot afford to leave the formation of public opinion to critics, especially when such arguments are sometimes based on inaccurate or misleading information. Proactively communicating the benefits, addressing misconceptions and correcting false claims will be essential for shaping a more informed public discourse, ultimately fostering consumer acceptance and ensuring the successful market integration of cultured meat.

Insights from consumer acceptance research offer crucial guidance in developing strategies that foster positive perceptions, address concerns and ultimately increase the likelihood of market success. Despite the limitations of existing research mentioned in the previous section, studies of consumer acceptance of cultured meat can provide businesses with some basic recommendations. They show that consumer acceptance is influenced by a complex interplay of factors, all of which businesses need to consider when developing their market strategies. Exclusive focus on

[64] Ajzen (1985).

[65] El Haffar et al. (2020).

technical product development, pricing or availability will not be sufficient to ensure success. Companies need to take a broader approach that includes consumer values, emotional responses, perceptions of the product and broader societal narratives surrounding cultured meat. While the findings should be interpreted with caution, existing research highlights key areas where strategic efforts can be directed, particularly in anticipating consumer concerns and framing the product within broader societal trends, such as sustainability and innovation. By acknowledging the multidimensional nature of consumer acceptance, businesses can develop more nuanced and robust strategies that are better aligned with the complexity of consumer expectations.

The complexity of consumer-acceptance dynamics also means that there is no universal approach to success. Each company operates under different conditions that shape its strategic priorities. Variations in target markets, cultural contexts, brand positioning, technological capabilities and regulatory environments create specific challenges and opportunities for the introduction of cultured meat. Companies also vary in the extent to which they can influence key product attributes such as price, taste and production transparency—critical factors for consumer acceptance. These distinctions require companies to adopt tailored strategies, focusing on the unique opportunities and constraints they face. By developing strategies that take these specific characteristics into account, companies can better navigate the complex landscape of consumer perceptions and improve their chances of achieving market acceptance.

Beyond these structural differences, initial studies indicate that organizational characteristics—which belong to external factors—can themselves significantly influence consumer acceptance.[66] For instance, whether cultured meat is marketed by a start-up or a multinational corporation may influence which acceptance factors are particularly influential. Start-ups, often associated with innovation and entrepreneurial spirit, tend to generate feelings of excitement and curiosity, which can increase acceptance among consumers who are more open to food innovations. In contrast, multinational corporations, known for their established reputation and resources, may activate perceptions of competence and product safety, appealing to consumers who are more risk-averse and cautious in adopting new foods. In addition, organizational attributes such as trustworthiness and Corporate Social Responsibility (CSR) affect consumer willingness to buy cultured meat. Trust in the producing or selling organization can help reduce uncertainty, a key barrier to the acceptance of radical food innovations, while strong CSR practices signal ethical commitment, thereby enhancing credibility. These findings highlight the need for businesses to focus not only on external market conditions but also on their internal strengths to shape consumer perceptions and improve acceptance.

In light of these insights, promoting consumer acceptance of cultured meat requires more than just product-level optimization or technological innovation. Companies need to include consumer expectations, organizational positioning and

[66] Lin-Hi et al. (2022, 2023).

cultural sensitivities in their strategies. Building on existing research, several strategic approaches can be identified for businesses. First, targeted communication and framing strategies are critical for addressing consumer concerns and emphasizing the benefits of cultured meat. Framing cultured meat in a way that resonates with prevailing societal values—such as environmental sustainability, animal welfare, and health—can positively influence perceptions and foster openness. Measures to build consumer trust are equally essential. Transparent communication on production processes, safety standards and ethical practices can help reduce uncertainty and increase consumer confidence. Furthermore, market strategies should be adapted to the specific needs and expectations of different consumer segments and cultural contexts. Recognizing that acceptance drivers differ across regions and demographics is a key to successfully navigating diverse markets. Finally, businesses should acknowledge the impact of their organizational identity and strategically manage their image, including demonstrating their commitment to social responsibility. By considering these factors, companies can develop comprehensive strategies that better align with the multifaceted nature of consumer acceptance and increase the likelihood of successfully introducing cultured meat to the market.

5 Final Remark

Consumer acceptance of cultured meat is a multifaceted and evolving challenge. While existing research has provided valuable insights into the factors that influence acceptance, it is clear that understanding consumer behaviour in this context requires a dynamic and comprehensive approach. Consumer perceptions are shaped by a complex interplay of personal, product-related and external factors, where perceptions often evolve as new information becomes available. It is therefore crucial for businesses and researchers to engage in a continuous dialogue about these factors and to adapt strategies to changing public perceptions and evolving societal narratives.

Bridging the gap between technical and social sciences could promote more comprehensive understanding of consumer acceptance. Collaboration between these fields is essential to advance a holistic understanding of consumer acceptance—one that combines scientific rigor with real-world concerns. Social science research, in particular, will benefit greatly from ensuring that the information provided about cultured meat in consumer studies is both accurate and up-to-date. Ensuring that the facts, benefits and challenges regarding cultured meat are communicated accurately will help reduce misconceptions, facilitate informed discussions and build public trust. Maintaining clarity and precision in the presentation of cultured meat in research studies is also essential to generate reliable findings useful for effective market strategies.

The influence of social media on public perceptions further highlights the importance of consistent and transparent communication. Misleading or incomplete information circulating in social channels can significantly hinder acceptance. On the other hand, well-crafted, proactive communication can promote understanding, correct misinformation and build trust. As consumer acceptance evolves, businesses must remain agile and engage social scientists and technologists to refine their strategies and adapt to the ever-changing landscape of public opinion.

The potential for cultured meat to contribute significantly to a more sustainable and ethical food system is immense. However, this potential can only be realized if widespread societal acceptance is achieved. Without public trust and understanding, even the most promising technological advancements are unlikely to gain traction or to spread. Therefore, fostering consumer acceptance is not only a strategic necessity for businesses but a fundamental prerequisite for cultured meat to fully unleash its transformative potential. By addressing consumer concerns, promoting informed discussions and facilitating collaboration across scientific disciplines, different stakeholders can work together to create the conditions necessary for cultured meat to advance sustainable development and reshape global food systems so that they meet the needs of future generations.

References

Ajzen I (1985) From intentions to actions: a theory of planned behavior. In: Kuhl J, Beckman J (eds) Action control: from cognition to behavior. Springer, Cham, pp 11–39

Arango L, Chaudhury SH, Septianto F (2023a) The role of demand-based scarcity appeals in promoting cultured meat. Psychol Mark 40(8):1501–1520

Arango L, Septianto F, Pontes N (2023b) Challenging cultured meat naturalness perceptions: the role of consumers' mindset. Appetite 190:107039

Asioli D, Fuentes-Pila J, Alarcón S, Han J, Liu J, Hocquette JF, Snell H, Nayga RM (2021) European consumer valuation for cultured beef burger: a multi-country investigation using choice experiments [Conference presentation]. Agricultural & Applied Economics Association Annual Meeting, Austin

Baum CM, Verbeke W, De Steur H (2022) Turning your weakness into my strength: how counter-messaging on conventional meat influences acceptance of cultured meat. Food Qual Prefer 97:104485

Baum CM, De Steur H, Lagerkvist CJ (2023) First impressions and food technology neophobia: examining the role of visual information for consumer evaluations of cultivated meat. Food Qual Prefer 110:104957

Baybars M, Ventura K, Weinrich R (2023) Can in vitro meat be a viable alternative for Turkish consumers? Meat Sci 201:109191

Boereboom A, Mongondry P, de Aguiar LK, Urbano B, Jiang Z, De Koning W, Vriesekoop F (2022) Identifying consumer groups and their characteristics based on their willingness to engage with cultured meat: a comparison of four European countries. Foods 11(2):197

Bryant C, Barnett J (2018) Consumer acceptance of cultured meat: a systematic review. Meat Sci 143:8–17

Bryant C, Barnett J (2020) Consumer acceptance of cultured meat: an updated review (2018–2020). Appl Sci 10(15):5201

Bryant C, Dillard C (2019) The impact of framing on acceptance of cultured meat. Front Nutr 6:103

Bryant CJ, Anderson JE, Asher KE, Green C, Gasteratos K (2019a) Strategies for overcoming aversion to unnaturalness: the case of clean meat. Meat Sci 154:37–45

Bryant C, Szejda K, Parekh N, Deshpande V, Tse B (2019b) A survey of consumer perceptions of plant-based and clean meat in the USA, India, and China. Front Sustain Food Syst 3:432863

Califano G, Furno M, Caracciolo F (2023) Beyond one-size-fits-all: consumers react differently to packaging colors and names of cultured meat in Italy. Appetite 182:106434

Chia A, Shou Y, Wong NMY, Cameron-Smith D, Sim X, Van Dam RM, Chong MFF (2024) Complexity of consumer acceptance to alternative protein foods in a multiethnic Asian population: a comparison of plant-based meat alternatives, cultured meat, and insect-based products. Food Qual Prefer 114:105102

Cornelissen K, Piqueras-Fiszman B (2023) Consumers' perception of cultured meat relative to other meat alternatives and meat itself: a segmentation study. J Food Sci 88(S1):A91–A105

Cox DN, Evans G (2008) Construction and validation of a psychometric scale to measure consumers' fears of novel food technologies: the food technology neophobia scale. Food Qual Prefer 19(8):704–710

de Oliveira GA, Domingues CHDF, Borges JAR (2021) Analyzing the importance of attributes for Brazilian consumers to replace conventional beef with cultured meat. PLoS One 16(5):e0251432

Deliza R, Rodríguez B, Reinoso-Carvalho F, Lucchese-Cheung T (2023) Cultured meat: a review on accepting challenges and upcoming possibilities. Curr Opin Food Sci 52:101050

Dupont J, Fiebelkorn F (2020) Attitudes and acceptance of young people toward the consumption of insects and cultured meat in Germany. Food Qual Prefer 85:103983

Dupont J, Harms T, Fiebelkorn F (2022) Acceptance of cultured meat in Germany—application of an extended theory of planned behaviour. Foods 11(3):424

El Haffar G, Durif F, Dubé L (2020) Towards closing the attitude-intention-behavior gap in green consumption: a narrative review of the literature and an overview of future research directions. J Clean Prod 275:122556

Engel L, Vilhelmsen K, Richter I, Moritz J, Ryynänen T, Young JF, Burton RJF, Kidmose U, Klöckner CA (2024) Psychological factors influencing consumer intentions to consume cultured meat, fish and dairy. Appetite 200:107501

Espinosa R, Treich N (2023) Eliciting non-hypothetical willingness-to-pay for novel products: an application to cultured meat. Environ Resour Econ 85(3):673–706

Fernandes AM, de Souza Teixeira O, Revillion JP, de Souza ÂRL (2022) Beef as a socio-cultural identity: rural and urban consumers' attitudes from Rio Grande do sul, Brazil, facing cultured beef. J Rural Stud 95:438–448

Fu W, Zhang H, Whaley JE, Kim YK (2023) Do consumers perceive cultivated meat as a sustainable substitute to conventional meat? Assessing the facilitators and inhibitors of cultivated meat acceptance. Sustainability 15(15):11722

Gonzalez-Arcos C, Joubert AM, Scaraboto D, Guesalaga R, Sandberg J (2021) How do I carry all this now? Understanding consumer resistance to sustainability interventions. J Mark 85(3):44–61

Hallman WK, Hallman WK II, Hallman EE (2023) Cell-based, cell-cultured, cell-cultivated, cultured, or cultivated. What is the best name for meat, poultry, and seafood made directly from the cells of animals? npj Sci Food 7(1):62

Heidenreich S, Kraemer T (2015) Passive innovation resistance: the curse of innovation? Investigating consequences for innovative consumer behavior. J Econ Psychol 51:134–151

Hibino A, Nakamura F, Furuhashi M, Takeuchi S (2023) How can the unnaturalness of cellular agricultural products be familiarized? Modeling public attitudes toward cultured meats in Japan. Front Sustain Food Syst 7:1129868

Ho SS, Ou M, Ong ZT (2023) Exploring the general public's and experts' risk and benefit perceptions of cultured meat in Singapore: a mental models approach. PLoS One 18(11):e0295265

Hwang J, You J, Moon J, Jeong J (2020) Factors affecting consumers' alternative meats buying intentions: plant-based meat alternative and cultured meat. Sustainability 12(14):5662

Kouarfaté BB, Durif FN (2023) A systematic review of determinants of cultured meat adoption: impacts and guiding insights. Br Food J 125(8):2737–2763

Laureati M, De Boni A, Saba A, Lamy E, Minervini F, Delgado AM, Sinesio F (2024) Determinants of consumers' acceptance and adoption of novel food in view of more resilient and sustainable food systems in the EU: a systematic literature review. Foods 13(10):1534

Lazou AE, Revelou PK, Kougioumtzoglou S, Strati IF, Kanellou A, Batrinou A (2024) Cultured meat: a survey of awareness among Greek consumers. AIMS Agric Food 9(1):356–373
Leite FP, Septianto F, Pontes N (2024) 'Meat' the influencers: crafting authentic endorsements that drive willingness to buy cultured meat. Appetite 199:10740
Lewisch L, Riefler P (2023a) Cultured meat acceptance for global food security: a systematic literature review and future research directions. Agric Food Econ 11(1):48
Lewisch L, Riefler P (2023b) Behavioural intentions towards cultured meat: the role of personal values, domain-specific innovativeness and distrust in scientists. Br Food J 125(5):1769–1781
Lin-Hi N, Schäfer K, Blumberg I, Hollands L (2022) The omnivore's paradox and consumer acceptance of cultured meat: an experimental investigation into the role of perceived organizational competence and excitement. J Clean Prod 338:130593
Lin-Hi N, Reimer M, Schäfer K, Böttcher J (2023) Consumer acceptance of cultured meat: an empirical analysis of the role of organizational factors. J Bus Econ 93(4):707–746
Liu J, Almeida JM, Rampado N, Panea B, Hocquette É, Chriki S, Ellies-Oury M, Hocquette JF (2023) Perception of cultured "meat" by Italian, Portuguese and Spanish consumers. Front Nutr 10:1043618
Onwezen MC, Dagevos H (2024) A meta-review of consumer behaviour studies on meat reduction and alternative protein acceptance. Food Qual Prefer 114:105067
Pakseresht A, Kaliji SA, Canavari M (2022) Review of factors affecting consumer acceptance of cultured meat. Appetite 170:105829
Pilařová L, Balcarová T, Pilař L, Kvasničková Stanislavská L, Rosak-Szyrocka J, Pitrová J, Moulis P, Kvasnička R (2023) Exploring ethical, ecological, and health factors influencing the acceptance of cultured meat among generation Y and generation Z. Nutrients 15(13):2935
Piochi M, Micheloni M, Torri L (2022) Effect of informative claims on the attitude of Italian consumers towards cultured meat and relationship among variables used in an explicit approach. Food Res Int 151:110881
Pivoraite G, Liu S, Roh S, Zhao G (2024) Framework for understanding consumer perceptions and attitudes to support decisions on cultured meat: a theoretical approach and future directions. In: Duarte SP, Lobo A, Delibašić B, Kamissoko D (eds) Decision support systems XIV. Human-centric group decision, negotiation and decision support systems for societal transitions. ICDSST 2024. Lecture notes in business information processing. Springer, Cham, pp 109–125
Pliner P, Hobden K (1992) Development of a scale to measure the trait of food neophobia in humans. Appetite 19(2):105–120
Pliner P, Salvy SJ (2006) Food neophobia in humans. In: Shepherd R, Raats M (eds) The psychology of food choice. CABI Digital Library, Wallingford, pp 75–92
Post MJ, Levenberg S, Kaplan DL, Genovese N, Fu J, Bryant CJ, Negowetti N, Verzijden K, Moutsatsou P (2020) Scientific, sustainability and regulatory challenges of cultured meat. Nat Food 1(7):403–415
Rehman N, Edkins V, Ogrinc N (2024) Is sustainable consumption a sufficient motivator for consumers to adopt meat alternatives? A consumer perspective on plant-based, cell-culture-derived, and insect-based alternatives. Foods 13(11):1627
Rogers EM (1983) Diffusion of innovations, 3rd edn. The Free Press, New York
Rolland NC, Markus CR, Post MJ (2020) The effect of information content on acceptance of cultured meat in a tasting context. PLoS ONE 15(4):e0231176
Rombach M, Dean D, Vriesekoop F, de Koning W, Aguiar LK, Anderson M, Mongondry P, Oppong-Gyamfi M, Urbano B, Gómez Luciano CA, Hao W, Eastwick E, Jiang ZV, Boereboom A (2022) Is cultured meat a promising consumer alternative? Exploring key factors determining consumer's willingness to try, buy and pay a premium for cultured meat. Appetite 179:106307
Rosenfeld DL, Tomiyama AJ (2022) Would you eat a burger made in a petri dish? Why people feel disgusted by cultured meat. J Environ Psychol 80:101758
Ruzgys S, Pickering GJ (2020) Perceptions of cultured meat among youth and messaging strategies. Front Sustain Food Syst 4:122
Schoormans JP, Orrt RJ, de Bont JP (1995) Enhancing concept test validity by using expert consumers. J Prod Innov Manag 12(2):153–162

Shang TT, Miao XM, Lu D, Xin XH (2020) The evolutionary process of disruptive green innovation in China: evidence from the electric bicycle sector. Technol Anal Strateg Manag 32(8):910–922
Shaw E, Mac Con Iomaire M (2019) A comparative analysis of the attitudes of rural and urban consumers towards cultured meat. Br Food J 121(8):1782–1800
Siddiqui SA, Khan S, Farooqi MQU, Singh P, Fernando I, Nagdalian A (2022) Consumer behavior towards cultured meat: a review since 2014. Appetite 179:106314
Siegrist M, Hartmann C (2020) Perceived naturalness, disgust, trust and food neophobia as predictors of cultured meat acceptance in ten countries. Appetite 155:104814
Sikora D, Rzymski P (2023) The heat about cultured meat in Poland: a cross-sectional acceptance study. Nutrients 15(21):4649
To KV, Comer CC, O'Keefe SF, Lahne J (2024) A taste of cell-cultured meat: a scoping review. Front Nutr 11:1332765
Tsvakirai CZ (2024) The valency of consumers' perceptions toward cultured meat: a review. Heliyon 10(6):e27649
van Der Weele C, Driessen C (2019) How normal meat becomes stranger as cultured meat becomes more normal; ambivalence and ambiguity below the surface of behavior. Front Sustain Food Syst 3:69
Vural Y, Ferriday D, Rogers PJ (2023) Consumers' attitudes towards alternatives to conventional meat products: expectations about taste and satisfaction, and the role of disgust. Appetite 181:106394
Wilks M, Phillips CJ (2017) Attitudes to in vitro meat: a survey of potential consumers in the United States. PLoS One 12(2):e0171904

Socio-Demographic and Behavioural Determinants of Consumer Attitudes to Cell-Based Meat and Insect-Flour-Based Products: An Empirical Study of Italians' Propensity to Accept New Foods for Human Nutrition and Pet Food

Giacomo Degli Antoni and Marco Faillo

Abstract The chapter analyses the socio-demographic and behavioural factors affecting Italian consumer attitudes to cell-based meat and insect-flour-based products. We surveyed a representative sample of the Italian adult population and found that the vast majority of respondents were reluctant to consume cell-based meat and insect-based foods. Consumption intentions were more favourable to cell-based meat. In terms of health consequences, environmental benefits, effects on animal welfare and taste, respondents had more positive perceptions of cell-based meat than insect-based foods. Alignment with tradition was perceived to be higher for cell-based meat than for insect-based products. In general, our findings support existing evidence on the role of socio-demographic factors, such as age, gender, education and political attitude, as drivers of acceptance. We also provide new evidence that technological beliefs and trust in science positively affect consumption intentions for cell-based meat, but not for insect-based products. Finally, we provide new insights into the drivers of acceptance of such products for pet food.

Keywords Cell-based meat · Insect-based food · Novel foods · Consumer perception · Technological beliefs

---

G. Degli Antoni (✉)
University of Parma, Department of Law, Politics and International Studies, Department of Excellence 2023–2027 "Food for Future", Parma, Italy
e-mail: giacomo.degliantoni@unipr.it

M. Faillo
University of Trento, Department of Economics and Management, Trento, Italy
e-mail: marco.faillo@unitn.it

G. Formici et al. (eds.), *Cell-Based Meat in the European Union and Beyond*,
https://doi.org/10.1007/978-3-032-14624-3_12

1 Introduction

Novel foods like insect-based foods and cell-based meat are increasingly seen as promising solutions to major global issues related to traditional food production.[1] This is particularly relevant in a scenario in which increased demand for meat, due to population and economic growth, especially in lower middle-income countries,[2] exposes the global population to potential aggravation of food insecurity, greenhouse gas emissions, land and water usage, soil contamination, zoonotic diseases and animal welfare. However, the contribution of novel foods to the transition to more sustainable production also depends crucially on consumer willingness to try and include them in their diets.

The empirical literature on attitudes to cell-based meat and insect-based foods has increased significantly in the last few years.[3] We now have fairly solid evidence of the major factors affecting consumer acceptance of these two novel food categories, although it comes mostly from Western countries.[4] In terms of socio-demographic characteristics, for example, there is robust evidence of higher acceptance of novel foods in general, among younger educated urban consumers.[5] Males are generally more willing to try cell-based meat and insects.[6] Food habits and preferences, like high levels of meat consumption, are positively associated with willingness to try and consume cell-based meat.[7] Impulsive reactions play a central role as determinants of consumer acceptance of novel foods and can significantly affect food choice cost and benefit assessment.[8] Disgust, fear of contamination and apprehension are frequently mentioned in the literature as associated with rejection of cell-based meat and insect-based foods.[9] Consumers' attitudes are also strongly influenced by expected or actual product properties, like taste, texture, healthiness, price and nutritional content.[10] Familiarity, in terms of acquaintance or exposure, increases acceptance.[11] Other strongly supported determinants are

---

[1] Mazac et al. (2022).

[2] FAO (2003) and OECD-FAO (2024).

[3] Hanan et al. (2024), Monaco and Giraldi (2024) and Tsvakirai et al. (2024).

[4] Siddiqui et al. (2023) provide a comprehensive world-wide comparative analysis on the issue. See also *Navigating Consumer Acceptance of Cultured Meat: Insights from Social Science Research* by Lin-Hi and Brudorf in this volume.

[5] Siegrist and Hartmann (2020), Padhilla et al. (2021), Pakseresht et al. (2022), Jacobs et al. (2024), Muiruri and Rickertsen (2024) and Giacalone and Jaeger (2023).

[6] Bryant and Barnett (2018), Andrić et al. (2023) and Siddiqui et al. (2022).

[7] Onwezen et al. (2021).

[8] Monaco and Giraldi (2024).

[9] Verbeke et al. (2015), Lammers et al. (2019), Onwezen et al. (2021), Kröger et al. (2022) and Siddiqui et al. (2023).

[10] Verbeke et al. (2015), Pakseresht et al. (2022), Wilks and Phillips (2019) and Tsvakirai et al. (2024).

[11] Hartmann and Siegrist (2017), Bryant and Barnett (2018), Onwezen et al. (2021), Pakseresht et al. (2022) and Monaco and Giraldi (2024).

information, ethical attitudes of consumers and cultural/social influences.[12] In the case of information, its impact depends not only on its content but also on how it is framed. In the case of cell-based meat, for example, information seems to increase resistance when descriptions are perceived to be too technical,[13] while it has a positive effect when it is interpreted as simple and transparent with respect to the regulation and cost/benefits of the product.[14] Ethical motivations and concern for animal welfare[15] and environmental sustainability have positive impacts on acceptance.[16] Cultural drivers are also important and may include the definition of food, religious beliefs and habits related to food preparation and consumption, and attachment to local traditions.[17] Social norms influence acceptance of novel foods by providing information on beliefs of what is appropriate in one's reference groups and may prevent consumers from trying alternative sources of proteins.[18] A negative impact on acceptance is associated with personal attitudes like neophobia, perceived unnaturalness and technophobia or distrust in the application of new technologies to food production.[19] Finally, distrust in science, technology, institutions and producers, together with conspiratorial ideation can act as barriers to acceptance.[20] Interesting evidence has also been collected on consumer acceptance of novel foods to feed their pets. For example, in a recent study, Fantechi[21] found that food neophobia and attention to environmental sustainability affect Italian consumers' attitudes to pet food.[22]

This chapter presents the results of an empirical analysis based on a dataset collected by means of an original questionnaire administered to a sample of 2102 respondents, representative of the adult population of Italy in terms of age groups, gender and macro-geographical area of residence. We analysed the effect of socio-demographic and behavioural factors on consumer attitudes to cell-based meat and insect-flour-based products. The main novelty of our approach resided in four aspects. First, we explored consumer acceptance of cell-based meat and insect-flour-based foods (specifically cricket or locust flour) by investigating not only respondents' consumption intentions, but also their perceptions in terms of safety for health, environmental impact, impact on animal welfare, consistency with local

---

[12] See *Who Hates Cultured Meat? Examining Public Opposition to Cellular Agriculture* by Broad and Glenna in this volume.

[13] Siegrist et al. (2018).

[14] Verbeke et al. (2015) and Laestadius and Caldwell (2015).

[15] See *Animal Welfare as a Constitutional Value: The Debate in Comparative Law and the Emerging Role of Cultivated Meat* by Piciocchi in this volume.

[16] Pakseresht et al. (2022).

[17] Siddiqui et al. (2022) and Gómez-Luciano et al. (2019). See also *Food Justice in Religious Laws: Hypotheses on Cultivated Meat* by Anello in this volume.

[18] Higgs (2015) and Onwezen et al. (2021).

[19] Hartmann and Siegrist (2017), Bryant and Barnett (2018) and Siegrist and Hartmann (2020).

[20] Wilks and Phillips (2019).

[21] Fantechi et al. (2024).

[22] See also Higa et al. (2021).

traditions and disgust. Second, we made a direct comparison of respondents' statements/attitudes about cell-based meat and insect-flour-based foods. Third, we considered behavioural drivers, such as impatience and risk attitude, which are underexplored in the literature on the topic. Fourth, we sought evidence of the socio-demographic and behavioural determinants of readiness to use cell-based meat and products made from insect flour to feed pets.

We found that the majority of respondents were unlikely to consume cell-based meat (74.69%) or increase their consumption, if any, of insect-based products (81.83%). We provide further support for existing evidence of the role of socio-demographic factors. We also found that acceptance of insect-flour-based products was lower among females, parents and risk-averse individuals, but these factors did not significantly impact attitudes to cell-based meat. Interestingly, trust in science and technology positively influenced attitude to cell-based meat consumption, but not to insect-flour-based foods. In terms of health consequences, environmental benefits, effects on animal welfare, taste and consistency with local food traditions, respondents have a more positive perception of cell-based meat than of insect-flour-based foods. Regarding novel ingredients in pet food, females, older individuals and right-wing respondents showed lower acceptance.

The chapter is organized as follows: in Sect. 2 we briefly review some of the recent evidence on Italian consumer acceptance of novel foods, with a focus on insect-based foods and cell-based meat. In Sect. 3 we describe the sample and the structure of the questionnaire and some descriptive statistics. In Sect. 4 we present our results on consumption intentions for cell-based meat and food products made from insect flour, and in Sect. 5 those on perceptions of the two food types. In Sect. 6 we explore the determinants of propensity to accept pet food containing cell-based meat or insect-based ingredients. Section 7 concludes.

2 Italian Consumer Acceptance of Cell-Based Meat and Insect-Based Foods: Recent Evidence

In the European Union (EU), the authorization procedure for Novel Foods (including insects and cell-based meat) is entirely centralised and binding for all Member States.[23] Despite this, in Italy, discussion on the commercialization of Novel Foods, especially cell-based meat and insect-based foods, is often at the centre of political debate and attracts public and media attention. The most striking case is the ban

[23] See the current Regulation (EU) 2015/2283 of the European Parliament and of the Council of 25 November 2015 on novel foods, amending Regulation (EU) no 1169/2011 of the European Parliament and of the Council and repealing Regulation (EC) no 258/97 of the European Parliament and of the Council and Commission Regulation (EC) no 1852/2001, in *Cell-Based Meat in the European Union: A Regulatory Crossroads* by Corazza and Formici in this volume.

endorsed by the Italian Parliament,[24] restricting the production, sale and import of cultivated cell-based meat in Italy, opening questions about its compliance with EU law.[25] A significant proportion of Italian consumers still seem to consider meat a fundamental and irreplaceable part of their diet and food culture, resulting in constant high per-capita consumption.[26] In this hostile context, understanding Italian consumers' opinions and attitudes to the two novel foods becomes even more urgent and interesting. In the last few years, we have witnessed an increase in studies, involving Italian respondents, aimed at measuring their willingness to try and/or consume novel foods and at identifying the main drivers of acceptance.[27]

Regarding insect-based foods, Menozzi,[28] for example, surveyed university students and reported little willingness to try insect-based products, females and respondents from southern Italy being less willing than the rest of the sample. Anticipated positive impacts on the environment and health were found to be the two main drivers of positive attitudes. In another study involving university students and faculty members, Sogari[29] confirmed the higher propensity of males to try these products, reporting a negative impact of food neophobia and a positive impact of familiarity on willingness to eat insect-based foods. La Barbera[30] conducted a laboratory experiment involving Italian students, providing further evidence of the negative impact of neophobia on willingness to eat insect-based products, though the variable's explanatory power was lower than that of disgust. Similar results were found by Iannuzzi[31] using a convenience sample of Italian citizens recruited online. Palmieri[32] conducted a study with a larger sample (n = 456) of Italian citizens living in four central-southern regions of the country, finding that only 17% of respondents were inclined to try insect-based foods. They also found that willingness to eat insects was positively associated with familiarity, concerns for environmental and

[24] Law 1 December 2023, No. 172. *Disposizioni in materia di divieto di produzione e di immissione sul mercato di alimenti e mangimi costituiti, isolati o prodotti a partire da colture cellulari o di tessuti derivanti da animali vertebrati nonchè di divieto della denominazione di carne per prodotti trasformati contenenti proteine vegetali.* (23G00188) (Official Gazette General Series No. 281 of 1 December 2023), https://www.gazzettaufficiale.it/eli/id/2023/12/01/23G00188/sg. Accessed 31 March 2025.

[25] See on this topic *Cell-Based Meat in the European Union: A Regulatory Crossroads* by Corazza and Formici in this volume.

[26] Mancini and Antonioli (2020) and Aureli et al. (2023).

[27] See for example Mancini and Antonioli (2020) for a review of studies on Italian consumers' attitude toward alternatives to meat, covering years 2016–2020.

[28] Menozzi et al. (2017).

[29] Sogari et al. (2018).

[30] La Barbera et al. (2018).

[31] Iannuzzi et al. (2019).

[32] Palmieri et al. (2019).

public health, and negatively correlated with neophobia,[33] fear of negative health consequences, aversion to new food technology, female gender and age.[34]

Regarding cell-based meat, Mancini and Antonioli[35] conducted a survey involving 525 Italian consumers, recruited through an Italian consumer association and provided with information on the laboratory production process of cell-based meat and on its positive externalities. Interestingly, the authors also collected data on respondents' perceptions on cell-based meat's intrinsic attributes (safety, taste and nutritional properties in comparison with conventional meat) and on its extrinsic attributes (impact on animal welfare, food safety and environment). Respondents did not expect cell-based meat to be as tasty or more nutritious or safer than a conventional burger and were more convinced of its extrinsic attributes. About half the respondents (54%) stated that they were willing to try cell-based meat. Females, respondents with a low level of education, those unfamiliar with the concept of cell-based meat and respondents who did not consume meat were less willing to try cell meat than younger respondents, well-educated males, those familiar with the idea of cell-based meat, and meat eaters.[36] Willingness to buy and to pay were also measured: 44% of respondents were willing to buy cell-based meat and 23% would pay a premium over the price of conventional meat. Mancini and Antonioli[37] showed that providing information on positive externalities of cell-based meat to the same sample of participants had a positive impact on willingness to buy when the information concerned safety and nutritional features. Using an online survey (n = 490), Palmieri[38] found that 78% of respondents were willing to try cell-based meat, willingness increasing with concerns for the environment,[39] food safety, animal welfare and curiosity about new food, and decreasing with technophobia and age. Piochi et al.[40] conducted a survey experiment involving students and faculty staff (n = 717) in which they manipulated the information provided on the extrinsic attributes of cell-based meat. The negative impact of neophobia, female gender and age on willingness to try cultivated meat was confirmed. In line with the literature, they also found that willingness to try was generally higher than willingness to substitute conventional with cell-based meat. Willingness to substitute seemed to be positively influenced by information, irrespective of the attribute considered. In a more recent survey with a representative sample of the Italian adult population, Aureli[41] found low acceptance of both insect-based foods and cultivated meat as a substitute for

[33] See also Merlino et al. (2024).

[34] See also Palmieri et al. (2023).

[35] Mancini and Antonioli (2019).

[36] See also Roma et al. (2020).

[37] Mancini and Antonioli (2020).

[38] Palmieri et al. (2020).

[39] See also Castellani et al. (2025).

[40] Piochi et al. (2022).

[41] Aureli et al. (2023).

conventional meat. De Martini[42] investigated the relation between dietary styles (omnivores, vegetarians, vegans, flexitarians) and acceptance of different alternatives to meat, finding low willingness to try insect products and cell-based meat across dietary styles.

3 Sample, Descriptive Statistics and Method

3.1 Sample and Survey Structure

The empirical analysis was based on original data collected between October and November 2024 through online surveys taken by 2102 subjects.[43] The sample was representative of the adult Italian population in terms of age groups, gender and macro-geographical area of residence (north-west, north-east, centre and south).

The survey was organized in six sections. The first section collected socio-demographic data such as age, gender, place of birth and residence, education level and financial condition. The second section focused on attitudes and habits, asking about risk preferences, impatience, technological beliefs regarding the effect of science and technology on our lives, trust in science, scientific research and various institutions, political orientation, eating habits and level of concern about environmental issues, health and food safety, animal welfare, and rising prices for food and basic necessities. The third section provided definitions and information regarding the concept of novel foods as defined by the European Union, and explored participant perceptions of same, particularly cell-based meats and food products made from insect flour. The fourth section delved deeper into eating attitudes and perceptions toward cell-based meats and insect-flour-based foods, asking respondents about their consumption intentions with regard to these products. The fifth section examined participant opinions regarding the normative market-authorization process, with particular reference to the European level. The sixth section concluded with two questions on whether the information about novel foods was sufficiently clear, and about respondents' perceptions regarding any political bias of the questions included in the survey. Between Sects. 3 and 4, surveys completed by half the sample included detailed information on the marketing authorization process in the European Union. This additional information aimed to investigate the effect of the information on attitudes to eating novel foods. However, in this chapter we do not examine the effect of information. But rather focus mainly on the questions in the

---

[42] De Martini et al. (2023).

[43] The survey was part of the research activities carried out within the project funded by the European Union under the Next Generation EU initiative, National Recovery and Resilience Plan (NRRP)—Mission 4, Component 2, Investment 1.3—Ministerial Call No. 341 of 15 March 2022 issued by the Ministry of University and Research; Project Code PE00000003, Ministerial Decree No. 1550 of 11 October 2022 granting the funding, project title: "ON Foods – Research and innovation network on food and nutrition Sustainability, Safety and Security – Working ON Foods."

first three sections. Only two variables based on questions from the fourth section of the questionnaire were used in the empirical analysis. When considering these two variables (Sect. 4), we restricted our attention to the 1051 participants who did not receive the above information, so that responses could be compared. Note that this sub-sample was also recruited to be representative of the adult Italian population in terms of age, gender and macro-geographical area of residence.

3.2 Descriptive Statistics

Descriptive statistics of all variables used in empirical analysis of the whole and the sub-sample of subjects (the latter did not receive information about the EU marketing-authorization process) are summarized in Table 8 of the Appendix.

In this section we describe the questions used to create the independent variables considered in the analysis and comment on the respective descriptive statistics. Dependent variables and respective descriptive statistics are analysed in detail in later sections.

Focusing on the whole sample, we had 51.43% of female respondents.[44] Average age was 50.93 years (min. = 18, max = 94; Standard Deviation (SD) = 16.602); 63.27% of respondents had children; 97.95% were Italian; 27.12% lived in NW Italy, 19.98% in the NE, 20.03% in the centre and 32.87% in the South (islands included); 34.95% of subjects had a university degree or higher qualification (Master or Ph.D.). Information on financial conditions was collected asking participants how well they were doing financially: 8.61% of participants replied comfortably; 67.65% acceptably; 21.03% that they barely get by and 2.71% really badly.

With respect to attitudes and preferences, we considered risk attitude, impatience, technological beliefs, trust in science and research, level of concern about various issues and political orientation (distribution of replies in Fig. 1a–f).

To measure risk aversion, we considered a validated measure[45] based on the answers to the question: "Are you generally a person who is fully prepared to take risks or do you try to avoid taking risks? Please tick a box on the scale, where 1 means: 'unwilling to take risks' and 10: 'fully prepared to take risks'." The average score was 4.919 (SD = 2.411), 54.54% of respondents indicating a value below 6 (Fig. 1a).

To measure impatience/patience, we asked "How would you describe yourself: would you say that you are more impatient or patient? Please respond using the scale 0 (highly impatient) to 10 (very patient)." This question aimed to capture people's attitude to the future in terms of the discount rate applied to potential future benefits compared to present benefits. The average score was 6.658 (SD = 2.258), 29.32% of respondents indicating a value below 6 (Fig. 1b).

[44] Two respondents classified themselves as "other" in response to the question "Are you: male; female; other".

[45] Based on Dohmen et al. (2011).

Technological beliefs (Fig. 1c) and trust in science and research (Fig. 1d) were measured with two questions. The first, on technological beliefs, was taken from the World Values Survey (Wave 7), and asked respondents to score their agreement (1–10) with the following statement: "Science and technology are making our lives healthier, easier and more comfortable." The average score was 6.642 (SD = 1.992). The second question was more general: "How much do you trust science and research?" with scores ranging from 1—I do not trust at all—to 5—I completely trust. The average score was 3.886 (SD = 0.779).

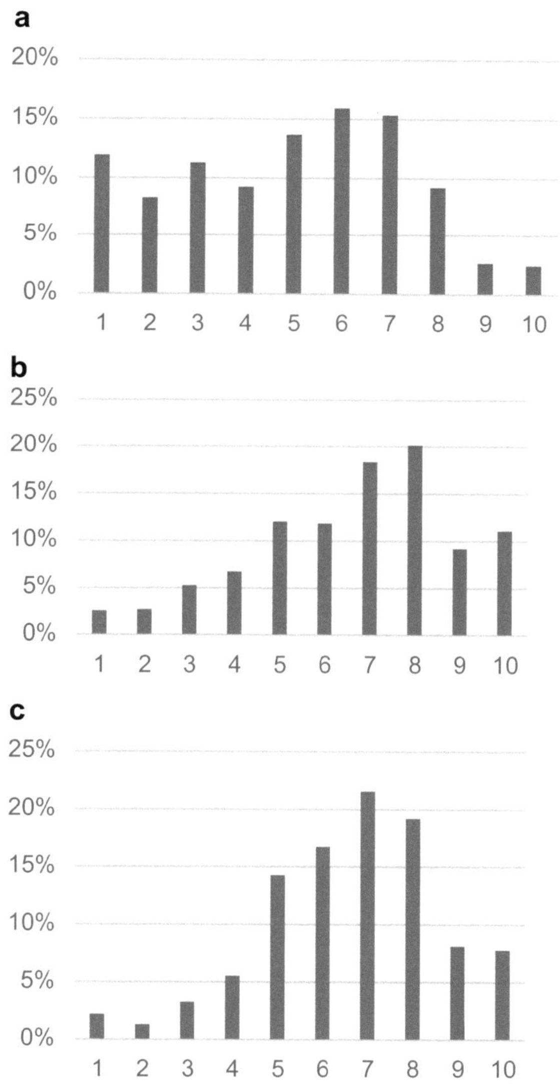

Fig. 1 (**a**) Risk propensity. (**b**) Patience. (**c**) Technological beliefs. (**d**) Trust in science and research. (**e**) Concerns. (**f**) Political orientation

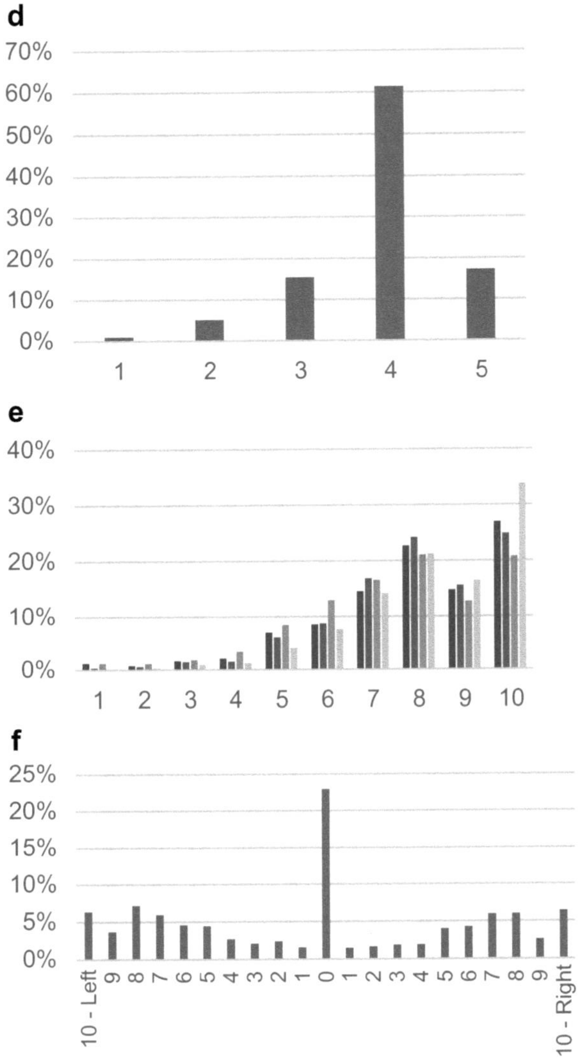

Fig. 1 (continued)

We also asked participants to score their concern about four issues on a scale from 1 (not worried at all) to 10 (completely worried): environmental issues—7.865 (SD = 1.980); health and food safety—7.939 (SD 1.781); animal welfare—7.484 (SD = 2.034); and rising prices of food and basic necessities—8.302 (SD = 1.680) (Fig. 1e, where the columns represent the same concerns, respectively, from left (dark grey) to right (light grey)).

Finally, to capture political orientation, we asked participants to mark their position on an integer scale (10—0—10), where 10 represented the two extremes of the political spectrum (Fig. 1f). We coded responses from 1 (10 on the right of the scale) to 21 (10 on the left of the scale). The average position was 10.715 (SD = 6.215).

3.3 *Plan of Analysis*

To investigate people's attitudes to novel foods, particularly cell-based meat and food products made from insect flour, we conducted a three-stage analysis. First, we examined subjects' propensity to consume cell-based meat if its commercialization is approved, and their willingness to increase consumption of insect-flour-based foods in the future, along with the determinants of these behaviours (Sect. 4). Second, we analysed people's level of agreement with five statements concerning cell-based meat and foods made from insect flour: "safety/health," "consistency with traditions," "environmental benefits," "animal welfare," and "disgust." We also investigated the factors influencing the answers to these questions, comparing attitudes to cell-based meat and insect food products and analysing the determinants of these differences (Sect. 5). Finally, we studied subjects' propensity to buy cell-based meat (if available) and insect-flour-based products for pet food (Sect. 6).

4 Consumption Intentions Regarding Cell-Based Meat and Food Products Made from Insect Flour

Two questions were specifically aimed at investigating future consumption intentions for cell-based meat and food made from insect flour, with specific reference to cricket or locust flour.[46]

Since cell-based meat is not yet available on the market in the EU, to study consumers' intentions regarding this product, we asked: "How likely are you to consume cell-based meat in future if it becomes commercially available?" Conversely, since food made from insect flour obtained from certain authorized species is already available on the market, we could not exclude the possibility that some respondents were already consuming it. We therefore also asked: "How likely are you to increase your consumption of products made from cricket or locust flour in the future?". The answer options were the same in both cases: 1 "completely unlikely"; 2 "very unlikely"; 3 "quite unlikely"; 4 "quite likely"; 5 "very likely"; 6 "completely likely".

Figure 2 shows clearly that the vast majority of participants considered unlikely an increase in their future consumption of cell-based meat (74.69% of respondents) or products made from cricket or locust flour (81.83%). In general, the distribution of responses revealed greater willingness to consume cell-based meat than to increase consumption of products made from insect flour: 4% of respondents stated that they were completely likely to consume cell-based meat in the future if it

[46] The independent variables used to study consumption intentions concerning products made from insect flour were based on questions specifically referring to cricket or locust flour. Conversely, when referring to perceptions, we opted for a broader definition of foods, asking about "insect-flour-based products" in general.

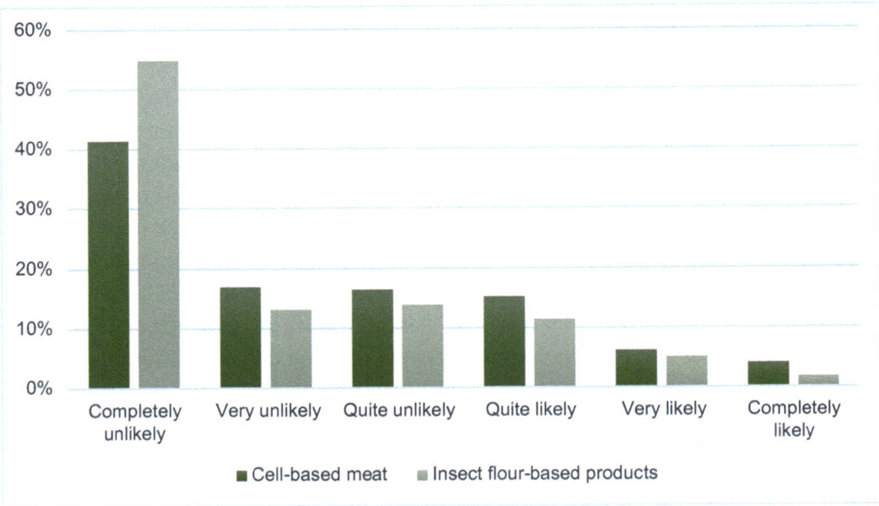

Fig. 2 Willingness to consume cell-based meat and to increase consumption of foods made from insect flour

became available, 6.09% that they were very likely to do so, and 15.22% that they were quite likely. The corresponding percentages related to the probability of increasing consumption of insect-based products were 1.71%, 5.04% and 11.42% (Fig. 2).

In the next section, we see that the gap in consumption intentions between the two novel foods appears to be linked to lower levels of liking and trust towards insect-based products than cell-based meat.

In line with the literature (Sect. 1) we assessed the effect of socio-demographic factors on consumption decisions regarding the two novel foods; we also added a set of measures of behavioural attitudes and technological beliefs, hitherto less investigated. We estimated the following equations using standard Ordinary Least Squares (OLS):

$$a_{i=}\alpha + x_i\beta + z_i\gamma + \varepsilon_i \tag{1}$$

$$b_{i=}\alpha + x_i\beta + z_i\gamma + \varepsilon_i \tag{2}$$

where a_i and b_i are the consumption intentions of individual i for cell-based meat and products made from cricket or locust flour, respectively, as captured by the above questions; x_i is a vector of socio-demographic controls and z_i is a vector of behavioural variables; ε_i is the independently distributed random term.

The socio-demographic variables included in vector x_i were: gender; age; having children; education level captured by a dummy equal to 1 if the respondents had a university degree or higher qualification; financial conditions; dummies related to the macro-region of residence[47] (see Sect. 2 and Table 8 of Appendix). Behavioural determinants and attitudes included in z_i were: risk preferences; impatience; technological beliefs and trust in science and research; level of concern about environmental issues, food safety/health; animal welfare; rising prices of food and basic necessities; political orientation.

Table 1 shows the results estimated by Eqs. (1) and (2). Interesting similarities and differences emerged between the determinants of future consumption intentions for the two novel foods.

Regarding similarities, we observed that persons less likely to consume cell-based meat or to increase their future consumption of products made from insect flour were older, less educated, more concerned about the rising price of food and basic necessities, and had a more right-wing political orientation. Impatient character and concern for animal welfare were not found to affect consumption intentions of either novel food.

Table 1 Determinants of consumption intentions regarding cell-based meat and foods made from insect flour

	(1)	(2)
	Consumption intentions: cell-based meat	Consumption intentions: insect-flour-based products
Method	OLS	OLS
Female	−0.154 (0.094)	−0.184* (0.088)
Age	−0.023** (0.003)	−0.017** (0.003)
Parents	−0.186 (0.109)	−0.259* (0.102)
Graduate	0.215* (0.099)	0.184* (0.093)
Financial condition	0.054 (0.080)	0.103 (0.075)
North-West	0.285* (0.119)	0.152 (0.111)
North-East	0.096 (0.129)	0.119 (0.121)

(continued)

[47] The empirical literature on acceptance of novel foods (Sect. 1) reveals a difference between urban and non-urban consumers. We did not explicitly seek this information in our questionnaire but verified the effect on our dependent variables of a dummy which takes the value of 1 for respondents living in towns with population < 10,000. This variable was not statistically significant [pr. = 0.502 and pr. = 0.759 in Eqs. (1) and (2) of Table 1, respectively]. Since being urban or non-urban consumers may be disconnected from the size of the town of residence, we opted for including only a control related to the macro-area of residence.

Table 1 (continued)

	(1)	(2)
Centre	0.144 (0.126)	0.046 (0.118)
Risk attitude	0.032 (0.021)	0.054** (0.019)
Patience	−0.006 (0.021)	−0.006 (0.020)
Technological beliefs	0.061* (0.026)	0.017 (0.025)
Trust in science	0.158* (0.071)	0.103 (0.066)
Political orientation	−0.043** (0.007)	−0.040** (0.007)
Concern_environment	0.071* (0.035)	0.080 (0.033)
Concern_health	−0.077* (0.035)	−0.052 (0.033)
Concern_animal welfare	0.054 (0.031)	0.003 (0.029)
Concern_prices	−0.129** (0.034)	−0.135** (0.032)
Constant	3.468** (0.480)	3.335** (0.448)
Observations	882	882
Adj R-squared	0.2365	0.1995
β(North-West) = β(North-East)	0.190 (0.136)	0.033 (0.127)
β(North-West) = β(Centre)	0.142 (0.134)	0.106 (0.125)
β(North-East) = β(Centre)	−0.048 (0.144)	0.073 (0.135)

Last rows show *lincom* (linear combinations of parameters) test of equality of coefficients
**$p < 0.01$; *$p < 0.05$; *$p < 0.1$

Regarding differences in terms of socio-demographic factors, females and persons with children were less likely to increase their consumption of insect-flour-based products. The two variables had no effect on cell-based meat consumption intentions. Respondents living in NW Italy showed higher intentions of consuming cell-based meat than those living in the South. No other regional differences emerged.

With regard to behavioural determinants and attitudes, we found that trust in science and research and positive technological beliefs had a positive impact on cell-based meat consumption intentions, but did not affect intention to increase consumption of insect-based products.

People more concerned about health problems related to food and less concerned about environmental issues were less likely to start consuming cell-based meat, were it available.

Finally, respondents who were more risk-averse tended to declare a lower probability of increasing their consumption of insect-based products, whereas no effect of risk aversion was observed for cell-based meat-consumption intentions.

5 Perceptions of Cell-Based Meat and Food Products Made from Insect Flour

In this section, we analyse the responses to five questions aimed at eliciting respondents' perceptions of cell-based meat and food made from insect flour. We asked participants to rate their level of agreement—from 0 (completely disagree) to 10 (completely agree)—with the following statements about the two products: "safe for health", "beneficial for the environment", "detrimental to animal welfare", "consistent with our traditions" and "disgusting".

5.1 Descriptive Statistics

Table 2 and Fig. 3 report the mean (and SD) and the distribution of responses to the ten questions, respectively. Two main pieces of evidence emerged from analysis of the responses.

First, apart from animal welfare which was of no concern for the vast majority of respondents, participants generally expressed concern about safety for health and environmental effects of both novel foods and perceived them as contrary to tradition and disgusting. The average responses regarding health/safety perceptions and environmental benefits were both below 6 for the two products. Regarding the distribution of responses (scored 0–10), a high percentage of respondents disagreed (scores <6) with the statement that cell-based meat and insect-flour-based products are safe for health (70.41% and 71.50%, respectively) and with the statement that they are beneficial for the environment (53.38% and 62.18%). Consistency with tradition was rated significantly low for both products (mean 1.980 for insect-based foods and 2.999 for cell-based meat), 76.93% and 85.82% of respondents, respectively, expressing agreement scores below 6. Respondents who declared that cell-based meat and insect-flour-based foods disgusted them represented the 52.90% and the 78.70% of the sample, respectively; their mean agreement scores were 5.939 and 7.044, respectively.

Table 2 Average responses concerning beliefs about cell-based meat and insect-flour-based products

	Safe for health	Beneficial for environment	Detrimental to animal welfare	Consistent with tradition	Disgusting
Cell-based meat	4.173 (2.865)	5.063 (2.988)	4.376 (3.076)	2.998 (3.083)	5.939 (3.362)
Insect-based foods	3.833 (3.004)	4.378 (3.088)	4.627 (3.192)	1.980 (2.771)	7.044 (3.396)
Paired t-tests for the null hypothesis that average responses regarding perception of the two novel foods is the same	$t = 6.380$ Pr $(T < t) = 1.0000$	$t = -3.596$ $Pr(T < t) = 1.0000$	$t = 11.608$ $Pr(T < t) = 1.0000$	$t = 17.066$ $Pr(T < t) = 1.0000$	$t = -14.452$ $Pr(T < t) = 1.0000$

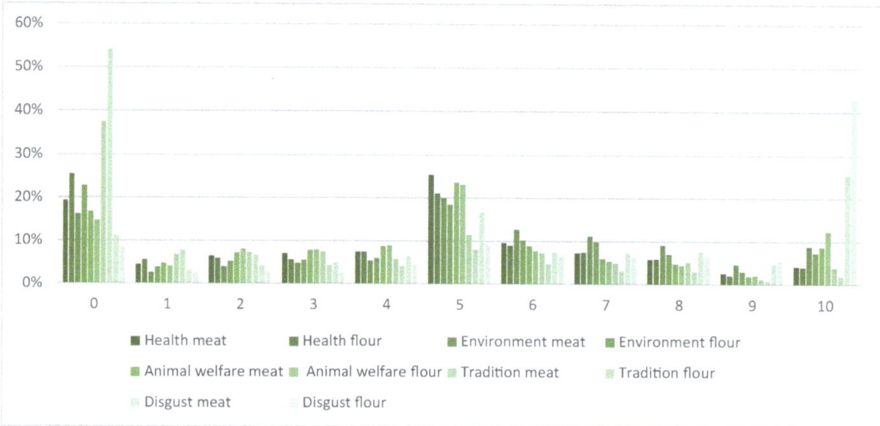

Fig. 3 Distribution of responses concerning perception of cell-based meat and insect-based products

Second, on average, perceptions were significantly more favorable to cell-based meat in terms of health/safety, environmental benefits, animal welfare and taste. Alignment with local tradition was perceived as higher for cell-based meat than for insect-based products. The paired t-tests in the last row of Table 2 reveal statistically significant differences in average responses for the two novel foods.

5.2 *Determinants of Perception of Cell-Based Meat and Insect-Based Products*

To investigate the determinants of people's perception of cell-based meat and foods made from insect flour, we performed multivariate analysis on the responses to each of the above *perception* questions using OLS according to the following equation:

$$c_{i=}\alpha + x_i\beta + z_i\gamma + \varepsilon_i \tag{3}$$

where c_i is individual i's level of agreement with the different statements about the two products; x_i is a vector of socio-demographic controls and z_i is a vector of behavioral variables as for Eqs. (1) and (2); ε_i is the independently distributed random term. The results are shown in Tables 3 and 4.

Analysis of the determinants of respondents' perceptions had two main goals. On the one hand, it aimed to explore similarities and differences across the various dimensions of perception (i.e. health/safety, environment, animal welfare, alignment with tradition and taste) of a given product category (cell-based meat or

Table 3 The determinants of perceptions about cell-based meat

	(1)	(2)	(3)	(4)	(5)
	Safe for health	Beneficial for the environment	Detrimental to animal welfare	Consistent with tradition	Disgusting
Method	OLS	OLS	OLS	OLS	OLS
Female	−0.341** (0.126)	−0.257 (0.134)	0.290* (0.142)	−0.160 (0.138)	0.389* (0.159)
Age	−0.024** (0.004)	−0.034** (0.005)	−0.003 (0.005)	−0.043** (0.005)	0.023** (0.005)
Parents	−0.334* (0.146)	−0.134 (0.156)	0.190 (0.166)	−0.161 (0.161)	0.208 (0.185)
Graduate	0.362** (0.132)	0.250 (0.141)	−0.098 (0.150)	0.330* (0.146)	−0.333* (0.167)
Financial condition	0.061 (0.110)	0.002 (0.118)	0.021 (0.125)	−0.010 (0.121)	−0.049 (0.139)
North-West	0.421** (0.159)	0.458** (0.169)	−0.260* (0.180)	0.434* (0.175)	−0.292 (0.200)
North-East	0.209 (0.173)	0.220 (0.184)	−0.485 (0.196)	0.019 (0.190)	−0.410 (0.218)
Centre	0.342* (0.170)	0.283 (0.182)	−0.089 (0.193)	0.388* (0.188)	−0.102 (0.215)
Risk attitude	0.136** (0.027)	0.110** (0.029)	0.058 (0.031)	0.106** (0.030)	−0.039 (0.035)
Patience	−0.011 (0.028)	−0.112** (0.030)	0.104** (0.032)	−0.005 (0.031)	0.057 (0.036)
Technological beliefs	0.136** (0.036)	0.165** (0.039)	0.012 (0.041)	0.153** (0.040)	0.011** (0.046)
Trust in science	0.535** (0.094)	0.686** (0.100)	−0.643** (0.106)	0.126 (0.103)	−0.586 (0.118)
Political orientation	−0.075** (0.010)	−0.073** (0.011)	0.067** (0.011)	−0.085** (0.011)	0.079** (0.013)
Concern_environment	0.137** (0.044)	0.239** (0.047)	−0.109* (0.050)	0.063 (0.048)	−0.116* (0.056)
Concern_health	−0.090* (0.046)	−0.034 (0.049)	−0.038 (0.052)	−0.076 (0.051)	0.118* (0.058)
Concern_animal welfare	0.074 (0.041)	0.106* (0.044)	0.143** (0.046)	0.133** (0.045)	0.005 (0.052)
Concern_prices	−0.258** (0.045)	−0.201** (0.048)	0.037 (0.051)	−0.286** (0.050)	0.182** (0.057)
Constant	3.641** (0.630)	3.129** (0.671)	4.914** (0.713)	5.417** (0.693)	4.478** (0.795)
No. observations	1767	1767	1767	1767	1767
Adj R-squared	0.2131	0.2281	0.0689	0.1809	0.0943
β(North-West) = β(North-East)	0.218 (0.180)	0.238 (0.192)	0.225 (0.204)	0.415* (0.198)	0.118 (0.228)
β(North-West) = β(Centre)	0.793 (0.180)	0.175 (0.192)	−0.171 (0.204)	0.046 (0.198)	−0.189 (0.227)
β(North-East) = β(Centre)	−0.133 (0.193)	−0.063 (0.205)	−0.396 (0.219)	−0.369 (0.212)	−0.308 (0.244)

Last rows show *lincom* (linear combinations of parameters) test of equality of coefficients
$^{**}p < 0.01$; $^{*}p < 0.05$; $^{*}p < 0.1$

Table 4 The determinants of perceptions about foods made from insect flour

	(1)	(2)	(3)	(4)	(5)
	Safe for health	Beneficial for the environment	Detrimental to animal welfare	Consistent with tradition	Disgusting
Method	OLS	OLS	OLS	OLS	OLS
Female	−0.268* (0.134)	−0.378** (0.142)	0.285 (0.149)	−0.207 (0.127)	0.683** (0.161)
Age	−0.023** (0.005)	−0.030** (0.005)	0.003 (0.005)	−0.031** (0.004)	0.006 (0.006)
Parents	−0.449** (0.156)	−0.237 (0.166)	−0.054 (0.173)	−0.059 (0.148)	−0.057 (0.188)
Graduate	0.391** (0.141)	0.164 (0.150)	−0.206 (0.157)	0.263* (0.134)	−0.295 (0.170)
Financial condition	0.041 (0.118)	0.178 (0.125)	−0.101 (0.130)	−0.039 (0.111)	−0.035 (0.141)
North-West	0.518** (0.169)	0.463* (0.179)	−0.353 (0.188)	0.097 (0.160)	−0.918** (0.203)
North-East	0.607** (0.185)	0.532** (0.196)	−0.352 (0.204)	0.171 (0.175)	−0.601 (0.222)
Centre	0.268 (0.182)	0.258 (0.193)	0.133 (0.201)	0.184 (0.172)	−0.267 (0.218)
Risk attitude	0.133** (0.029)	0.123** (0.031)	0.022 (0.032)	0.190** (0.028)	−0.124** (0.035)
Patience	−0.005 (0.030)	−0.017 (0.032)	0.097** (0.033)	0.042 (0.028)	0.081* (0.036)
Technological beliefs	0.126** (0.039)	0.135** (0.041)	−0.033 (0.043)	0.045 (0.036)	−0.002 (0.046)
Trust in science	0.392** (0.100)	0.404** (0.106)	−0.492** (0.111)	0.108 (0.094)	−0.219 (0.120)
Political orientation	−0.084** (0.011)	−0.083** (0.011)	0.038** (0.012)	−0.052** (0.010)	0.065** (0.013)
Concern_environment	0.190** (0.047)	0.279** (0.050)	−0.217** (0.052)	0.056 (0.044)	−0.074 (0.056)
Concern_health	−0.120* (0.049)	0.016 (0.052)	0.027 (0.054)	−0.075 (0.047)	0.079 (0.059)
Concern_animal welfare	0.015 (0.044)	−0.003 (0.046)	0.140** (0.048)	0.062 (0.041)	−0.005 (0.052)
Concern_prices	−0.240** (0.048)	−0.334** (0.051)	0.044 (0.053)	−0.331** (0.046)	0.270** (0.058)
Constant	4.058** (0.673)	3.782** (0.713)	5.934** (0.745)	4.624** (0.635)	5.107** (0.807)
No. observations	1767	1767	1767	1767	1767
Adj R-squared	0.1866	0.1939	0.0521	0.1594	0.0760
β(North-West) = β(North-East)	−0.089 (0.193)	−0.069 (0.204)	−0.001 (0.213)	−0.075 (0.181)	−0.316 (0.231)
β(North-West) = β(Center)	0.250 (0.192)	0.205 (0.204)	−0.486* (0.213)	−0.088 (0.182)	−0.651** (0.231)
β(North-East) = β(Centre)	0.339 (0.206)	0.274 (0.218)	−0.485* (0.228)	−0.013 (0.195)	−0.335 (0.247)

Last rows show *lincom* (linear combinations of parameters) test of equality of coefficients
**p < 0.01; *p < 0.05; *p < 0.1

insect-based products). On the other hand, it aimed to highlight similarities and differences in determinants of like perceptions of the two novel foods.

We first examine the determinants of perceptions of cell-based meat.

5.2.1 Cell-Based Meat

Safety for Human Health

We found that females, older individuals, those with children and without a university degree believed that cell-based meat is less safe for human health. Risk propensity, positive technological beliefs and trust in science and research increased perceptions that cell-based meat is safe; likewise for leftist political orientation. Regarding the effects of macro-area of residence, we found that people living in southern Italy perceived cell-based meat as less safe for health than those living in NW or central Italy. Finally, concerns about environmental issues positively affected perceptions on the health/safety of cell-based meat, while concerns about health effects and rising prices of food and basic necessities had a negative impact.

Environmental Effects

With respect to socio-demographic factors, we found that only age and area of residence affected the perception of beneficial environmental effects of cell-based meat, older people and those living in southern Italy believing it to be less beneficial for the environment than did younger people and those living in NW Italy. Like for perceptions about health/safety, we found that less risk-averse and left-leaning individuals, those with more positive technological beliefs and a higher level of trust in science and research were more likely to perceive cell-based meat as beneficial in terms of environmental effects. Concerns about environmental issues and animal welfare positively affected perceptions about the environmental benefits of cell-based meat, while concerns about rising prices had a negative effect.

Animal Welfare

Regarding the impacts of cell-based meat on animal welfare, we found that females, right-wing politically oriented individuals and those living in southern Italy were more likely than those living in the NW to consider them detrimental. Neither risk attitude nor technological belief in the positive effects of science and technology on people's lives affected perceptions about the effect of cell-based meat on animal welfare, whereas respondents having a higher level of trust in science and research believed the effects of cell-based meat on animal welfare to be less detrimental than those with a lower level of trust. Respondents who were less (more) concerned about environmental issues and animal welfare believed cell-based meat to be more (less) detrimental for animal welfare.

Tradition

Younger and more educated respondents tended to believe cell-based meat to be more aligned with tradition, as did left-leaning individuals, risk lovers and people with more positive technological beliefs. On the other hand, individuals living in southern Italy tended to believe it to be less aligned with tradition than those living in other regions of Italy. Concerns about animal welfare and rising prices of food and basic necessities increased and decreased, respectively, perceptions that cell-based meat is in line with tradition.

Disgust

We found that females and older, right-leaning and less educated individuals were more likely to feel disgusted by cell-based meat. Trust in science and research and concerns about environmental issues tended to reduce feelings of disgust, while positive technological beliefs, concern about effects of diet and rising prices of food and basic necessities tended to increase them.

5.2.2 Insect-Flour-Based Foods

The determinants of perceptions about insect-based foods were analysed by highlighting the differences with respect to the determinants of perceptions about cell-based meat.

Regarding perceptions about health/safety and animal welfare, we found essentially the same determinants as for cell-based meat, except for a lack of significance of animal welfare concerns, a lack of gender effect and more negative perceptions of those living in central than NW Italy in relation to the detrimental effects on animal welfare.

With respect to perceptions on the environmental effects of insect-based products, we noted three differences: (1) the emergence of a gender effect, non-females believing more than females that these products harm animal welfare; (2) a lack of significance of impatience, that showed a negative correlation with perceptions of the environmental benefits of cell-based meat; and (3) a lack of significance of animal welfare concerns.

Perceptions of the compatibility of these products with traditions were not affected by geographical area of residence, technological beliefs or animal welfare concerns, all of which proved to have a positive correlation with perceptions about cell-based meat's compatibility with tradition.

Finally, many differences characterized the determinants of taste perception of insect-based products compared to those of cell-based meat. Significant factors were the female dummy variable (positive effect); geographical area of residence, respondents living in southern and central Italy more often declaring disgust than those in the NW; the level of risk aversion and impatience, less (more) risk averse (patient) respondents feeling more disgust; concern about rising prices was correlated with a higher perception of disgust; political orientation with right-leaning persons finding insect-based foods more disgusting than left-leaning respondents.

5.3 Determinants of Differences in Perception of Cell-Based Meat and Insect-Flour-Based Products

To analyse the determinants of differences in perceptions of the two types of product, we created a categorical variable ranging from −1 to 1 associated with different levels of agreement with the five statements, as they pertained to cell-based meat and insect-based products: "safe for health", "beneficial for the environment",

Table 5 Perceptions about cell-based meat (P_m) and foods made from insect flour (P_f)

	Safe for health (%)	Beneficial for the environment (%)	Detrimental to animal welfare (%)	Consistent with tradition (%)	Disgusting (%)
$P_m = P_f$	41.67	39.01	35.78	49.62	39.53
$P_m < P_f$	23.93	21.03	33.97	10.75	44.20
$P_m > P_f$	34.40	39.96	30.26	39.63	16.27

"detrimental to animal welfare", "consistent with traditions" and "disgusting". The variable was: equal to 1 for respondents who were more in agreement with a statement when it pertained to cell-based meat than insect-flour-based products; equal to 0 when agreement with a statement was exactly the same for the two products; equal to −1 for respondents who were less in agreement with a statement when it pertained to cell-based meat than insect-based products. Therefore, the variable took the value of 1 (or 0 or −1) for respondents who believed that cell-based meat was: (a) safer (equally safe; less safe) for health; (b) more (equally; less) beneficial for the environment; (c) less (equally; more) detrimental to animal welfare; (d) more (equally; less) aligned with tradition; (e) more (equally; less) disgusting than the insect product. Table 5 shows the percentage of 1, 0 and −1 values for each statement. The percentage of respondents providing the same replies (variable = 0) for the two products ranged from 35.78% for animal welfare to 49.62% for compatibility with tradition. In the case of respondents providing different replies, we found that perceptions regarding health/safety, beneficial environmental effects, detrimental effects on animal welfare and disgust were more favorable for cell-based meat than for insect-flour-based products. Cell-based meat was perceived as more consistent with tradition than insect-flour-based foods by 39.63% of respondents, while the opposite was true for 10.75% of respondents.

Table 6 presents ordered logit estimates investigating factors that increased the probability of replying to the five questions with a higher level of agreement for cell-based meat than for insect-flour-based foods. The dependent variables in the analysis were the five categorical variables described above.[48]

Only a few variables had a significant effect in explaining differences in respondents' perceptions of the two novel foods, and only two variables seemed to reveal effects that characterize multiple dimensions captured by our perception questions. The first was the level of trust in science and research which increased the probability of having a more favorable perception of cell-based meat in terms of environmental effects and also the probability that respondents, ceteris paribus, considered cell-based meat relatively more compatible with tradition and less disgusting than insect-flour-based products. The second variable was age, older individuals being more likely to perceive cell-based meat as less consonant with tradition, more disgusting and less detrimental for animal welfare than insect-flour-based products. Finally, respondents living in southern Italy were more likely than those living in

[48] Note that since our dependent variable takes three values (1; 0; −1), the comparison between P_m and P_f regards three alternatives: (a) $P_m = P_f$ vs $P_m < P_f$; (b) $P_m > P_f$ vs $P_m = P_f$; (c) $P_m > P_f$ vs $P_m < P_f$.

Table 6 The determinants of differences in perceptions of cell-based meat and foods made from insect flour

	(1)	(2)	(3)	(4)	(5)
	Differences in perceptions				
	Safe for health	Beneficial for the environment	Detrimental to animal welfare	Consistent with tradition	Disgusting
Method	Ologit	Ologit	Ologit	Ologit	Ologit
Female	−0.050 (0.093)	0.134 (0.093)	−0.027 (0.091)	0.017 (0.096)	−0.159 (0.094)
Age	−0.002 (0.003)	−0.003 (0.003)	−0.008* (0.003)	−0.008* (0.003)	0.011** (0.003)
Parent	0.139 (0.107)	0.132 (0.107)	0.165 (0.106)	−0.057 (0.111)	0.065 (0.110)
Graduate	−0.098 (0.098)	0.015 (0.098)	0.052 (0.097)	−0.015 (0.102)	0.047 (0.099)
Financial condition	0.000 (0.080)	−0.048 (0.081)	−0.019 (0.080)	0.002 (0.084)	0.004 (0.082)
North-West	−0.109 (0.117)	−0.083 (0.117)	0.058 (0.115)	0.227 (0.121)	0.165 (0.118)
North-East	−0.350** (0.127)	−0.322* (0.127)	−0.048 (0.126)	−0.298* (0.131)	0.051 (0.130)
Centre	−0.097 (0.125)	−0.052 (0.125)	−0.174 (0.125)	−0.010 (0.131)	0.024 (0.129)
Risk attitude	0.004 (0.020)	−0.021 (0.020)	−0.001 (0.020)	−0.032 (0.021)	0.029 (0.020)
Patience	−0.011 (0.021)	−0.043 (0.021)	−0.005 (0.020)	−0.013 (0.021)	−0.027 (0.021)
Technological beliefs	−0.006 (0.026)	0.029 (0.026)	0.013 (0.026)	0.039 (0.027)	0.001 (0.027)
Trust in science	0.127 (0.068)	0.159* (0.068)	−0.044 (0.068)	0.153* (0.072)	−0.185** (0.069)
Political orientation	0.001 (0.007)	0.000 (0.007)	0.009 (0.007)	−0.024** (0.008)	0.007 (0.007)
Concern_ environment	−0.041 (0.032)	−0.042 (0.032)	0.023 (0.032)	0.010 (0.033)	−0.054 (0.032)
Concern_health	0.040 (0.034)	−0.044 (0.034)	−0.030 (0.034)	0.003 (0.035)	0.031 (0.034)
Concern_animal welfare	0.019 (0.030)	0.038 (0.030)	−0.005 (0.029)	0.028 (0.031)	0.017 (0.030)
Concern_prices	−0.061 (0.033)	0.076* (0.034)	−0.007 (0.033)	−0.005 (0.035)	−0.013 (0.034)
Cut 1	−1.316 (0.469)	−0.929 (0.456)	−1.169 (0.458)	1.977 (0.489)	−0.427 (0.471)
Cut 2	0.513 (0.468)	0.802 (0.456)	0.305 (0.457)	0.560 (0.486)	1.438 (0.473)
Observations	1767	1767	1767	1767	1767
Pseudo R2	0.0053	0.0082	0.0038	0.0172	

(continued)

Table 6 (continued)

	(1)	(2)	(3)	(4)	(5)
$\beta(North\text{-}West) = \beta(North\text{-}East)$	0.241 (0.132)	0.239 (0.133)	0.106 (0.131)	0.525** (0.138)	0.115 (0.135)
$\beta(North\text{-}West) = \beta(Center)$	−0.012 (0.132)	−0.031 (0.132)	0.232 (0.132)	0.238 (0.139)	0.141 0.136
$\beta(North\text{-}East) = \beta(Center)$	−0.253 (0.141)	−0.271 (0.141)	0.126 (0.141)	−0.288 (0.148)	0.026 0.146

Last rows show *lincom* (linear combinations of parameters) test of equality of coefficients
**p < 0.01; *p < 0.05; * p < 0.1

the NW to perceive cell-based meat as more in line with tradition, beneficial for the environment and safer for health than insect-based products.

6 Propensity to Accept Cell-Based Meat and Insect-Based Foods as Pet Food

Pet owners made up 59% (n = 1242) of our sample. Among them, 57.98% had a dog, 58.62% had a cat, 7.17% had fish, 5.93% turtles/tortoises, 2.90% rabbits, 1.69% hamsters or other rodents, and 3.30% other pets, mainly birds.

We analysed respondents' propensity to feed their pets cell-based meat and insect flour-based products by considering replies to the following questions: "Would you be inclined to give your pet: cell-based meat/products made from insect flour?" with answer options: "absolutely not", "more no than yes", "more yes than no", "absolutely yes".

When we aggregated responses "absolutely not" and "more no than yes", we found that 72.14% and 72.79% of respondents were not willing to accept cell-based meat or insect flour-based products, respectively, as pet food (44.04% and 46.62% of responses being "absolutely not", respectively) (Fig. 4).

Table 7 shows the results of OLS analysis of determinants of willingness to feed pets the two novel foods. The OLS estimate is specified by Eq. (1).

We found that females, older individuals, those with right-wing political orientations and those living in southern Italy were less willing to use cell-based meat and insect-based products as pet food than non-females, younger individuals, those with left-wing political orientations, and those living in NW Italy. Additionally, concerns about the rising prices of food and basic necessities reduced the propensity to accept either product as pet food. An interesting difference regarding the two novel foods concerned the role of technology and related aspects. While trust in science and technology positively affected the propensity to accept both cell-based meat and insect flour-based products as pet food, positive technological beliefs about the role of science and technology in making our lives healthier, easier and more comfortable positively affected the former but had no impact on the latter.

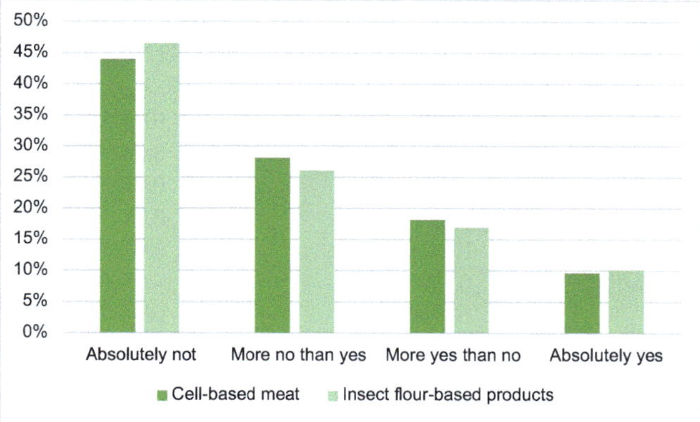

Fig. 4 Propensity to feed pets cell-based meat and insect-flour-based products

Table 7 The determinants of accepting novel foods for pet food

	(1)	(2)
	Pet food with cell-based meat	Pet food with insect-based ingredients
Method	OLS	OLS
Female	−0.127* (0.057)	−0.148* (0.060)
Age	−0.012** (0.002)	−0.011** (0.002)
Parent	0.002 (0.067)	−0.053 (0.071)
Graduate	0.105 (0.061)	0.076 (0.064)
Financial condition	−0.026 (0.051)	0.042 (0.054)
North-West	0.221** (0.074)	0.170* (0.078)
North-East	0.100 (0.078)	0.090 (0.083)
Centre	0.131 (0.078)	0.118 (0.082)
Risk attitude	0.015 (0.012)	0.005 (0.013)
Patience	0.001 (0.013)	−0.001 (0.014)
Technological beliefs	0.041* (0.016)	0.020 (0.017)
Trust in science	0.190** (0.042)	0.142** (0.044)

(continued)

Table 7 (continued)

	(1)	(2)
Political orientation	−0.033** (0.005)	−0.027** (0.005)
Concern_environment	0.021 (0.020)	0.062** (0.021)
Concern_health	0.034 (0.021)	0.014 (0.022)
Concern_animal welfare	0.003 (0.019)	−0.034 (0.020)
Concern_prices	−0.072** (0.021)	−0.064** (0.022)
Constant	1.925** (0.283)	2.094** (0.300)
Observations	1082	1082
Adj R-squared	0.1801	0.1309
β(North-West) = β(North-East)	0.121 (0.082)	0.080 (0.087)
β(North-West) = β(Center)	0.090 (0.082)	0.052 (0.087)
β(North-East) = β(Center)	−0.030 (0.086)	−0.028 (0.091)

Last rows show *lincom* (linear combinations of parameters) test of equality of coefficients
**$p < 0.01$; *$p < 0.05$; *$p < 0.1$

7 Discussion and Conclusions

This chapter aimed to explore acceptance of cell-based meat and insect-flour-based foods by investigating consumers' intentions for human nutrition and feeding pets. We analysed their perceptions with regard to health/safety, environmental impact, animal welfare, consistency with local traditions and disgust, and by directly comparing respondents' perceptions about the two foods.

Our findings can be summarized in different ways. In terms of consumption intentions, the great majority of respondents were reluctant to consume cell-based meat and insect-flour-based foods, favouring the former. We found further support for a role of socio-demographic factors, like age, gender, education and political orientation as drivers of acceptance, and new evidence that acceptance of insect-based products was lower for parents and risk-averse individuals. An interesting finding regarded technological beliefs and trust in science, which positively affected consumption intentions for cell-based meat, but not for insect-based products. This may suggest that cell-based meat is perceived as a more technological product than foods made from insect flour, so that its consumption is favoured by positive beliefs that science and technology make our lives healthier, easier and more comfortable and by greater trust in science and research.

In terms of responder perceptions, we found that those regarding health/safety, environmental effects, animal welfare and taste were more favorable to cell-based meat than insect-based foods. Consistency with tradition emerged as higher for cell-based meat than for insect-flour-based products.

Regarding the determinants of perceptions of cell-based meat and insect-based foods, we obtained general confirmation that age, being female, having children, living in southern Italy, being concerned about health/food safety and the rising of price of food and basic necessities have a negative effect on acceptance. Conversely, educational level, risk propensity, technological beliefs, trust in science and research, concern for the environment and left-wing political orientation showed a positive effect (Table 9 Appendix summarizes the effects of the different determinants).

Focusing on differences in the perceptions of respondents with respect to the two novel foods, we found that the level of trust in science and research positively affected the probability of having a more favourable perception of cell-based meat than insect-flour-based products in terms of environmental effects, consistency with tradition and taste. Age was negatively correlated with the likelihood of perceiving cell-based meat as more in line with tradition, less disgusting and more detrimental for animal welfare than insect-flour-based products. Respondents living in southern Italy were more likely than those in the NW to perceive cell-based meat as more aligned with tradition, beneficial for the environment and safer for health than insect-based products.

Finally, regarding propensity to feed pets with cell-based meat or insect-flour-based products, we found support for lower propensity on the part of female, older and right-wing respondents. Regarding the determinants of propensity to feed pets with the two novel foods, we found an interesting difference that concerned the role of technology and related aspects. In fact, while trust in science and technology positively affected the propensity to feed both cell-based meat and insect-flour-based products to pets, positive technological beliefs only favoured the former product.

Our analysis confirmed that different factors influenced consumer perceptions of different novel foods, as well as their consumption intentions, in different ways. To support the design of targeted policies, future research should therefore also aim to identify the determinants of acceptance of different novel foods in a comparative perspective.

Our analysis suggests that policies aimed at generating a positive attitude towards science and technology may also have a positive effect on consumption intentions regarding cell-based meat, but not insect-based products.

Appendix: Variable Legend and Descriptive Statistics

Table 8 Descriptive statistics and legend

Name	Description	Full sample n = 2102	Half sample with no information n = 1051
Consumption intentions	Variables ranging from 1 "Completely unlikely" to 6 "Completely likely" based on the questions:		
Consumption intentions—cell-based meat	"If it were commercially available, how likely would you be to consume cell-based meat?"	2.505 (1.513)	2.399 (1.489)
Consumption intentions—insect-based products	"How likely are you to increase your consumption of products made from cricket or locust flour?"	2.177 (1.441)	2.039 (1.368)
Perceptions	Variables ranging from 0 (completely disagree) to 10 (completely agree) depending on the level of agreement with the following statements:		
Safe for health_meat	"Cell-based meat is safe for health"	4.173 (2.865)	4.127 (2.889)
Safe for health_flour	"Insect-based products are safe for health"	3.833 (3.004)	3.750 (3.005)
Beneficial for the environment_meat	"Cell-based meat is beneficial for the environment"	5.063 (3.076)	4.993 (3.095)
Beneficial for the environment_flour	"Insect-based products are beneficial for the environment"	4.378 (3.192)	4.296 (3.193)
Detrimental to animal welfare_meat	"Cell-based meat is detrimental to animal welfare"	4.376 (2.988)	4.340 (2.959)
Detrimental to animal welfare_flour	"Insect-based products are detrimental to animal welfare"	4.627 (3.088)	4.718 (3.084)
Consistent with tradition_meat	"Cell-based meat is consistent with our traditions"	2.999 (3.083)	2.937 (3.055)

(continued)

Table 8 (continued)

Name	Description	Full sample n = 2102	Half sample with no information n = 1051
Consistent with tradition_flour	"Insect-based products are consistent with our traditions"	1.980 (2.771)	1.919 (2.713)
Disgusting_meat	"Does cell-based meat disgust you"	5.939 (3.362)	5.973 (3.369)
Disgusting_flour	"Do insect-based products disgust you"	7.044 (3.396)	7.064 (3.393)
Pet food	Variable ranging from 1 "absolutely not" to 4 "absolutely yes", based on the questions:		
Feeding pets with cell-based meat	"Would you be inclined to give your pet cell-based meat?"	1.935 (1.203) n = 1242	1.906 (0.983) n = 627
Feeding pets with insect-based products	"Would you be inclined to give your pet products made from insect flour?"	1.908 (1.019) n = 1242	1.900 (1.001) n = 627
Age	Age	50.934 (16.602)	50.834 (16.609)
Female	DV = 1 if female.	51.43%	51.38%
Graduate	DV = 1 if the respondent had a university degree or higher qualification	34.82%	34.54%
Parent	DV = 1 if respondent has children	63.27% n = 1843	62.77% n = 924
Financial condition	1 = "Really bad"; 2 = "Barely getting by"; 3 = "Living in an acceptable way"; 4 = "Living in a comfortable way"	2.822 (0.611)	2.794 (0.630)
North-West	DV = 1 if respondents were born in NW Italy	27.12%	27.21%
North-East	DV = 1 if respondents were born in NE Italy	19.98%	19.98%
Centre	DV = 1 if respondents were born in Central Italy	20.03%	19.98%
South and islands	DV = 1 if respondents were born in southern Italy or the islands	32.87%	32.83%

Risk propensity	Variable ranging from 1 to 10 based on the question: "Are you generally a person who is fully prepared to take risks or do you try to avoid taking risks? Please score as follows: 1 means: 'unwilling to take risks' and 10 'fully prepared to take risk'."	4.919 (2.411) n = 2059	4.819 (2.351) n = 1023
Patience	Variable ranging from 1 to 10 based on the question: "Would you describe yourself as impatient or patient? Please score as follows, where 1 means highly impatient and 10 means very patient."	6.658 (2.258) n = 2077	6.570 (2.226) n = 1037
Technological beliefs	Variable ranging from 1 (completely disagree) to 10 (completely agree) depending on the level of agreement with the following statement: "Science and technology are making our lives healthier, easier and more comfortable."	6.842 (1.992) n = 2055	6.630 (1.979) n = 1032
Trust in science and research	Variable ranging from 1 (I do not trust them at all) to 5 (I completely trust them) depending on your reply to the question: "How much do you trust science and research?"	3.886 (0.779) n = 2077	3.898 (0.765) n = 1031
Political orientation	Variable reflecting political orientation on an integer scale (10—0—10) ranging from 1 (10 on the right of the scale) to 21 (10 on the left of the scale) where 10 represented the two extremes of the political spectrum.	10.715 (6.215)	10.904 (6.251)
Concerns	Variable ranging from 1 (not worried at all) to 10 (completely worried) in relation to level of concern about:		
Concern_environment	Environmental issues	7.865 (1.980)	7.849 (1.985)
Concern_health	Health problems related to food	7.939 (1.781)	7.934 (1.812)
Concern_animal welfare	Animal welfare	7.484 (2.034)	7.514 (2.041)
Concern_prices	Rising prices of food and basic necessities	8.302 (1.680)	8.326 (1.684)

Note: Standard deviations in brackets. Number of observations shown when different from the maximum
DV dummy variable

Table 9 Determinants of perceptions of cell-based meat and insect-based foods [a]

Factor	Safe for health	Beneficial for environment	Detrimental to animal welfare	Consistent with tradition	Disgusting
Female	−,−	,−	+,		+,+
Older	−,−	−,−		−,−	+,
Parent	−,−				
Graduate	+,+			+,+	−,
Higher risk propensity	+,+	+,+		+,+	,−
Patience		−,	+,+		,+
Positive technological beliefs	+,+	+,+		+,	+,
Trust in science	+,+	+,+	−,−		
Left-wing orientation	+,+	+,+	−,−	+,+	−,−
Southern Italy vs. NW (and/or NE or Centre)	−,−	−,−	+,−	−,	,+
Centre vs. NW			,+		,+
NE vs NW				−,	
Environmental concerns	+,+	+,+	−,−		−,+
Health/safety concerns	−,−				+,
Animal welfare concerns	−,	+,	+,+	+,	
Rising food prices	−,−	−,−		−,−	+,+

a + and − before and after the comma refer to perception regarding cell-meat and insect-based foods, respectively

References

Andrić A, Miličić M, Bojanić M, Obradović V, ŠaŠić Zorić L, Petrović M, Gadjanski I (2023) Survey on public acceptance of insects as novel food in a non-EU country: a case study of Serbia. J Insects Food Feed 10(1):91–106

Aureli V, Nardi A, Palmieri N, Peluso D, Di Veroli JN, Scognamiglio U, Rossi L (2023) Sustainability perception of Italian consumers: is it possible to replace meat, and what is the best alternative? Nutrients 15(18):3861

Bryant C, Barnett J (2018) Consumer acceptance of cultured meat: a systematic review. Meat Sci 143:8–17

Castellani P, Cassia F, Vargas-Sánchez A, Giaretta E (2025) Food innovation towards a sustainable world: a study on intention to purchase lab-grown meat. Technol Forecast Soc Chang 211:123912

De Martini E, Marescotti ME, Amato M, Corradini A, Verneau F, Gaviglio A (2023) Acceptance of alternative meats among different dietarian styles: an explorative analysis in Italy. Food Qual Prefer 113:105060

Dohmen T, Falk A, Huffman D, Sunde U, Schupp J, Wagner GG (2011) Individual risk attitudes: measurement, determinants and behavioural consequences. J Eur Econ Assoc 9(3):522–550

Fantechi T, Califano G, Caracciolo F, Contini C (2024) Puppy power: how neophobia, attitude towards sustainability, and animal empathy affect the demand for insect-based pet food. Food Res Int 177:113879

Food and Agriculture Organization (2003) World agriculture: towards 2015/2030: a FAO perspective. Earthscan Publications Ltd., London
Giacalone D, Jaeger SR (2023) Consumer acceptance of novel sustainable food technologies: a multi-country survey. J Clean Prod 408:137119
Gómez-Luciano CA, de Aguiar LK, Vriesekoop F, Urbano B (2019) Consumers' willingness to purchase three alternatives to meat proteins in the United Kingdom, Spain, Brazil and The Dominican Republic. Food Qual Prefer 78:103732
Hanan FA, Karim S, Aziz YA, Ishak FAC, Sumarjan N (2024) Consumer's cultured meat perception and acceptance determinants: a systematic review and future research agenda. Int J Consum Stud 48(5):e13088
Hartmann C, Siegrist M (2017) Consumer perception and behaviour regarding sustainable protein consumption: a systematic review. Trends Food Sci Technol 61:11–25
Higa JE, Ruby MB, Rozin P (2021) Americans' acceptance of black soldier fly larvae as food for themselves, their dogs, and farmed animals. Food Qual Prefer 90:104119
Higgs S (2015) Social norms and their influence on eating behaviours. Appetite 86:38–44
Iannuzzi E, Sisto R, Nigro C (2019) The willingness to consume insect-based food: an empirical research on Italian consumers. Agric Econ 65(10):454–462
Jacobs AK, Windhorst HW, Gickel J, Chriki S, Hocquette JF, Ellies-Oury MP (2024) German consumers' attitudes toward artificial meat. Front Nutr 11:1401715
Kröger T, Dupont J, Büsing L, Fiebelkorn F (2022) Acceptance of insect-based food products in western societies: a systematic review. Front Nutr 21(8):759885
La Barbera F, Verneau F, Amato M, Grunert KG (2018) Understanding westerners' disgust for the eating of insects: the role of food neophobia and implicit associations. Food Qual Prefer 64:120–125
Laestadius LI, Caldwell MA (2015) Is the future of meat palatable? Perceptions of in vitro meat as evidenced by online news comments. Public Health Nutr 18(13):2457–2467
Lammers P, Ullmann LM, Fiebelkorn F (2019) Acceptance of insects as food in Germany: is it about sensation seeking, sustainability consciousness, or food disgust? Food Qual Prefer 77:78–88
Mancini MC, Antonioli F (2019) Exploring consumers' attitude towards cultured meat in Italy. Meat Sci 150:101–110
Mancini MC, Antonioli F (2020) To what extent are consumers' perception and acceptance of alternative meat production systems affected by information? The case of cultured meat. Animals 10(4):656
Mazac R, Meinilä J, Korkalo L, Järviö N, Jalava M, Tuomisto HL (2022) Incorporation of novel foods in European diets can reduce global warming potential, water use and land use by over 80. Nat Food 4:286–293
Menozzi D, Sogari G, Veneziani M, Simoni E, Mora C (2017) Eating novel foods: an application of the theory of planned behaviour to predict the consumption of an insect-based product. Food Qual Prefer 59:27–34
Merlino VM, Mosca O, Fornara F, Roma R, Bonerba E, Schiavone A, Passaro RL, Tarantola M (2024) Which factors affect the Italian consumer's intention to eat insects? An application of an integrated attitude-intention-eating model. Food Qual Prefer 113:105040
Monaco A, Giraldi L (2024) Heuristics and biases in consumer decision-making for cultured meat. J Behav Decis Mak 37(1):45–62
Muiruri SW, Rickertsen K (2024) Norwegian consumers' willingness to try cultured meat. Future Foods 10:100409
OECD and FAO (2024) OECD-FAO Agricultural Outlook 2024–2033. OECD Publishing/FAO, Paris/Rome
Onwezen MC, Bouwman EP, Reinders MJ, Dagevos H (2021) A systematic review on consumer acceptance of alternative proteins: pulses, algae, insects, plant-based meat alternatives, and cultured meat. Appetite 159:105058
Padhilla LG, Malek L, Umberger WJ (2021) Food choice drivers of potential lab-grown meat consumers in Australia. Br Food J 123(7):2354–2371

Pakseresht A, Ahmadi Kaliji S, Canavari M (2022) Review of factors affecting consumer acceptance of cultured meat. Appetite 170:105829
Palmieri N, Perito MA, Macrì MC, Lupi C (2019) Exploring consumers' willingness to eat insects in Italy. Br Food J 121(11):2937–2950
Palmieri N, Perito MA, Lupi C (2020) Consumer acceptance of cultured meat: some hints from Italy. Br Food J 123(1):109–123
Palmieri N, Nervo C, Torri L (2023) Consumers' attitudes towards sustainable alternative protein sources: comparing seaweed, insects, and jellyfish in Italy. Food Qual Prefer 104:104735
Piochi M, Micheloni M, Torri L (2022) Effect of informative claims on the attitude of Italian consumers towards cultured meat and relationship among variables used in an explicit approach. Food Res Int 151:110881
Roma R, Ottomano Palmisano G, De Boni A (2020) Insects as novel food: a consumer attitude analysis through the dominance-based rough set approach. Foods 9(4):387
Siddiqui SA, Khan S, Farooqi MQU, Singh P, Fernando I, Nagdalian A (2022) Consumer behavior towards cultured meat: a review since 2014. Appetite 179:106314
Siddiqui SA, Tettey E, Yunusa BM, Ngah N, Debrah SK, Yang X, Fernando I, Povetkin SN, Shah MA (2023) Legal situation and consumer acceptance of insects being eaten as human food in different nations across the world - a comprehensive review. Compr Rev Food Sci Food Saf 22(6):4786–4830
Siegrist M, Hartmann C (2020) Perceived naturalness, disgust, trust and food neophobia as predictors of cultured meat acceptance in ten countries. Appetite 155:104814
Siegrist M, Sütterlin B, Hartmann C (2018) Perceived naturalness and evoked disgust influence acceptance of cultured meat. Meat Sci 139:213–219
Sogari G, Menozzi D, Mora C (2018) The food neophobia scale and young adults' intention to eat insect products. Int J Consum Stud 43(1):68–76
Tsvakirai CZ, Nalley LL, Tshehla M (2024) What do we know about consumers' attitudes towards cultured meat? A scoping review. Future Foods 9:100279
Verbeke W, Marcu A, Barnett J (2015) Would you eat cultured meat? Consumers' reactions and attitude formation in Belgium, Portugal, and the United Kingdom. Meat Sci 102:49–58
Wilks M, Phillips CJ (2019) Testing potential psychological predictors of attitudes towards cultured meat. Appetite 136:137–145

Concluding Observations

Bernd van der Meulen

The book you are now holding in your hands provides a worthy conclusion to the Research Project of Relevant National Interest (PRIN) "The Future of Food, the Food of the Future. Novel Food, Innovation, Sustainability and Legal Issues".

Advancing technologies raise questions that have never been raised before as to the essence of what food of animal origin really is. What is it that makes meat? Is animal suffering unavoidable in a human diet that includes protein of animal origin? What about the culinary pleasures?[1]

Although the science to grow animal tissue out of cells still has many obstacles to overcome, already now the market access of products faces organised opposition. Paradoxically, this opposition implies a believe that the science is viable and the product likely to access the market soon. After all, if people spend valuable time in opposing cell-based production, they must be convinced that the product has the capacity to spread on the market. If people considered cell-based meat a product unlikely to be accepted by consumers, they probably would not have given priority to opposing it.

1 The Long and Winding Road

This book traces and analyses steps aiming to clear a road that ultimately may lead from vision to realisation.

[1] See also *Cultivating a Revolution: Legal and Policy Trajectories for a Sustainable Protein Transition* by Scaffardi in this volume.

B. van der Meulen (✉)
European Institute of Food Law, Amstelveen, The Netherlands
e-mail: bernd.vandermeulen@food-law.nl

G. Formici et al. (eds.), *Cell-Based Meat in the European Union and Beyond*,
https://doi.org/10.1007/978-3-032-14624-3_13

The vision is to create a way towards providing animal protein that could be high in nutritional value, mild in cruelty, moderate in thirst for energy, limited in hunger for land space, small sized in environmental footprint and low in exposure to food safety hazards. This is to be achieved by growing tissue out of animal cells in controlled environments.[2]

What would the realisation of that vision entail? Judging by the severity of some of the opposition, detractors expect cell-based products to cause a disruption by replacing substantial parts of animal husbandry and profoundly affecting existing food cultures.[3] Also many proponents speak in terms of 'replacing'. In this sense, battle lines are being formed and preparations are being made for hostilities.

Interestingly, the opponents seem to be the strongest believers that all obstacles that still stand in the way of realising the vision can be overcome. Probably things will not progress as fast as that. Would it not already be quite an accomplishment if upscaling of production and downscaling of resource intensity were achieved to such an extent that economically viable options could be offered to consumers, and perhaps part of the global growth in demand for animal protein could be captured?

This book addresses among many other important topics the question whether consumers would be willing to accept cultured meat, *foie gras*, dairy or other cell-based products.[4] In the political debate, another form of acceptance lies hidden. This is not the willingness of consumers to consume such products themselves, but a willingness to allow others the option to consume in case they would want to.

A famous quote in Dutch literature from the Flemish author Willem Elsschot goes something like this in English: "between dream and deed, laws and practical issues stand in the way."[5] In realising the vision of cultured meat as set out above, many practical issues need to be overcome. Several chapters in the book address the laws that "stand in the way".[6] This element, in particular, deserves a final reflection in these conclusive remarks.

---

[2] On the vision, see several chapters of this book: *What Is Cell-Based Meat? Food Safety and Sustainability Aspects* by Cheli et al.; *Cell-Based Meat and Food Policy* by Mancini; and, as far as animal welfare is concerned, *Animal Welfare as a Constitutional Value: The Debate in Comparative Law and the Emerging Role of Cultivated Meat* by Piciocchi; more specifically as seen from religious and ethical perspective, *Food Justice in Religious Laws: Hypotheses on Cultivated Meat* by Anello.

[3] See *Who Hates Cultured Meat? Examining Public Opposition to Cellular Agriculture* by Broad and Glenna in this volume.

[4] In particular, see *Navigating Consumer Acceptance of Cultured Meat: Insights from Social Science Research* by Lin-Hi and Burdorf, and *Socio-Demographic and Behavioural Determinants of Consumer Attitudes to Cell-Based Meat and Insect-Flour-Based Products: An Empirical Study of Italians' Propensity to Accept New Foods for Human Nutrition and Pet Food* by Degli Antoni and Faillo, both in this volume.

[5] A translation is available here: https://puntkomma.wordpress.com/2011/04/13/the-marriage-by-willem-elsschot/. Accessed 10 August 2025.

[6] In particular, for an in-depth scientific presentation of the mechanisms through which issues of scientific risk assessment, food safety and broader ethical, economic and social considerations are formally addressed in the path towards market approval, see *Cell-Based Meat in the European Union: A Regulatory Crossroads* by Corazza and Formici; *The Legal Debate on Cultured Meat*

2 Science Based Food Law

As of the early 1960s, the supra-national organisation that today is known as the European Union has employed a legislative strategy of prohibiting certain categories of products or substances from use *as* or *in* food. This prohibition can then be lifted through an authorisation on a case-by-case basis.[7] In German doctrine this legislative strategy is considered a matter of principle "Verbotsprinzip mit Erlaubnisvorbehalt".[8] Principle or not, the strategy has been expanded over time. Initially it applied to colourants and other food additives.[9] It was later expanded to substances such as (not exhaustive and not in chronological order) intelligent packaging materials,[10] food enzymes,[11] genetically modified foods,[12] and other Novel Foods.[13] The structure is always the same. Every product or substance that fits the legal definition is categorically denied market access except when specifically authorised. To acquire authorisation, an applicant must provide scientific evidence that its product is safe to consume, does not mislead consumers and is not

Around the World: Comparative Insights from the United States, Singapore and Israel by Pitto; *The Regulatory Landscape of Cultured Meat and Fish: Terminology and Consumer Information Provisions* by Sirsi and Rolandi; and *Cell-Based Meat and Food Policy* by Mancini.

[7] Again, see *Cell-Based Meat in the European Union: A Regulatory Crossroads* by Corazza and Formici for a deep analysis of the authorisation process in the EU and the different views and evaluations expressed by the Authors on the EU regulatory framework.

[8] In one of the oldest publications on comparative food law, it is expressed as follows: "According to the principle of prohibition, everything is forbidden which is not expressly permitted. The burden of proof is on the subjected to the law, who must in each case supply the proof that it is essential for a given situation to be expressly authorized and that for this reason this situation must be incorporated in the list of those things which are permissible. Following the principle of abuse, everything is [allowed] which is not expressly prohibited. The burden of proof is on the authorities which must demonstrate, in each case, that a de facto situation not expressly forbidden constitutes an 'abuse' and must for this reason be prohibited." Frenzel (1965), p. 134.

[9] Some early examples include the following. Directive of 23 October 1962 on the approximation of the rules of the Member States concerning the colouring matters authorized for use in foodstuffs intended for human consumption; Council Directive 64/54/EEC of 5 November 1963 on the approximation of the laws of the Member States concerning the preservatives authorised for use in foodstuffs intended for human consumption, Council Directive 70/357/EEC of 13 July 1970 on the approximation of the laws of the Member States concerning the antioxidants authorized for use in foodstuffs intended for human consumption.

[10] Regulation (EC) no 1935/2004 of the European Parliament and of the Council of 27 October 2004 on materials and articles intended to come into contact with food.

[11] Regulation (EC) no 1332/2008 of the European Parliament and of the Council of 16 December 2008 on food enzymes.

[12] Regulation (EC) no 1829/2003 of the European Parliament and of the Council of 22 September 2003 on genetically modified food and feed.

[13] Initially Regulation (EC) no 258/97 of the European Parliament and of the Council of 27 January 1997 concerning novel foods and novel food ingredients. Now replaced by Regulation (EU) 2015/2283 of the European Parliament and of the Council of 25 November 2015 on novel foods.

nutritionally disadvantageous.[14] As of 2002 the European Food Safety Authority (EFSA)[15] is the agency responsible for risk assessment.[16] In the case of authorisation procedures, risk assessment means scrutinising the completeness and reliability of the science provided with the application.[17]

It is mostly this system of required evidence of safety that lies at the heart of EU food law's claim to be science based.[18]

The burden of proof in authorisation schemes is high. Although it seems fair that businesses that want to gain from innovations bear the burden of providing the scientific evidence, there seem to be some disadvantages, indeed flaws in the system.

The authorisation system is resource and time intensive. This functions as a funnel towards innovations likely to profit corporations that have the resources to invest and the time to wait until approval which is rarely received within 3 years and may take a lot longer or never come at all. It goes without saying that smaller businesses and academia play a negligible role in this profit driven type of innovation. The same is true, for example, for health claims that also require authorisation. By consequence, most foods in the database of authorised health claim are industrial foods.[19]

---

[14] For Novel Foods these criteria are provided in Article 7 of Regulation (EU) 2015/2283 on Novel Foods.

[15] Established by Article 22 of Regulation (EC) no 178/2002 of the European Parliament and of the Council of 28 January 2002 laying down the general principles and requirements of food law, establishing the European Food Safety Authority and laying down procedures in matters of food safety.

[16] For the concepts of risk analysis and risk assessment see Articles 3 and 6 of Regulation (EC) no 178/2002.

[17] On the role of EFSA and the scientific challenges in assessing cultured meat, see *Safety Assessment of Cell-Culture-Derived Food: The Role of the European Food Safety Authority* by Dall'Asta and Gjoni in this volume.

[18] Article 6(1) of Regulation (EC) no 178/2002: "(...) food law shall be based on risk analysis (...)". Recital 16 of the same regulation: "Measures adopted by the Member States and the Community governing food and feed should generally be based on risk analysis except where this is not appropriate to the circumstances or the nature of the measure. Recourse to a risk analysis prior to the adoption of such measures should facilitate the avoidance of unjustified barriers to the free movement of foodstuffs."

[19] See for the database: https://food.ec.europa.eu/food-safety/labelling-and-nutrition/nutrition-and-health-claims_en. Accessed 10 August 2025. Also the following considerations may contribute to this effect. When one has raw materials or food coming directly from nature, the characteristics may vary depending on place of growth, soil, climate conditions, *et cetera*. Industrial foods are more standardised so their characteristics are steady, which makes it is easier to provide evidence of their effects on health. Further, if operators ask for authorisation on the use of claims for non-industrial products, everybody will be entitled to use that claim. This may discourage a food business operator to invest in a claim that will then be used by all the competitors.

Within the broader scope of authorisation requirements in EU food law, the regulatory context for cultured meat in the EU is mainly the Novel Foods Regulation.[20] This book has discussed pros and cons in detail as well as experiences in regulatory frameworks in other jurisdictions.[21]

3 Political Preferences

Apart from the commercial bias just mentioned, there is another aspect in the EU system of authorisation schemes such as the Novel Foods Regulation that may influence which innovations become available to consumers and which innovations do not. This is the political element. In many countries the emphasis of political involvement is on setting out the general rules and policies. The implementation is then rules based. Decisions can be measured up for compliance to these rules and can be struck down by courts in case of noncompliance. In theory this applies to the EU as well. However, the EU has to content with the subtleties of the relation between the Union and the Member States where politics is never far away and always ahead of the rules.

After the application has successfully navigated the scientific part of the authorisation process, the European Commission has to take a decision. In most situations, this decision takes the form of a regulation updating the list of authorised substances. In taking the decision the European Commission is bound by the law. It has to take account of the risk assessment—i.e. EFSA's favourable[22] opinion—the precautionary principle and other legitimate factors.[23] The precautionary principle does not and cannot play a role in authorisation procedures. At least not the precautionary principle as codified in Article 7 of the General Food law.[24] To put it simply: this provision allows risk managers to work from a worst-case scenario in situations of

---

[20] As analysed in *Cell-Based Meat in the European Union: A Regulatory Crossroads* by Corazza and Formici in this volume.

[21] See *The Legal Debate on Cultured Meat Around the World: Comparative Insights from the United States, Singapore and Israel* by Pitto; *Cell-Based Meat and Food Policy* by Mancini, both in this volume.

[22] See *Safety Assessment of Cell-Culture-Derived Food: The Role of the European Food Safety Authority* by Dall'Asta and Gjoni in this volume.

[23] Article 6(3) of Regulation (EC) no 178/2002: "Risk management shall take into account the results of risk assessment, and in particular, the opinions of the Authority referred to in Article 22, other factors legitimate to the matter under consideration and the precautionary principle where the conditions laid down in Article 7(1) are relevant, in order to achieve the general objectives of food law established in Article 5."

[24] Article 7(1) of Regulation (EC) no 178/2002: "In specific circumstances where, following an assessment of available information, the possibility of harmful effects on health is identified but scientific uncertainty persists, provisional risk management measures necessary to ensure the high level of health protection chosen in the Community may be adopted, pending further scientific information for a more comprehensive risk assessment."

scientific uncertainty. In other words, on a temporary basis, risk managers may treat scientific uncertainty the same way as they would treat the availability of evidence of a risk. In an authorisation procedure there cannot be scientific uncertainty. In case the evidence provided by the applicant does not prove conclusively that the product or substance for which authorisation has been asked, is safe, the application must be rejected. Not for reason of scientific uncertainty, but simply because the burden of proof was not fulfilled. Admittedly, the expression 'precautionary principle' is also used in a meaning that is different from the General Food Law.[25] When legislators impose an obligation to provide scientific evidence—as they do in authorisation requirements[26]—this in itself is sometimes also called 'precautionary principle'. On the basis of this meaning, it has to be assured that no unsafe products are placed on the market.[27] It is clear however that the precautionary principle that the Commission

[25] Recital 8 of Regulation (EC) no 1107/2009 of the European Parliament and of the Council of 21 October 2009 concerning the placing of plant protection products on the market for example reads: "(...) The precautionary principle should be applied and this Regulation should ensure that industry demonstrates that substances or products produced or placed on the market do not have any harmful effect on human or animal health or any unacceptable effects on the environment." What the legislature has in mind here, is clearly different from Article 7 of Regulation (EC) 178/2002. It is not about decision making in situations of scientific uncertainty but about placing the burden of proof on applicants. This dichotomy in meanings of precautionary principle was already present in the Communication from the Commission on the precautionary principle, COM(2000) 1 final, p. 20: "Community rules and those of many third countries enshrine the principle of prior approval (positive list) before the placing on the market of certain products, such as drugs, pesticides or food additives. This is one way of applying the precautionary principle, by shifting responsibility for producing scientific evidence. This applies in particular to substances deemed "a priori" hazardous or which are potentially hazardous at a certain level of absorption. In this case the legislator, by way of precaution, has clearly reversed the burden of proof by requiring that the substances be deemed hazardous until proven otherwise. Hence it is up to the business community to carry out the scientific work needed to evaluate the risk. As long as the human health risk cannot be evaluated with sufficient certainty, the legislator is not legally entitled to authorise use of the substance, unless exceptionally for test purposes."

[26] Not only for categories of innovative foods, but also for other substances such as veterinary drugs, medicines and pesticides.

[27] See for example Court of Justice of the European Union (CJEU) Judgement (25 July 2018) Case C-528/16, *Confédération paysanne and Others v Premier ministre and Ministre de l'agriculture, de l'agroalimentaire et de la forêt*, para. 53: "As laid down in Article 4(1) of Directive 2001/18, it is for the Member States to ensure, in accordance with the precautionary principle, that all appropriate measures are taken to avoid adverse effects on human health and the environment which might arise from the deliberate release or placing on the market of GMOs. This implies, in particular, that such deliberate release or the placing on the market may take place only on completion of procedures of assessment of the risks referred to in part B and part C of that directive respectively. However, as set out in paragraph 48 of the present judgment, the risks for the environment or human health linked to the use of new techniques/methods of mutagenesis to which the referring court refers might be similar to those which result from the production and release of a GMO through transgenesis. It follows that an interpretation of the exemption in Article 3(1) of Directive 2001/18, read in conjunction with point 1 of Annex I B thereto, which excludes organisms obtained by means of techniques/methods of mutagenesis from the scope of that directive, without any distinctions, would compromise the objective of protection pursued by the directive and would fail to respect the precautionary principle which it seeks to implement."

has to take into account when taking risk management decisions in accordance with Article 6 of Regulation (EC) no 178/2002, is the one of Article 7 of Regulation (EC) no 178/2002.

This leaves the European Commission with the Other Legitimate Factor (OLFs) to take into account. Some examples exist where the European Commission actually overruled EFSA's opinion on the basis of OLFs.[28] Literature suggests that there is some limitation included in the concept of OLFs.[29] In other words, not every factor can be considered as legitimate. However this may be, there are in fact no limitations to the other political element in the decision-making process. The reason for this is that the decision by the European Commission is taken in comitology.[30] This means that the Commission is "assisted" (as the EU euphemism goes) by the Standing Committee on Plants Animals Food and Feed (PAFF Committee).[31] This in turn means that the Commission needs the endorsement of a majority in the PAFF Committee to be able to take its intended decision. One might argue that the criteria provided by the law would also apply to the PAFF Committee. However, there is no functional mechanism to hold the PAFF Committee members accountable for the way they vote and the factors—legitimate or otherwise—that to them make the difference. By consequence, they are unrestrained in following the political priorities of the moment of the Member States they represent.

4 But They Won't Do That?

An example from another corner of food law may illustrate the scope of the impact this may have. This is decontaminants. Decontaminants are substances that are used to remove contamination from products of animal origin. These substances are also within the scope of an authorisation requirement.[32]

At the moment of writing this piece, the news informs us that in France two people lost their lives as a consequence of the contamination of pasteurised soft cheeses with *Listeria monocytogenes*.[33] Many others fell ill. The impact of the issue

---

[28] See CJEU Judgement (16 March 2016) Case T-100/15, *Dextro Energy GmbH & Co. KG v European Commission*.

[29] See for example, Szajkowska (2012).

[30] For details of the procedure, see: Regulation (EU) no 182/2011 of the European Parliament and of the Council of 16 February 2011 laying down the rules and general principles concerning mechanisms for control by Member States of the Commission's exercise of implementing powers.

[31] Established by Article 58 of Regulation (EC) no 178/2002.

[32] Article 5(3) of Regulation (EC) no 853/2004 of the European Parliament and of the Council of 29 April 2004 laying down specific hygiene rules for food of animal origin.

[33] See https://www.rfi.fr/en/france/20250813-deadly-listeria-outbreak-in-france-tied-to-nationwide-cheese-recall, https://www.connexionfrance.com/news/listeria-fears-prompt-global-recall-of-french-cheeses/740099. Accessed 15 August 2025; https://www.reuters.com/sustainability/boards-policy-regulation/france-recalls-soft-cheeses-after-two-suspected-deaths-listeriosis-2025-08-13/. Accessed 15 August 2025.

was global with recalls taking place all over the EU,[34] in the UK,[35] the USA,[36] Hong Kong,[37] Singapore[38] and New Zealand.[39] *Listeria monocytogenes* is a big and life-threatening problem. At the beginning of this Millennium, a start-up business made an invention. They found ways to isolate, produce, store and apply to products the natural enemies of *Listeria monocytogenes*. These enemies are viruses, bacteriophages, that 'feed' on *Listeria monocytogenes* killing the bacteria in the process. In the USA this process was recognised as safe for cheeses in 2006[40] and for all foods in 2007.[41] The inventors received awards for demonstrably saving human lives. After protracted debates in the EU on what would be the appropriate framework to assess whether this method could be allowed—are the phages food additives, processing aids, novel foods or something else?—in 2015 an application was submitted for authorisation as decontaminant under Article 3(2) of Regulation (EC) no 853/2004. In 2017 EFSA gave a favourable opinion.[42] According to EFSA, phages can effectively protect the safety of food by killing *Listeria monocytogenes* while at the same time being safe for human consumers. Taking account of EFSA's favourable opinion, the European Commission presented a draft decision authorising the product as decontaminant to the PAFF Committee. The PAFF Committee, however, did not provide its needed support and the Commission did not further pursue the matter. All this is evident from a ruling of the Court of Justice of the European

[34] Belgium https://www.euractiv.com/section/agriculture-food/news/listeria-outbreak-in-cheese-kills-two-in-france-spreads-to-belgium/; https://www.brusselstimes.com/1702063/listeria-case-in-belgium-linked-to-deadly-french-cheese-contamination; Ireland https://www.fsai.ie/news-and-alerts/food-alerts/recall-of-various-branded-goat-s-cheese-products; https://www.independent.ie/irish-news/fsai-recalls-two-types-of-cheese-due-to-presence-of-listeria/a1516230164.html; Netherlands https://nltimes.nl/2025/08/19/listeria-infection-netherlands-linked-soft-cheese-two-killed-france. All accessed 28 August 2025.

[35] https://www.mirror.co.uk/news/world-news/two-dead-dozens-ill-listeria-35757780. Accessed 20 August 2025.

[36] https://www.eatingwell.com/wegmans-cheese-recall-listeria-11790295. Accessed 15 August 2025.

[37] https://www.timeout.com/hong-kong/news/hong-kong-citizens-urged-not-to-consume-two-popular-cheeses-due-to-listeria-contamination-082625. Accessed 27 August 2025.

[38] https://www.channelnewsasia.com/singapore/three-french-cheeses-recalled-sfa-listeria-food-poisoning-5294551. Accessed 15 August 2025.

[39] https://www.1news.co.nz/2025/08/19/cheeses-recalled-due-to-possible-presence-of-listeria/. Accessed 20 August 2025.

[40] https://wayback.archive-it.org/7993/20171031020542/https:/www.fda.gov/Food/IngredientsPackagingLabeling/GRAS/NoticeInventory/ucm154675.htm. Accessed 10 August 2025.

[41] https://wayback.archive-it.org/7993/20171031015752/https:/www.fda.gov/Food/IngredientsPackagingLabeling/GRAS/NoticeInventory/ucm153865.htm; https://www.foodonline.com/doc/fda-extends-gras-approval-listex-to-all-food-0001. All accessed 10 August 2025.

[42] EFSA Panel on Biological Hazards (2016), https://www.efsa.europa.eu/en/efsajournal/pub/4565. Accessed 10 August 2025.

Union.[43] From this ruling it is further explained that despite all possible efforts undertaken by the applicant, no remedy was available in EU law. Thus, the food sector in the EU cannot have access to phages as weapon in their struggle with *Listeria monocytogenes*.

From the opinion of EFSA and the draft decision of the European Commission it becomes sufficiently clear that the legal requirements had been met. In the absence of a written opinion from the PAFF Committee it cannot be known which factors prompted their lack of support. As there is not a procedure where the position of the PAFF Committee can be contested, it cannot be ascertained whether these factors could have been considered as legitimate within the meaning of Article 6 of Regulation (EC) no 178/2002.

5 Politics Needs Accountability

What do phages have to do with cultured meat? Not much as such. They do, however, provide an example of how the political element in the authorisation procedures in EU food law is unrestrained. It is unrestrained by legal criteria on which to base decisions—even though the law does provide such criteria—and it is unrestrained by procedures of accountability—because the law does not provide for checks and balances in comitology.[44] One could therefore assume that the political majority can do anything it likes. This makes the procedure lopsided. Applicants have to provide scientific evidence that all the criteria have been fulfilled and in particular that consumer safety is assured. After this evidence has been shown to be conclusive, it is a political decision whether market access is granted or not granted.

Under EU authorisation schemes such as the Novel Foods Regulation, fulfilling all the legal obligations and requirements does not give the applicant access to the market. Whether, on the basis of the scientific evidence, authorisation will follow, ultimately depends on what the members of the PAFF Committee politically appreciate or do not appreciate.

It is not a forgone conclusion how the majority may vote on cultured meat. However, a massif lobby for the Member States to vote against has already been mobilised. Not the science, not the safety, not the sustainability, not the care for animal welfare[45] but the political preferences of a majority of the 27 members of the PAFF Committee representing the preferences in 27 Member States is the last and largest hurdle on the path towards the realisation of the vision in the European

[43] CJEU Judgement (22 February 2024) Case C-745/22, *Micreos Food Safety BV v Eniaios Foreas Elenchou Trofimon (EFET)*.

[44] Apart from a recently introduced option for the European Commission to appeal to a committee composed of other Member State representatives, who enjoy the same liberties as the original PAFF Committee members.

[45] See *Animal Welfare as a Constitutional Value: The Debate in Comparative Law and the Emerging Role of Cultivated Meat* by Piciocchi in this volume.

Union. Well, maybe not the last. In case the PAFF Committee would support the authorisation of cultured meat, there is still the opposition in Member States such as Austria, France, Italy and Hungary[46] to take into account.

6 Informed Choice

The impact of a refusal of authorisation of products found safe in scientific risk assessment is considerable. The product at issue cannot become available for any consumers. Not for those who are not interested in the product, but also not for those who would make a choice in favour of the product.[47]

Way back in the 1970s in its Cassis de Dijon case law,[48] the Court of Justice of the European Union addressed the question when it might be justified for EU Member States to ban products from their markets. In line with the proportionality principle,[49] the Court queried whether less intrusive measures could not solve the problem. The Court introduced the model of the average consumer who is reasonably well-informed,[50] reasonably observant and circumspect. This consumer is willing and able to take responsibility and capable of making informed decisions.[51] If such consumers can protect themselves when properly informed by making their own choices, then Member States should not ban products. From the case law follows that safety issues should not be left to the choice of consumers, but that with regard to other quality aspects informing consumers should be preferred over banning products. It seems to work well for example for genetically modified organisms (GMOs). It is allowed to use authorised GMOs in foods, but they have to be labelled as such. As a consequence, several GMOs are legal on the EU market, but are rare to actually find there. Given the choice, consumers make the choice. This

---

46 See *Cell-Based Meat in the European Union: A Regulatory Crossroads* by Corazza and Formici in this volume.

47 For an insight about these two categories, see *Navigating Consumer Acceptance of Cultured Meat: Insights from Social Science Research* by Lin-Hi and Burdorf, and *Socio-Demographic and Behavioural Determinants of Consumer Attitudes to Cell-Based Meat and Insect-Flour-Based Products: An Empirical Study of Italians' Propensity to Accept New Foods for Human Nutrition and Pet Food* by Degli Antoni and Faillo, both in this volume.

48 CJEU Judgement (20 February 1979) Case 120/78, *Cassis de Dijon: Rewe-Zentral AG v Bundesmonopolverwaltung für Branntwein.*

49 See also CJEU Judgement (23 September 2003) Case C-192/01, *Commission v Denmark*; CJEU Judgement (5 February 2004) Case C-95/01, *Greenham v Abel*; CJEU Judgement (2 December 2004) Case C-41/02, *Commission v Netherlands.*

50 *The Regulatory Landscape of Cultured Meat and Fish: Terminology and Consumer Information Provisions* by Sirsi and Rolandi in this volume, addresses the question how to inform consumers, and possible limitations set on words used in this context.

51 CJEU Judgement (20 February 1979) Case 120/78, *Cassis de Dijon: Rewe-Zentral AG v Bundesmonopolverwaltung für Branntwein.*

See also See also CJEU Judgement (16 July 1998) Case C-210/96, *Gut Springenheide GmbH and Rudolf Tusky v Oberkreisdirektor des Kreises Steinfurt – Amt für Lebensmittelüberwachung.* See also, Prechal et al. (2010).

choice can be in favour of the product at offer, or—as the GMO example shows—not in favour.

When it comes to cultured meat, will consumers be able to solve the problems perceived by politics by making choices? What are these problems? There are several concerns. Will cultured meat be able to live up to the expectation of sustainability? If not, it will have no unique selling point and only low price might convince some consumers to consume. However, low price so far does not seem to come in sight. Will cultured meat change our food culture? Food culture is under constant development. Traditions inspire each other.[52] Daring chefs push boundaries and turn haute cuisine into something different every year. Movements such as Slow Food make great efforts to preserve and to promote culinary treasures. Among the many options, consumers make their choices. Is there a problem here that needs solving? Is it the government's place to decide on the food culture a country should have?

Finally, there is the position of the meat industry. Should governments want to protect the scale and methods of the current animal production sector? And if they do, should they want to protect it from competition from alternative sources of animal protein or can it be left it to consumers to make informed choices?

7 Live and Let Live

This book provides a wealth of thoughts and information on cultured meat.

In research and development many steps still need to be taken to unlock the promises these technologies hold.

These final remarks reflected on who holds the key to allowing or disallowing cultured meat to take its chances with consumers if and when the technical challenges are overcome. To whom should the decision to accept cultured meat on the market or not to accept it be entrusted and on the basis of which criteria should the decision be taken? In principle, three models present itself. The democratic model, where the decision is taken on the basis of political considerations backed up by political majority support; the economic model, where consumer preferences decide through spending decisions; and the scientific model, where decisions are taken by experts based on predetermined objective scientific criteria.

The EU is based on the democratic decision of its constituting Member States to grant the lead to market mechanisms in many walks of life. Part of the deal is the commitment of public authorities to interfere only in the market when such interference is necessary. To assess the justification of interference—i.e. its necessity—the Court of Justice of the European Union developed the requirement to provide scientific evidence for the problem they perceive before they can impose their national desires on imported products.[53]

[52] See also *Food Justice in Religious Laws: Hypotheses on Cultivated Meat* by Anello in this volume.

[53] CJEU Judgement (23 September 2003) Case C-192/01, *Commission v Denmark*; CJEU Judgement (5 February 2004) Case C-95/01, *Criminal proceedings against John Greenham and Léonard Abel*; CJEU Judgement (2 December 2004) Case C-41/02, *Commission v Netherlands*.

The Novel Foods Regulation lifts this approach to the Union level. It requires pre-market authorisation of foods that do not have a history of safe use in the European Union prior to 1997 and that do fall under at least of the categories listed in the Regulation. The purpose of this Regulation is to ensure food safety. However, the instrument of authorisation provides a tool that can also be used to block market access for other reasons than food safety. The decision to use this option or not to use it, should be based as much as possible on a thorough understanding of the issue at hand, i.e. the vision and the promise of cultured meat. Such understanding might be gained from reading this book.

References

EFSA Panel on Biological Hazards (2016) Evaluation of the safety and efficacy of Listex™ P100 for reduction of pathogens on different ready-to-eat (RTE) food products. https://www.efsa.europa.eu/en/efsajournal/pub/4565. Accessed 10 Aug 2025

Frenzel H (1965) Basic principles of food laws divergences and harmonization. Food law in the Europe of to-morrow. Institut d' études Européennes: 127–150

Prechal A, de Graaf KJ, Jans JH, Widdershoven RJGM (2010) European administrative law: top down and bottom up. proceedings of the first REALaw research forum. Europa Law Publishing, Groningen

Szajkowska A (2012) Different actors, different factors. Science and other legitimate factors in the EU and national food safety regulation. Eur J Risk Regul 2:523–539

MIX
Papier aus verantwortungsvollen Quellen
Paper from responsible sources
FSC® C105338

If you have any concerns about our products,
you can contact us on
ProductSafety@springernature.com

In case Publisher is established outside the EU,
the EU authorized representative is:
Springer Nature Customer Service Center GmbH
Europaplatz 3, 69115 Heidelberg, Germany

Printed by Libri Plureos GmbH
in Hamburg, Germany